DISCRETE MATHEMATICS WITH CRYPTOGRAPHIC APPLICATIONS

DISCRETE MATHEMATICS WITH CRYPTOGRAPHIC APPLICATIONS

A *Self-Teaching Introduction*

Alexander I. Kheyfits, PhD

MERCURY LEARNING AND INFORMATION
Dulles, Virginia
Boston, Massachusetts
New Delhi

Publisher: David Pallai
MERCURY LEARNING AND INFORMATION
22841 Quicksilver Drive
Dulles, VA 20166
info@merclearning.com
www.merclearning.com
800-232-0223

Alexander I. Kheyfits. *Discrete Mathematics with Cryptographic Applications.*
ISBN: 978-1-68392-763-1

Library of Congress Control Number: 2021942465

CONTENTS

*P*REFACE

Mathematics and sciences, including political science have always involved many problems requiring large computations and sometimes, thousands of people. A century ago, the word "computer" meant a human being performing those computations. The top-level scientists, like Gottfried Leibniz, Lady Ada Lovelace, Alan Turing, and others were thinking how to automatize those computations. The breakthrough happened in the thirties-forties of the previous century, when Alan Turing proved the theorem about the universal program and John von Neumann developed the scheme of the universal computer machine, which all of us have used since then – see Chapter 23. The technology was ready, and the very first electronic computer "ENIAC" executed its initial programs in 1945. Since then, "computer" has meant an electronic device.

ENIAC declassifying in 1946 led to the explosive growth of the number of electronic computers, and correspondingly, to the growth of the number of people working with computers. Those people were mostly mathematicians, but very soon mathematics departments could not provide enough cadres, and the colleges and universities had to start teaching different courses to the new "computer scientists," whose first mathematical course was often called *Discrete Mathematics* or *Finite Mathematics*. Within a few years, discrete mathematics has become an independent mathematical discipline. Paul Halmos predicted [25, p. 19] that in the foreseeable future ". . . discrete mathematics will be an increasingly useful tool in the attempt to understand the world, and ... analysis will therefore play a proportionally smaller role."

Textbooks for the new discipline quickly followed, from concise lecture notes to Kenneth Rosen's monumental treatise [43], whose 8th edition (2019) amounts to more than 1000 pages. But every new generation meets their own challenges and requires new textbooks, see, e.g., [42], whatever good are the previous ones. A quarter century back, it was important to emphasize the algorithmic nature of discrete mathematics [37]. Now the 20-30-year-old discussions – see, e.g., [36], are mostly forgotten. The Internet security issues are more important than ever before, and this textbook was written with strengthened attention to these topics. A short version of these lectures was initially taught in 1971 and has evolved during the following half-century, being greatly influenced by the times and people – by my students, friends, and colleagues on both sides of the Atlantic.

To describe the contents of this book in more detail, let us notice that our thoughts are initially appeared as electrical-chemical potentials in the "grey matter" in our brains. To communicate our thoughts to other people, those potentials must be converted into electrical potentials governing oscillations of our vocal cords, and the latter create the fluctuations of the air pressure, carrying our speech. When these acoustical oscillations reach our ears, all the transformations are done in the opposite direction. Thus, the information that we create and obtain, is to be encoded and decoded in many different ways. These transformations are studied by coding theory, which is an important part of the discrete mathematics. However, to perform those conversions, we must represent the signals in a suitable form, i.e., as Boolean functions, and the book starts with an exposition of elementary logic and Boolean calculus. What is more, we often do not want other people to know the outcomes of our dealings with information, hence giving rise to the cybersecurity issues.

Boolean functions are maps, and chapters devoted to the functions, sets, and relations follow. We also consider in more detail predicates and quantifiers, which are necessary in many applications. When it is appropriate, we include expositions of some applications of these mathematical questions to computer-related issues, such as, e.g., relational databases or hashing functions. Several classical discrete mathematics topics, namely combinatorics, graph theory, complete systems of functional elements, etc., and some applications like finite automata, necessary for the potential users of the book, are included. These developments are based upon some algebraic structures, like groups, rings, fields, and Boolean Algebras. That allowed us to consider certain cryptography issues, in particular, the Discrete Logarithm Problem.

The book also contains chapters about number theory and game theory. No discussion of cryptography is possible these days without number-theoretical issues, like clock arithmetic and CRT. That background allows us to consider affine ciphers and some procedures of sharing secrets. The book also includes a chapter about game theory, the topic, whose inclusion in a discrete mathematics text is long overdue.

It is not the first text, which uses cryptology to enhance the teaching of mathematics – see, e.g., an interesting note [7]. Due to the volume restrictions, specifically crypto-questions are not covered here as much as they deserve. If the discussion of crypto–issues would deviate too far from the discrete mathematics, we are to stop there. To proceed farther, we would recommend the excellent books [2, 13, 14, 40]. The table of contents shows in more detail, what is in the book.

Many sections contain more material than can be reasonably taught at a two-hour class. That allows a lecturer some freedom of choice, and also provides the material for the student's individual work.

No mathematical exposition is possible without mathematical induction, which is unknown to most high school students. The author's experience shows though, that the method of mathematical induction, being properly explained, is well accessible to college freshmen. The very first chapter of the book is devoted to mathematical induction and to a very brief discussion of elementary functions, necessary in the textbook. That sets out the prerequisite level for the whole book. It is supposed that most college freshmen, independently upon their specialization, will benefit from this textbook. However, those lower prerequisites brought an unexpected problem. Certain language, which the college sophomores usually know quite well, can be unfamiliar to some potential readers. Thus, the book explains in more detail than usual, certain parlance, e.g., "necessary and sufficient conditions" or the like. This is especially necessary and useful now due to proliferation of on-line courses, where the student often cannot ask immediately certain "simple" questions. The reader, who is fluent with that material, can skip this material with no harm.

Of course, this is a mathematics textbook, and its reading requires concentration. To develop this culture, the reader is supposed to solve at least some of the included problems, and to analyze the suggested solutions. This is especially important in the times of online education. We remark in passing that it is not unusual, when the students find new solutions of old problems or suggest new problems. The author always welcomes any such input. The book contains more than 600 problems of various levels of difficulty. All of the exercises for the students' individual work are placed in the end of chapters. Many other problems can be found in the cited literature, e.g., in [19].

We thank many people for help. Parts of the book were discussed at the DIMACS Center at Rutgers University during the *Reconnect* conferences. The Reconnect-2019 workshop at the Champlain College in Burlington,

VT, was invaluable in finalizing this project. The present author, as anyone else, has its favorite books; I cannot list all of them, but I want to mention delightful books by M. Schroeder [45] and by L. Lovasz, J. Pelikan and K. Vesztergombi [35]. We want to personally thank the following people: Midge Cozzens, Fred Roberts, David Pallai and the staff of Mercury Learning and Information, who made this project possible.

Alexander I. Kheyfits, PhD.
September 2021

CHAPTER 1

A BRIEF SURVEY OF ELEMENTARY FUNCTIONS

1.1   MATHEMATICAL INDUCTION

The Principle, or Axiom, or Postulate of Mathematical Induction, is one of the cornerstones of any mathematical reasoning and, in particular, of our course. It is claimed that the outstanding mathematician Leopold Kronecker (1823–1891) said that "God created natural numbers, all else is humans' business." As with any trivial truth, it can be wrong. Indeed, many thousand years ago, together with learning to talk, people started to count, and eventually, the names for small numbers had appeared, like "one," "two," "three," etc., different for various languages. Moreover, we know, for example, from observations of Russian ethnographer and traveler Nikolas Miklukho-Maklai (1846–1888) over the indigenous people of Papua-New Guinea, that people had initially developed several specialized versions of the word "one," such that the phrase "one tree" sounded initially differently than "one boat," or than "one kid." Only gradually, during millennia, those various versions of "one something" merged in the abstract word "one," which means the natural number without any specific meaning attached.

During the centuries, arithmetic has been developed together with the human society, and now there is the highly sophisticated mathematical discipline, the *Number Theory*, which studies, in particular, the properties of the *natural numbers*[1] $\{0,1,2,3,...\}$, of the (positive, negative, and zero) *integers* $\{...,-3,-2,-1,0,1,2,3,4,...\}$, of the *prime numbers*, etc. We accept as the known facts that the natural numbers and the integers satisfy the four standard arithmetic operations. In particular, if any three integers are connected as $a = b \times c$, which can be written as $a = b \cdot c$, then the integers b and c are called *factors* or *multipliers*, and a is called the *product*. If we rewrite the equation as $a \div b = c$, then a is called the *dividend*, b the *divisor*, and c the *quotient*. It is also said that b (and c as well) divides a, or that a is divisible by b and by c, or that b (and c) *divides into a*.

Definition 1 *A natural number $p > 1$ is called prime, iff* [2] *it has only two positive divisors, 1 and itself. For example, 3 is an improper divisor of 3 and a proper divisor of 6, but it is not a divisor of 2. A natural non-prime number $p > 1$ is called composite. The number 1 is neither prime nor composite, it is an improper divisor of any integer.*

Thus, the sequence of primes begins with $\{2,3,5,7,11,13,...\}$, while the initial composite numbers are $\{4,6,8,9,10,12,...\}$.

The integers can also be classified in terms of their *parity*, that is, their divisibility by 2. If we divide an integer by 2, there are only two possibilities, and the remainder is either 0 or 1. The integers of the first kind, with the zero remainder, are called *even* numbers, the other integers are called *odd* numbers; thus, the integers $\{...,-6,-4,-2,0,2,4,6,8,...\}$ are even, and $\{...,-5,-3,-1,1,3,5,7,...\}$ are odd. Of course, the integers can be classified in terms of their divisibility over any other integer.

Problem 1 *Prove that 2 is the only even prime number.*

Proof. If p is an even prime, then $p = 2p_1$. But if $p_1 > 1$, then p has two factors, thus, it is not a prime number, and we arrived at the *contradiction*. So that, $p = 2 \times 1 = 2$.

This method of proof is called The Proof By Contradiction (Reductio Ad Absurdum) and is systematically used in mathematical reasoning; see

---

[1] The zero is included in the set of natural numbers, as is the common agreement in discrete mathematics and computer science. We do not give an axiomatic description of the set of natural numbers. The most common, Peano system of axioms, includes The Axiom of Mathematical Induction, which is discussed in detail below.

[2] The abbreviation "iff" stands for "if and only if" and means "necessary and sufficient condition(s)"; we discuss it in detail in the next Lecture.

Problem 52. When one wants to prove a statement by contradiction, she assumes an opposite statement and uses logical methods to deduce a statement, which contradicts the axioms or some known statements. More regarding the logic behind this and similar laws will be said in lecture 19.

We will continue our study of the integers in Chapter 8, devoted to the Number Theory.

Because we usually want to be correct in our statements, we must study certain properties of our reasoning, making us believe that our proofs are correct. All the more this is crucial now when we delegate some proofs to the computers and have to trust them. We discuss now only the Natural Numbers. In particular, a profound property called *the Axiom or Principle of Mathematical Induction* is considered at the beginning because we cannot proceed further without it.

During millennia, people established two important, in a sense, opposite methods for proving their claims, called *Deduction* and *Induction*. Both methods are widely employed in mathematics. Employing the *deduction* means that we have a general statement, maybe without regard to the source of the statement, and then apply it, *deduce* the properties of the specific entities. Consider, for example, a plane right triangle with the acute angles 20° and 70°. Since it is said that the triangle is right, we *deduce* that the triangle satisfies the Pythagorean theorem, that is, its sides satisfy $a^2 + b^2 = c^2$, where a and b are the lengths of the legs and c the length of the hypotenuse. When we study mathematics, all the time, we make such deductions.

Employing the induction method, though, we move in the opposite direction – we consider certain particular cases, analyze them, and draw a conclusion about the general case. Consider an example. Remind that the integers $\{2,4,6,8,10,\ldots\}$ are called *even* numbers, while the *odd* numbers are $\{1,3,5,\ldots\}$.[3]

Problem 2 *Find the sum of n consecutive odd numbers, starting with 1.*

Solution. Mathematics is, in a sense, an experimental science. Let us perform a few *numerical experiments* and compute these sums. Indeed,

$$1+3=4, 1+3+5=9, 1+3+5+7=16.$$

---

[3] These two methods have occurred not only in mathematics and sciences, but also, for example, in historiography: cf. the discussion of Gesetzwissenschaft versus Geschichtswissenschaft in [23, p. 251].

We observe that all these sums are perfect squares, $4 = 2^2$, $9 = 3^2$, $16 = 4^2$, and even the very first $1 = 1^2$. The evidence is compelling, that if we add first n natural numbers, the sum is the square of their number n. To do the mathematics, let us express all that in symbols. In particular, let us notice that any odd number can be denoted as $2j+1$; thus, if $j = 1$, then $2j+1 = 2 \times 1 + 1 = 3$, if $j = 5$, then $2j+1 = 2 \times 5 + 1 = 11$, etc. Vice versa, say $7 = 2 \times j + 1$ implies $j = 3$. One can also write the odd numbers as $2 \times k - 1$, but here the parameter $k = j + 1$.

There is a convenient notation for the summation formulas above, called the *sigma-notation*. "Sigma" is the name of the Greek capital letter Σ traditionally used to denote various *sums*; the small sigma is σ. Thus, we can write

$$1 + 3 + 5 + \cdots + (2n-1) = \Sigma_{j=1}^{n}(2j-1).$$

Here, j is called the index of summation, or the summation index, or *dummy index*; $j = 1$ is the lower index, $j = n$ is the upper index. As another example, consider $\sum_{k=1}^{-1} k = 1 + 0 + (-1) = 0$. With the same success, instead of j we can choose any other convenient symbol; the only restriction is that we must avoid the *collision of variables*, that is, the same letter cannot stand for different variables.

Thus, the sums of odd numbers above can be written as

$$1 + 3 = (1 \times 0 + 1) + (1 \times 1 + 1) = \Sigma_{j=0}^{1}(2j+1),$$

or as

$$1 + 3 + 5 = (1 \times 0 + 1) + (1 \times 1 + 1) + (1 \times 2 + 1) = \Sigma_{j=0}^{2}(2j+1),$$

or as

$$1 + 3 + 5 + \cdots + (2n-1) = (1 \times 0 + 1) + (1 \times 1 + 1) + (1 \times 2 + 1) + \cdots + (2n-1),$$

etc. As an example, we use this notation to prove that

$$\Sigma_{j=1}^{n}(2j-1) = n^2. \tag{1.1}$$

The actual difficulty is that we do not want to establish these equations only for $n = 2$ or $n = 3$, or say, for $n = 77$. That would be easy, just do the computations long enough. Indeed, suppose we have just finitely many statements to prove, say, we want to prove the statements $S_0, S_1, S_2, \ldots, S_{N_0}$, where N_0 is, maybe very big, but *finite natural number*. We remember that the true conditional with the true premise *must have the true conclusion*. So that, $S_1 \equiv T$, since $S_0 \equiv T$ and $S_0 \to S_1 \equiv T$. But now from $S_1 \equiv T$ and $S_1 \to S_2 \equiv T$ we conclude that $S_2 \equiv T$. By the same token, we imply next that $S_3 \equiv T$, then $S_4 \equiv T$, etc. Thus, after the finitely many steps, we conclude that $S_{N_0} \equiv T$. The actual magnitude of the integer number N_0 in this reasoning is immaterial; only its finiteness is important.

The situation is quite different if we have *infinitely many* statements. In this case, we physically cannot finish this domino-game and must claim certain new properties of the natural series. During the years, people separated that key property of the set of natural numbers. We must prove equation (1.1) for every natural n, that is, we have to establish it for **infinitely** many values of n. But our life is **finite**, and independently upon the speed of our computations, we cannot perform all of them in finite time. The method of proof, avoiding that infinity, is called the Principle (or the Axiom or the Postulate) of Mathematical Induction and goes in two steps as follows.

The Axiom of Mathematical Induction. Consider a set of statements, maybe formulas $S_{n_1}, S_{n_1+1}, S_{n_1+2}, \ldots$, numbered by all sufficiently large integers $n \geq n_1$. Usually $n_1 = 1$ or $n_1 = 0$, but it can be any integer. The statement S_n is called the *Induction Hypothesis* or *Inductive Assumption*.

(1) First, suppose that the statement S_{n_1}, called the *basis step of induction*, or just *the base* is valid. In applications of the method of mathematical induction, the verification of the basic step is an independent problem. In some problems, this step may be trivial, but it can never be skipped altogether.

(2) Second, suppose that for each integer $n \geq n_1$, that is, bigger than or equal to the basis value, we can prove a conditional statement $S_n \Rightarrow S_{n+1}$, that is, we can prove the validity of the hypothesis S_{n+1} for each specified natural $n > n_1$ assuming the validity of S_n, and this conditional statement is valid for all natural $n \geq n_1$. This part of the method is called the *inductive step*.

(3) If we can independently show these two steps, then the Principle of Mathematical Induction claims that all *infinitely many* of the statements S_n, for all integer $n \geq n_1$ are valid.

This method of proof is accepted as an axiom because nobody can actually verify infinitely many statements S_n, $n \geq n_1$; the method cannot be justified without using some other, maybe even less intuitively obvious, properties of the set of natural numbers. Mathematicians have been using this principle for centuries and never arrived at a contradiction. Therefore, we accept the method of mathematical induction without a proof, as a postulate, and believe that this principle properly expresses a certain fundamental property of the infinite set $\mathcal{N}$ of natural numbers.

We will apply the postulate (the axiom) of mathematical induction many times in the sequel chapters; the method will be employed in many proofs in this chapter; however, sometimes the method presents itself only implicitly, through some known results that have been already proved by the use of the mathematical induction. In the following problem, we give a detailed example of an application of the method of mathematical induction. We believe that this postulate is an intrinsic property of the set of natural numbers. When we apply this principle, we, for short, say that the proof was done *by induction*. First, we show how the method works in solution below, and than give the exact statement.

Solution of Problem 2. We show now how the method works. We have already formulated the inductive assumption:

$$S_n = \Sigma_{j=1}^n (2j-1) = 1+3+5+\cdots+(2n-1); n = 1,2,\ldots.$$

We can choose the obvious statement $S_1 : 1 = 1^2$ as a basis of induction. That establishes the first step, the basis of induction.

To make the inductive step, it is often useful, but not necessary! to denote the parameter as $k > 1$ (or any other convenient character) and write down

$$S_k : 1+3+\cdots+(2k-1) = k^2 \tag{1.2}$$

and

$$S_{k+1} : 1+3+\cdots+(2k-1)+(2k+1) = (k+1)^2. \tag{1.3}$$

To apply the *inductive reasoning*, we assume that the statement S_k holds good for any natural $k \geq 1$, and have to prove the validity of the statement S_{k+1}. It is crucial in the proof of the implication $S_k \Rightarrow S_{k+1}$ that we do not use any additional properties of k: for instance, we cannot suppose that k is an *odd* number; the proof must work for all the natural numbers.

Let us compare the left sides of equations (1.2) and (1.3): we observe that the former is a part of the latter. Moreover, its value is known by the inductive step; thus, we can rewrite the latter as

$$\{1+3+\cdots+(2k-1)\}+(2k+1) = k^2 +(2k+1).$$

Since we know that $k^2 + (2k+1) = (k+1)^2$, we arrive at (1.3).

We want to emphasize that we did not perform infinitely many computations; we replaced them by referring to the Axiom of Mathematical Induction. In what follows, we often say that we proceed "by induction." Important properties of the prime numbers, in particular, their applications to cryptography, will be studied in the following chapters. Now, we consider more examples, starting with the simplest.

Problem 3 *Prove that for every natural n, n = 1, 2, 3, ...,*

$$1^2 + 2^2 + \cdots + n^2 = \frac{1}{6}n(n+1)(2n+1).$$

Proof. Here $n_1 = 1$ and S_n stands for the equation above; thus, S_1 denotes the equation $1^2 = \frac{1}{6}1(1+1)(2 \cdot 1 + 1)$, which is certainly true, S_2 denotes $1^2 + 2^2 = \frac{1}{6}2(2+1)(2 \cdot 2 + 1)$, which is true as well; S_3 is also a valid statement $1^2 + 2^2 + 3^2 = \frac{1}{6}3(3+1)(2 \cdot 3 + 1)$. Therefore, we have the basis of induction; of course, it was enough to verify only one statement S_1. Now we have to validate the inductive step, that is, to prove the equation S_{n+1}, *assuming* that S_n is valid for some unspecified but fixed natural $n = k$. In this problem, we must prove the statement (the equation) S_{k+1}, which reads

$$S_{k+1}: \quad 1^2 + 2^2 + \cdots + k^2 + (k+1)^2 = \frac{1}{6}(k+1)(k+2)(2k+3)$$

assuming that S_k is valid, that is, using the equation

$$S_k: \quad 1^2 + 2^2 + \cdots + k^2 = \frac{1}{6}k(k+1)(2k+1)$$

as if it were correct. Its validity is unknown yet; however, in the procedure, we suppose it to be true. It is worth repeating that our reasoning must be valid for any natural number $n \geq 1$, that is, the reasoning can use only properties common to all natural numbers. For instance, we cannot assume that n is an even number.

Observe that the left-hand side of the equation

$$S_{k+1}: 1^2 + 2^2 + \cdots + k^2 + (k+1)^2$$

contains the left-hand side of S_k, and the latter in our inductive reasoning is considered to be known. This observation gives us the idea of the proof. Since we assume that the equation

$$S_k: \quad 1^2 + 2^2 + \cdots + k^2 = (1/6)k(k+1)(2k+1)$$

holds true, we employ S_k to transform the left-hand side of S_{k+1} as follows,

$$S_{k+1}: \quad \left(1^2 + 2^2 + \cdots + k^2\right) + (k+1)^2 =$$

$$= \left((1/6)k(k+1)(2k+1)\right) + (k+1)^2 = (1/6)(k+1)(k+2)(2k+3).$$

Thus, we have derived the statement S_{k+1} from S_k for an arbitrary fixed natural $k \geq 1$. Since we completed both steps of the Principle of Mathematical Induction, we claim that S_n is valid for all natural n.

Problem 4 *Set* $S_n = \dfrac{1}{1 \times 2} + \dfrac{1}{2 \times 3} + \dfrac{1}{3 \times 4} + \cdots + \dfrac{1}{n \times (n+1)}$. *Prove that for any* $n \geq 1$, $S_n = \dfrac{n}{n+1}$.

Proof. We leave a standard inductive proof to the reader, and give even simpler elementary proof without the induction to show that however powerful, the Method of Mathematical Induction is not the panacea. Indeed, since $\dfrac{1}{1 \cdot 2} = \dfrac{1}{1} - \dfrac{1}{2}$, $\dfrac{1}{2 \cdot 3} = \dfrac{1}{2} - \dfrac{1}{3}$, etc., the sum is telescoping, that is,

$$S_n = \frac{1}{1 \times 2} + \frac{1}{2 \times 3} + \cdots + \frac{1}{n \times (n+1)} =$$

$$= \frac{1}{1} - \frac{1}{2} + \frac{1}{2} - \frac{1}{3} + \cdots + \frac{1}{n} - \frac{1}{n+1} = 1 - \frac{1}{n+1} = \frac{n}{n+1}.$$

Problem 5 *Let the integer* $q = 12\,055\,735\,790\,331\,359\,447\,442\,538\,767$. *Prove that the integer* $991q^2 + 1$ *is a perfect square, and q is the smallest natural number with this property.*

This is a confirmation that the incomplete induction, that is, the reasoning, based on finitely many confirmed cases, does not work in mathematics; it can result in a wrong conclusion. Such reasoning, called *incomplete induction,* can lead to mistakes; it does not matter how many particular cases we have verified.

Problem 6 *Let $P(x)$ be a quadratic polynomial $P(x) = x^2 + x + 41$. Verify that the integers $P(0), P(1), P(2),\ldots, P(39)$ are the prime numbers, but $P(40)$ is composite.*

Proof. Now, $P(0) = 41$, which is prime. For a proof by induction, it is enough to have the base of one element 0, but in this example it is interesting to check a few more cases. Thus, $P(1) = 43$, $P(2) = 47$, $P(3) = 53$, and all these values are prime numbers. It is quite natural to assume that $P(k)$ is a prime number for every natural k. We continue the trials, $P(38) = 1523$ and $P(39) = 1601$, which are both prime numbers, thus confirming our guess. However, after that we get a surprise, since $P(40) = 1681 = 41^2$, thus, it is not prime. The latter means that the hypothesis "The number $P(n)$ is prime for every natural n" is wrong. The number, 40 in the example, is called a *counter-example* to our hypothesis.

Problem 7 *Find the polynomial $P(x)$ with integer coefficients, such that the values $P(x)$ are prime for $x = 0, 1, 2$. What is the degree of this polynomial?*

Solution. There are infinitely many such polynomials, for example, $p(x) = 2x^2 - 2x + 3$.

The following example shows that while performing the inductive step $S_k \Rightarrow S_{k+1}$, one cannot omit even *one* step, even one value k.

Example 1 *Prove that all ladies have blue eyes.*

Proof. Of course, everyone knows at least one lady with blue eyes – we fix her, and this is the base of induction. To make the inductive step, we have to prove that any group of $k+1$ ladies has the blue eyes, assuming that any group of k ladies has the same blue eyes. To do that, we select any group of $k+1$ ladies, select any lady G_1 in this group, and remove this lady from the group. By inductive assumption, the remaining k ladies all have the blue eyes. Now we return the lady G_1 into the group and separate another lady $G_2 \neq G_1$. All the ladies except G_2 also have blue eyes. Therefore, G_1 and G_2 also have the blue eyes, and all ladies have blue eyes.

Certainly, this is a joke,[4] and the actual problem is to find at what point this reasoning goes wrong.

The Principle of Mathematical Induction can be stated in several equivalent forms, such as the following one.

---

[4] Such a mathematical joke with a wrong proof is called a *sophism* or a *fallacy*.

The Axiom of Mathematical Induction in equivalent form. *Let S be some set of natural numbers. Let $0 \in S$ and if some natural number $k \in S$, then also the following natural number $k + 1 \in S$. Then S is the set of all natural numbers $\{0, 1, 2, \ldots\}$.*

Problem 8 *Prove that in every problem above one can apply any of the equivalent forms of the axiom of induction.*

Problem 9 *Prove that $n - 3$ diagonals divide an $n - gon$, that is, the polygon with n sides (not necessarily convex) into $n - 2$ parts.*

Proof. It is convenient in this problem to start at $n = 3$. Thus, a polygon is a triangle, which has no diagonal, and consists of $3 - 2 = 1$ part, that establishes the basis of induction. Now assume that for all the $k - gons$ the statement is true, and consider any polygon P with $k + 1$ sides. Any its diagonal d splits P into two smaller polygons with a k_1 and a k_2 sides, respectively, where $k_1 + k_2$ counts d twice. Thus, $k_1 + k_2 = k + 1 - 1 = k$. On the other hand, the total number of parts is $k_1 - 1 + k_2 - 1 + 1 = k_1 + k_2 - 1 = k - 1 = (k + 1) - 2$.

The following example shows that the induction can be employed to prove inequalities as well.

Problem 10 *Prove that $\dfrac{1}{\sqrt{1}} + \dfrac{1}{\sqrt{2}} + \dfrac{1}{\sqrt{3}} + \cdots + \dfrac{1}{\sqrt{n}} > \sqrt{n}$.*

Problem 11 *The next is called The Strong Mathematical Induction Principle. Let S be some set of natural numbers and $0 \in S$. Moreover, if for any natural number k, the natural numbers $0, 1, 2, \ldots, k - 1, k \in S$, then also the following natural number $k + 1 \in S$. Then S is the set of all natural numbers $\{0, 1, 2, \ldots\}$.*

The assumptions here seem to be stronger than in the standard principle above, since we assume something not only about k, but also about all smaller natural numbers $j \leq k$. Thus the statement looks weaker. But both principles are equivalent.

Indeed, prove that the Strong Principle of Mathematical Induction is equivalent to the Principle of Mathematical Induction in standard form.

The natural numbers satisfy the Axiom of Mathematical Induction. Moreover, they have a more general property, the Well-Ordering Principle; it considered in more detail, for example, in [31, p. 20].

1.2   FACTORIALS AND THE STIRLING FORMULA

In the previous computation, we had to multiply *consecutive natural numbers*. This procedure occurs so often that it deserved its own name and symbol.

Definition 2 *The product of n consecutive natural numbers from 1 through n inclusive is called the n factorial and is denoted as n!.*

For example, $1! = 1, 2! = 2 \times 1 = 2, 3! = 3 \times 2 \times 1 = 6$; if $k < n$, then $\frac{n!}{k!} = (k+1)(k+2)\cdots n$. If we want to preserve this property for $k = 0$, it is natural to *define* $0! = 1$.

The symbol $n!$ does not look very impressive for small n but let us study it more carefully. Already $10! = 3,628,800$, and we observe that the factorials grow very fast. James Stirling (1692–1770), the younger contemporary of Newton (1643–1727), showed the asymptotic formula to be proved in Lemma 2,

$$n! \approx \left(\frac{n}{e}\right)^n \sqrt{2\pi n}, n \to \infty. \tag{1.4}$$

The letter e is a standard symbol for the famous real number $e \approx 2.7$, which will be discussed below. The symbol $\approx$ and name here mean that the ratio of the left- and right-hand sides of the formula tend to 1 as $n \to \infty$. This formula, without the precise value of the constant, was known to Abraham de Moivre (1667-1754) even before Stirling.

Problem 12 *Approximate* 10! *by formula (1.4) and compare with the exact value given above.*

Solution. By Stirling's formula, $10! \approx 3.6 \times 10^6$, with the relative error less than 1%. If n is increasing, the relative error is even smaller.

Problem 13 *Compute* $\frac{(n+1)!}{(n-1)!}$. *For what natural numbers is this expression defined?*

Solution. For $n - 1 \geq 0$, hence $n \geq 1$.

Problem 14 *How many digits are in the number* 100!?

Solution. Of course, we are not going to compute the huge number 100! precisely and count the digits. It is much easier to use a calculator and get $\log_{10}(100!) \approx 157.97$. Hence,

$$10^{157} < 100! < 10^{158},$$

so that the decimal number 100! has 158 decimal digits.

1.3   RECURSIVE DEFINITIONS

The definition of the factorial function can be written as

$$n! = n \times (n-1)!.$$

Such a definition is called *recursive* because at the second step, it returns to the same definition, but with a smaller value of the parameter. Indeed, we compute the $n-$ factorial through the $(n-1)-$ factorial. Recursive definitions are often used in computer science and mathematics. As another example, let us consider a recursive definition of integer powers $pow(a,n)$, $a \neq 0$, that can be defined for any natural n as $pow(a,0) = 1$ and $pow(a, n+1) = a \times pow(a,n)$.

The definitions of well-known arithmetic and geometric progressions (sequences), namely,

$$a_{n+1} = a_n + d,$$

where a_0 is the initial term and d, called the difference, are given numbers, and

$$a_{n+1} = a_n \times q; a_0 \text{ and } q \text{ are given,}$$

are also recursive definitions.

Problem 15 *List the first five terms of the arithmetic progression with the first term $a_0 = 1$ and the difference $d = -2$. Prove that the terms of any arithmetic progression satisfy $2a_{n+1} = a_n + a_{n+2}$*

$$a_{n+k} = a_n + k \cdot d$$

$$\sum_{k=0}^{r} a_{n+k} = (r+1)a_n + \frac{1}{2}r(r+1)d.$$

Solution. To prove the first equation, it is enough to employ the definition $a_{n+1} = a_n + d$ and iterate it, $a_{n+2} = a_{n+1} + d = a_n + 2d$.

Problem 16 *Given $a_0 = A, a_1 = B$, where A and B are some constants, and $a_n = 5a_{n-1} - 6a_{n-2}$ where $n \geq 2$; find an explicit expression for $a_n, n = 2,3,\ldots$.*

Solution. Look for the solution as $a_n = \alpha\, 2^n + \beta\, 3^n, n = 0,1,2,\ldots$, where the constants α and β are the roots of the quadratic equation $t^2 - 5t + 6 = 0$. The conditions for a_0 and a_1 give $\alpha = 3A - B$ and $\beta = B - 3A$.

This solution can be easily generalized for any homogeneous *difference* equations with constant coefficients and with a certain non-homogeneity. The method is similar to the case of linear differential equations with constant coefficients, see, for example, [28, Sect. 4.4].

Problem 17 *Find explicitly the Fibonacci numbers, which satisfy the difference relation $f_{n+2} = f_{n+1} + f_n$ and the initial conditions $f_0 = f_1 = 1$.*

Solution. The relation and the initial conditions lead to the quadratic equation $t^2 = t + 1$ with the irrational roots $\frac{1}{2}(1 \pm \sqrt{5})$ and the formula for the Fibonacci numbers

$$f_n = \frac{3 + \sqrt{5}}{2\sqrt{5}}\left(\frac{1}{2}(1 + \sqrt{5})\right)^n + \frac{\sqrt{5} - 3}{2\sqrt{5}}\left(\frac{1}{2}(1 - \sqrt{5})\right)^n, n = 0,1,2,\ldots.$$

This formula, containing radicals, gives for every natural n integer values. For example, $f_0 = 1, f_1 = 1, f_2 = 2, f_3 = 3, f_4 = 5$, etc.

To prove that a recursive definition returns the quantity it is supposed to, one *must* use the mathematical induction. For instance, in the example above, the zeroth part of the definition, $pow(a,0) = 1$, must be used as a basis of induction. Then, the inductive step is justified by the second part of the recursive definition, $pow(a, n+1) = a \times pow(a, n)$. Any rigorous exposition, which includes a recursive definition or a recursive procedure, must be accompanied by an inductive proof of its validity.

Problem 18 *Find the recursive formula for the general term of a sequence, if it starts with $a_0 = 1$, and every subsequent term is 3 more than twice the previous term.*

Solution. $a_0 = 1$, $a_n = 3 + 2a_{n-1}$ for $n \geq 1$.

Problem 19 *Does the sequence $a_n = n^2 + 4, n = 0,1,\ldots$, satisfy the recurrence relation $a_n = 4a_{n-1} - 2$?*

Solution. No, since in general $4[(n-1)^2 - 22 \neq n^2 + 4$.

Functions of two or more variables can also be defined recursively. A common example is the Ackermann function $A(m,n)$, defined as follows:

$$\begin{cases} A(0,n) = n+1 & \text{if} \quad m = 0 \\ A(m,0) = A(m-1,1) & \text{if} \quad m \neq 0 \text{ and } n = 0 \\ A(m,n) = A(m-1, A(m,n-1)) & \text{if} \quad m \cdot n \neq 0. \end{cases}$$

Problem 20 *Compute the values* $A(0,0), A(0,1), A(1,0), \ldots, A(10,10)$.

Problem 21 *Give a recursive definition of the set of pairs of positive integers whose sum is odd.*

1.4   ELEMENTARY FUNCTIONS

The next few pages contain a very brief survey of the basic elementary functions[5] – Power, Exponential, Logarithmic, and Trigonometric Functions. If the reader is familiar with that material, she can safely skip it and go to the next lecture. However, we know from the experience that many students, especially at the community colleges, know (if any) this stuff insufficiently, that is why it is included here.

Consider a quadratic equation $x^2 = 3$. It has two real roots, $\pm\sqrt{3}$. A similar equation $x^2 = -3$ has no real solution, but if we consider it over the larger set of complex numbers, the equation has two roots. The reason for that is that the map $y = x^2$ for real x is not a surjection, that is, given a y, we not always can return to x. This is a very common problem, and we address it now.

First, we consider bijective maps and let $f: X \to Y$ be bijective. This means that for every element $y \in Y$ there exists one and only one element $x = x_y \in X$ such that $y = f(x)$. Now we construct the map $g: Y \to X$ as follows. For every $y \in Y$ we set $g(y) = x_y$, where x_y has been just defined. Since the element y_x was defined uniquely, we uniquely defined the map $g: Y \to X$. By our construction, the map g has the following properties.

The domain of g is Y and the co-domain is X. For each $x \in X$,

---

[5] Here "elementary functions" only means that some features of these functions are studied at high school.

$$g \circ f(x) = g(f(x)) = g(y) = x \text{ and } f \circ g(y) = f(g(y)) = f(x_y) = y, \qquad (1.5)$$

therefore,

$$g \circ f = I_X \text{ and } f \circ g = I_Y. \qquad (1.6)$$

The map g is called the *inverse map* of the map f (because it is unique) and is denoted $g = f^{-1}$. If the inverse map exists, f is called *invertible*. Let us repeat that the domain of f coincides with the co-domain of g and vice versa; the co-domain of f coincides with the domain of g. It is clear also that in this case, the inverse map, g is also bijective and invertible, and we can write $f = g^{-1}$. We restate the conclusion of this argument in the following theorem, where the uniqueness obviously follows from the construction of the inverse map.

Theorem 1 *Any bijective map has the unique inverse map, which is also bijective and satisfies the equations (1.5)–(1.6).*

Theorem 1 says, in particular, that a non-bijective function cannot have the inverse function. However, there may exist one-sided (from left or from right) invertible functions. We proceed as one often does in mathematics: namely, we use the properties, which have been proven in a particular case (in this case, equations (1.5) – (1.6) valid for bijective maps) as the definitions in general situation.

Definition 3 *Consider two arbitrary sets X and Y, and an arbitrary map $f : X \to Y$; f is called left-invertible if there exists a map $f_l^{-1} : Y \to X$, such that $f_l^{-1} \circ f = I_X$; the map f_l^{-1} is called a left-inverse map of f.*

A map f is called right-invertible if there exists a map $f_r^{-1} : Y \to X$, which is called a right-inverse map for f, such that $f \circ f_r^{-1} = I_Y$. The name (left or right) is given according to the position with respect to the given map f. Map f itself is a right-inverse for f_l^{-1} and is a left-inverse for f_r^{-1}.

Theorem 2 *Criterion of the unilateral invertibility. A map $f : X \to Y$ is left-invertible iff it is injective. A left-inverse map exists iff $|X| \leq |Y|$. It does not have to be unique. It is unique if either f is bijective or the domain X contains only one element x_0.*

A map $f : X \to Y$ is right-invertible iff it is surjective. It exists iff $|X| \geq |Y|$. A right-inverse map may not be unique. It is unique if f is bijective or the co-domain Y is a one-element set.

A map $f : X \to Y$ is invertible (from both sides) iff it is bijective. In this case, f is both left- and right-invertible or $|X| = |Y|$, the left-inverse map and the right-inverse map are unique, coincide with one another and both are equal to the inverse map f^{-1}.

Proof. Let f be left-invertible and $Y_1 \subset Y$ be its range. The map $f_1 : X \to Y_1$ is bijective and $\forall x \in X (f(x) = f_1(x))$. Therefore, f is injective. Vice versa, if f is injective, the function $f : X \to Rang(f)$ is a bijection; thus, it has an inverse map. Extending its definition onto $Y \setminus Rang(f)$ in either way, for example, defining $f_l^{-1}(y) = x_0$ for every $y \in Y \setminus Rang(f)$, where x_0 is an arbitrary element of X, we derive a left-inverse map to f. It is obvious that any such extension generates a left-inverse of f.

Problem 22 *Finish the proof of Theorem 2*

1.4.1 Quadratic Functions

Let $X = Y = \mathcal{R}$, and $f : X \to Y, y = f(x) = x^2$. We know already that this f is neither injective nor surjective, thus, it has no unilateral inverse, on ether side. To proceed, we should reduce either its co-domain or domain, or both. First we reduce the co-domain by removing the negative numbers and get a surjective map $g : \mathbb{R} \to \mathbb{R}_+$, where $\mathbb{R}_+ = [0, \infty)$. This g has at least one, maybe many right-inverse maps. One of them is $g_1 : \mathbb{R} \to \mathbb{R}_+$, $g_1(y) = \sqrt{y} = x$, where $y = x^2$. We can also define $g_2(y) = -\sqrt{y} = x$, where $y \ge 0$ and, as before, $y = x^2$. We can also scramble these two branches, say, use one for $0 \le x \le 2$ and use another for $x > 2$.

Problem 23 *Verify that for all right-inverses g_r of g, $g \circ g_r = I_Y$.*

Next, we consider function $h : \mathbb{R}_+ \to \mathbb{R}$ given by the same formula $h(x) = x^2$. It is an injection, and therefore it has left-inverse maps. One of them is $h_1(y) = \sqrt{y}, y \ge 0$, but we also must define $h_1(y)$ for $y < 0$, for example, as $h_1(y) = -5$.

Problem 24 *Define another left-inverse function for $h(x)$ and verify that for all left-inverses h_l, $h_l \circ h = I_X$.*

The map $f(x) = x^2$ for $X = Y = \mathbb{R}_+$ is a bijection, so that it is invertible. Its inverse is a bijection too, and we have $\sqrt{x} \ge 0$ for all $x \ge 0$. This function is the main ingredient of all unilateral inverses above. It is called the *principal square root* of a positive number x.

1.4.2 Exponential and Logarithmic Functions

If three real numbers b, x, y satisfy the equation $b^x = y$, then b is called the *base*, x is the *exponent*, and y the *power*. For example, $2^3 = 8$, while $3^2 = 9$. If the base b is fixed, $y = y(x)$ becomes a function of x, it is called the *exponential* function (to the) base b. Experimenting with convenient values, for example, with $b = 2$, we conclude that x can be any real number, but y is always positive, even if x is negative, say, $2^{-3} = \dfrac{1}{2^3} = \dfrac{1}{8}$. This statement is deep; its proof requires some mathematical analysis. Actually, for different bases b there are *different* exponential functions $y = b^x$, however, their properties are very similar, that is why you can often read about "the exponential function."

The next statement summarizes the basic properties of the exponential function. We consider them only when the base $b > 1$ or $0 < b < 1$; when $b = 1$, we have the constant function $1^x \equiv 1$, which is in a sense trivial. What is more, if $b = 1$, the values of $y = 1^x = 1$ are the same for different values of x, thus, the function is not injective. The case $b < 0$ necessitates consideration of the *complex numbers*, which is beyond this book.

Theorem 3 *Let the base* $b \in (0,1) \cup (1,\infty)$. *The domain of the exponential function* $y = b^x$ *consists of all real numbers, and its range consists of all the positive real numbers. If* $b > 1$, *then it is a continuous positive monotonically increasing function, if* $0 < b < 1$, *it is monotonically decreasing. Therefore, as a function from* $\mathbb{R}$ *to itself, it is injective. Moreover, as a function from* $\mathbb{R}$ *to strictly positive real numbers,* $\mathbb{R}_+ = (0, +\infty)$ *it is a bijection and, therefore, it has the inverse function. The following properties follow from the properties of the exponents,*

$$b^{x_1 + x_2} = b^{x_1} \times b^{x_2}; \quad b^{x_1 - x_2} = b^{x_1} \div b^{x_2}; \quad (b^x)^k = b^{kx}.$$

Problem 25 *Graph the 16 exponential functions* $y(x) = \pm 3 \pm 2 \cdot 5^{\pm(1/4)x \pm 1}$ *and* $y(x) = \pm 3 \pm 2 \cdot (1/5)^{\pm(1/4)x \pm 1}$.

Consider the exponential function $y(x) = b^x$ with the positive base $b \neq 1$. Therefore, this function is monotone, and so that, it is *injective* if the domain consists of all real numbers, or it is even bijective, if we consider only the positive real numbers as its domain. Either way, this exponential function has the (left) inverse, called the *logarithmic* function and denoted as

$$y(x) = \log_b(x)$$

or just $\log_b x$.

The definition and Theorem 3 imply the following features of the logarithmic function.

Theorem 4 *The domain of logarithmic function $y = \log_b x$ consists of positive real numbers, and the range consists of all real numbers. If $b > 1$, then, it is a continuous monotonically increasing function, for $0 < b < 1$ it is monotonically decreasing. It is (left) inverse of the exponential function, namely, $\log_b b^x = x$ for all real x, while $b^{\log_b x} = x$ for $x > 0$. Next,*

$$\log_b(x_1 \cdot x_2) = \log_b x_1 + \log_b x_2$$

$$\log_b(x_1 \div x_2) = \log_b x_1 - \log_b x_2$$

$$\log_b(x^k) = k \log_b x.$$

Lemma 1 *Change of Base formula. For any positive $a, b, c, b \neq 1, c \neq 1$ it is hold*

$$\log_b a = \frac{\log_c a}{\log_c b}.$$

Problem 26 *Prove the statements of this section.*

Next, we show a few simple applications of these functions to *compound interest* problems.

Problem 27 *On January 1, you have the balance of P dollars in your bank account; this amount is called the principal P. Since you keep your money in this account and allow the bank to use this money, once or several times a year, the bank pays you some small amount, called the interest p, and deposits this interest into your account, thus adding to the current amount. If the interest equals p% of the current balance, this method is called the compound interest. In other words, the bank pays "interest on interest." Here p is the annual interest, and if the bank pays the interest k times per year, each payment is equal $(p/k)\%$ of the annual interest. Let $P = \$500$, $p = 5\%$, and $k = 2$, that is, the interest is paid semi-annually. What is the balance on the account after the five full years will expire?*

Problem 28 *A bank pays compound interest k times per year. How much time is it necessary to double the balance in your account?*

Problem 29 *Let x be a real number and k be an integer, such that $2^k \leq x < 2^{k+1}$. Prove that $[\log_2 x] = k$.*

Problem 30 *If $n > 1$ is an odd number, then $[\log_2(n-1)] = [\log_2(n)]$.*

1.4.3 Trigonometric Functions

Consider the function $y = f(x) = \sin x$, where $X = Y = \mathbb{R}$. This is a periodical function with the smallest positive period 2π and with the range $[-1,1]$, therefore, it is not injective nor surjective and has neither left, nor right inverse. However, if we restrict it to the set $[-\pi/2, \pi/2]$, the restricted function is monotone and takes on in this interval every its possible value from -1 to 1 exactly once. Thus, we consider the restricted function $\sin_0 : [-\pi/2, \pi/2] \to [-1,1]$. This is a bijective function, so that it has the inverse map, which is denoted as $y = \arcsin x$, or (in particular, on calculators) as $y = \sin^{-1} x$. This function is the main part of all the inversions of the sin function.

Since the function $y = \arcsin x$ is bijective, when $x \in [-\pi/2, \pi/2]$ and $y \in [-1,1]$, it has both left- and right-inverse maps, and we have

$$\arcsin(\sin x) = x, \forall x \in [-\pi/2, \pi/2] \tag{1.7}$$

$$\sin(\arcsin x) = x, \forall x \in [-1,1]. \tag{1.8}$$

By tradition, we denoted the argument in the second equation as x, even though visually it runs over the y – axis.

Consider a surjective function $y = \sin_0(x) = \sin x, ; x \in \mathbb{R}, -1 \leq y \leq 1$. Again, there are infinitely many different right-inverse functions, one of them is $y = \arcsin x : x \in \mathbb{R}, y \in [-1,1]$. We have $\sin(\arcsin x) = x$ – but only for $x \in [-1,1]$. Outside of this set the expression $\arcsin x$ is undefined (as a real number).

Finally, we consider an injective function $f(x) = \sin x, x \in [-\pi/2, \pi/2], y \in \mathbb{R}$. It is injective but not surjective, thus, it is left-invertible, and we define one of its left-inverses as $\phi(x) = \begin{cases} \arcsin x, & x \in [-1,1] \\ 0, & |x| > 1 \end{cases}$. Equation $\phi \circ f = I_X$ becomes now (1.7). It is worth noticing that the left-hand side of (1.7), namely, $\arcsin(\sin x)$ is defined now for all real x, and not only for $[\pi/2, \pi/2]$, however, for $|x| > \pi/2$ formula (1.7), in general, fails. For instance, if $[x \in [\pi/2, 3\pi/2],$ then $\arcsin(\sin x) = \pi - x.$

Problem 31 *Compute* $\arcsin(\sin x)$ *for all real* x.

Solving various equations is one of the most common mathematical tasks, and the results of this section clarify what can be expected while solving the equations. In general, consider an equation $y = f(x)$, $x \in X$, $y \in Y$. If f is surjective, then the equation has a solution for every $y \in Y$, and this solution is generally not unique. If f is injective, then the equation has the unique solution for every $y \in Rang(f)$; however, for $y \in Y \setminus Rang(f)$ the equation has no solution. If f is bijective, then the equation has the unique solution for every $y \in Y$. Obviously, these theorems are pure *existence theorems* – they contain no algorithms for constructing a solution in particular cases.

Problem 32 *Are the following functions bijective? Injective? Surjective? Neither? If possible, construct left- or right- or inverse functions.*

1. $f : \mathbb{R} \to \mathbb{R}, f(x) = \exp x$

2. $f : \mathbb{R} \to (0, \infty), f(x) = \exp x$

3. $f : (0, \infty) \to (0, \infty), f(x) = \exp x$

4. $f : (0, \infty) \to \mathbb{R}, f(x) = \log x$

5. $f : (e, \infty) \to \mathbb{R}, f(x) = \log(1 + x)$.

Problem 33 *Prove that the set of polynomials with integer coefficients is countable, that is, can be put in a one-to-one correspondence with the set of natural numbers.*

1.5   STIRLING ASYMPTOTIC FORMULA

We consider now quantities that depend upon integer parameters and can take on natural values, that is, the functions from the set of the integers to the set of natural numbers. We study these functions or maps (mappings), often called *sequences,* in detail later on. The reader can now skip the end of this lecture, which includes a small foray into analysis, and return to it later, as necessary.

If there are two variables, say two sequences $a(n)$ and $b(n)$, depending upon a parameter[6] $n \to \infty$, then it is written $a(n) \sim b(n)$ as $n \to \infty$. The

---

[6] The symbol ∞ is not a number, the writing $n \to \infty$ means that the parameter n becomes and then remains bigger than any positive number.

notation[7] $a(n) = O(b(n))$, $n \to \infty$, means that $|a(n)/b(n)|$ is bounded for all $n \geq n_0$. The notation[8] $a(n) = o(b(n))$, $n \to \infty$, means that $|a(n)/b(n)| \to 0$ as $n \to 0$.

It is also convenient to define two integer-valued functions from $\mathcal{R}$ to $\mathcal{Z}$, where $\mathcal{R}$ and $\mathcal{Z}$ are the sets of the real numbers and the integers, respectively.

Definition 4 *The floor function $\lfloor x \rfloor$, called also the integer part $[x]$ of x, or $E(x)$ (from French "Entier"), denotes the largest integer which is smaller than or equal to x, that is, $[x]$ is always an integer and $[x] \leq x$; for example, $[-3] = -3$ and $[-\pi] = -4$. Similarly, the smallest integer, which is no less than x, is called the ceiling function and is denoted as $\lceil x \rceil$; for example, $\lceil -3 \rceil = -3$ and $\lceil -e \rceil = -2$, where $e \approx 2.71828$ is the Napierian number.*

Problem 34 *Compute $\lfloor 3.5 \rfloor$, $\lfloor -\pi \rfloor$, $\lceil 3.5 \rceil$, $\lceil -\pi \rceil$.*

Solution. $\lfloor 3.5 \rfloor = 3$, $\lfloor -\pi \rfloor = -4$, $\lceil 3.5 \rceil = 4$, $\lceil -\pi \rceil = -3$.

Problem 35 *(1) Graph the functions $[x]$ and $\lfloor x \rfloor$.*

(2) Prove that $x - 1 < \lceil x \rceil \leq x \leq \lfloor x \rfloor < x + 1$. Study the equality cases.

(3) Prove that $\lfloor 2x \rfloor = \lfloor x \rfloor + \lfloor x + \frac{1}{2} \rfloor$.

(4) Prove that $\lfloor x \rfloor = -\lceil -x \rceil$ for every real x.

(5) Prove that $\lceil x \rceil = \lfloor x \rfloor + 1$ if x is not integer, and $\lceil x \rceil = \lfloor x \rfloor$ if x is an integer.

(6) Prove that the number of multiples of a natural number k among the integers $\{1, 2, \ldots, n\}$ is given by the floor function $\lfloor n/k \rfloor$. Let us notice that the integer part $\lfloor n/k \rfloor = 0$ if $0 \leq n < k$. What is this quantity among the integers $\{m, m+1, \ldots, n\}$, where $m < n$ is any integer?

Problem 36 *How many integers are between 1 and 6 inclusive?*

Solution. One immediately sees that the answer is 6.

Problem 37 *How many integers are between −8 and 6 inclusive?*

Solution. Now the answer consists of three addends: 6 integers for the positive numbers, 8 for the negative numbers, and 1 for the 0, thus the final answer is $8 + 1 + 6 = 15$, which can be written as $6 - (-8) + 1 = 15$.

---

[7] Called "Big O."

[8] Called "Small o."

Problem 38 *How many multiples of a natural number p are between m and n, inclusive, where $m \leq n$ are the integers?*

Solution. The fraction n/p is not always integer, hence the answer includes the floor and/or ceiling function. With this observation, you can immediately check that the answer is

$$\left\lfloor \frac{n}{k} \right\rfloor - \left\lfloor \frac{m}{k} \right\rfloor.$$

Problem 39 *How many integers between 4 and 44 are prime?*

Problem 40 *Check whether the following equations are true or false.*

1) $2 \cdot \lfloor 3/2 \rfloor = 3$; *2)* $\lceil 5/3 \rceil \cdot \lceil 3/5 \rceil = 1$; *3)* $[3.5+1.5] = [3.5]+[1.3]$.

Lemma 2 $n! \approx \left(\dfrac{n}{e} \right)^n \sqrt{2\pi n}, n \to \infty.$

Proof. Hereafter, the symbol log always means *natural logarithms*, that is, logarithms are to the base e, the Napier number $e \approx 2.7182818....$ We have

$$\log n! = \log 1 + \log 2 + \cdots + \log n =$$

$$= \int_{1/2}^{n} \log x d[x] = \int_{1/2}^{n} \log x d([x] - x - \frac{1}{2}) + \int_{1/2}^{n} d(x+1/2).$$

Integrating by parts and considering that $|x - [x] - 1/2| \leq 1/2$, we finish the proof.

Problem 41 *Stirling formula can be made more precise and include the decaying terms of orders $n^{-1}, n^{-2}, \ldots, n \to \infty$. Prove that*

$$n! \approx \left(\frac{n}{e} \right)^n \sqrt{2\pi n} \left(1 + \frac{1}{12(n+1)} + O(n^{-2}) \right), n \to \infty.$$

and

$$n^n \sqrt{2\pi n} \exp\left(-n + \frac{1}{12n} - \frac{1}{360n^3} \right) \leq n! \leq n^n \sqrt{2\pi n} \exp\left(-n + \frac{1}{12n} \right).$$

Problem 42 *Compute the precise values of 5! and 10!, and compare with their Stirling approximations in Lemma 2.*

Solution. $5! = 1 \cdot 2 \cdot 3 \cdot 4 \cdot 5 = 120$ and $10! = 3,628,800$. By Lemma 2, $5! \approx 118.02$ and $10! \approx 3.599 \cdot 10^6$.

Definition 5 *The "double factorial" is the product of consecutive natural numbers from 1 to n of the same parity as n. For example,*

$$7!! = 7 \times 5 \times 3 \times 1 = 105 \text{ and } 8!! = 8 \times 6 \times 4 \times 2 = 384.$$

Problem 43 *Define the double factorial as recursive function.*

Problem 44 *Modify the Stirling asymptotic formula for the double factorials.*

1.6   EXERCISES

Exercise 1.1 *Give a formal definition of the odd and of the even integers in terms of their remainders after divisibility over 2.*

Exercise 1.2 *Prove that there are infinitely many prime numbers, as well as infinitely many composite numbers.*

Exercise 1.3 *What is the parity of the result, if one adds, subtracts, multiplies or divides two even integers, or two odd integers, or an even and an odd integer numbers?*

Exercise 1.4 *Prove that* $\Sigma_{k=0}^{m} k = \dfrac{m(m+1)}{2}.$

Exercise 1.5 *Prove that* $\Sigma_{k=1}^{p} k(2k-1) = \dfrac{p(p+1)(4p-1)}{6}.$

Exercise 1.6 *Find the polynomial $P(x)$ with integer coefficients, which values are prime for $x = 0,1,2,\ldots,999,1000$, but $P(1001)$ is a composite number. What is the degree of this polynomial?*

Exercise 1.7 *Find the polynomial $P(x)$ with integer coefficients, which is prime for $x = 0,1,2,\ldots,L$, where L is any natural number, but $P(L+1)$ is composite. What is the degree of this polynomial?*

Exercise 1.8 *Prove that* $1^3 + 2^3 + 3^3 + \cdots + n^3 = \left(\dfrac{n(n+1)}{2} \right)^2.$

Exercise 1.9 *Prove that* $1 \cdot 2 \cdot 3 + \cdots + n(n+1)(n+2) = \dfrac{1}{4} n(n+1)(n+2)(n+3).$

Exercise 1.10 *Prove that* $\dfrac{1}{1 \cdot 3} + \dfrac{1}{3 \cdot 5} + \dfrac{1}{5 \cdot 7} + \cdots + \dfrac{1}{(2n-1)(2n+1)} = \dfrac{n}{2n+1}.$

Exercise 1.11 *Is the well-ordering property valid for the set of integers?*

Answer. No.

Exercise 1.12 *Estimate approximately 30! by formula (1.4).*

Solution. By Stirling's formula, $30! \approx 2.64 \times 10^{32}$.

Exercise 1.13 *There are five bags and five keys, but nobody knows from what bag is a certain key. It takes one minute to try a key with all bags. How long will it take, in the worst case, to find the key to every bag?*

Solution. $5! = 120$ min $= 2$ h.

Exercise 1.14 *If an integer n is a multiple of 4, than $2^n \equiv 1 (\mathrm{mod}\, 1)5$.*

Solution. $2^n = 2^{4m} = (15+1)^m = 15l + 1$.

Exercise 1.15 *Compute precisely $20! \div 5$ and compare the calculations with that given by the Stirling formula.*

Exercise 1.16 *Verify that 10! has 7 decimal digits.*

Solution. A "mechanical" solution is to calculate 10! precisely, but that is impossible for large umbers. Another approach is outlined in Problem 14.

Exercise 1.17 *List the first five terms of the geometric progression with the initial term $a_0 = 1$ and the common ratio $q = 2$. Prove that the terms of any geometric progression satisfy $a_{n+1}^2 = a_n \cdot a_{n+2}$, $a_{n+r} = a_n \cdot q^r$, and $\sum_{k=0}^r a_{n+k} = a_n \dfrac{q^{r+1}-1}{q-1}$ if $q \neq 1$. Derive the formula for the sum of geometric progression if $q = 1$.*

Exercise 1.18 *Give the recursive definitions of the following sequences.* $(a)\, 1,3,5,1,3,5,1,\ldots$ $(b)\, 1,4,9,16,25,\ldots$ $(c)\, 1,2,3,5,8,13,21,\ldots$

Solution. If we start the numbering with 0, then $a_{3k} = 1$ for $k = 0,1,2,3,\ldots$, $a_{3k+1} = 3$, and $a_{3k+2} = 5$. If we start the numbering with 1, then there must be an obvious shift.

Exercise 1.19 *Compute a_0, a_{13}, a_{k-1} and a_{k^2}, if a_n are terms of the sequence with the general term $a_n = 4n - 3$.*

Solution. $a_0 = -3, a_{13} = 49, a_{k-1} = 4k - 7, a_{k^2} = 4k^2 - 3$.

Exercise 1.20 *Does the sequence $a_n = 2^{n+1} + (-1)^n$, $n = 0,1,\ldots$, satisfies the conditions $a_0 = a_1 = 3$ and the recurrent relation $a_n = a_{n-1} + 2a_{n-2}$?*

Exercise 1.21 *Define the functions $f_+(n) = n+1 : \mathbb{N} \to \mathbb{N}$ and*

$f_-(n) = \max\{0; n-1\} : \mathbb{N} \to \mathbb{N}$. Calculate them for $0 \leq n \leq 5$. Show that f_+ is injective, f_- is surjective, but none of them is bijective.

Exercise 1.22 *Find the inverses of the functions*

$f(x) = 1/x : (0,\infty) \to (0,\infty), f(x) = 1/x$ and $g(x) = 2 - x : \mathbb{R} \to \mathbb{R}.$

Solution. $f^{-1}(x) = 1/x; g^{-1}(x) = 2 - x.$

2

PROPOSITIONAL ALGEBRA

2.1   PROPOSITIONS

From an "applied" point of view, mathematics develops mathematical models of various phenomena. For example, if you want to re-carpet a room, with its length and width, in feet, being l and w, and the price of one square foot of new carpet to be $\$(p)$, then the cost of a new carpet is $C = \$(p \cdot l \cdot w)$. This equation, however simple it is, gives the *mathematical model* of the problem. In this lecture, we develop a mathematical model of a simple human language. The model, which describes only some basic formal features of the language, was developed by George Boole (1815–1864) more than a century and a half ago, and for a century, it was thought to be useful only for solving puzzles. But now, every computer employs this mathematical theory, called Algebra of Logic or algebra of propositions.

Mathematical logic is the art and the skill of correct, convincing reasoning. The name tells that we use it to construct our arguments using some formal methods. Reasoning first appears in our brain, and even there, we quickly resort to some usual human language, for example, to English. What is more, to communicate the ideas and arguments to other people, we have no choice but to use a live human language, oral or written, so that the very first, and one of the important goals of mathematical logic is to build a formal model of a human language.

Our speech consists of sentences. In grammar, we consider declarative (narrative), interrogative, and exclamatory sentences. Nonetheless, if a reader has any experience in studying mathematical texts, she has already noticed that mathematical books have very few interrogative or exclamatory sentences. Of course, mathematics gives rise to strong emotions and impulses to our feelings, but these emotions are mostly hidden behind words and formulas. Thus, in this textbook, we use only declarative sentences and called them *propositions*. We denote propositions by lower-case Latin letters $p, q, r, \ldots, a, b, \ldots$, sometimes with subscripts, and use these symbols to develop the *algebra of propositions*.

We do not study here the internal structure of the sentences, how they are built from nouns, verbs, and other words. Moreover, we do not consider the meaning of the sentences. As in grammar, we consider simple and compound sentences. The latter are formed out of the simple sentences by making use of different *connectives*, such as NO, OR, AND, IF ... THEN, XOR, etc. We assume that the simple sentences have no internal structure, and construct the mathematical models of these and some other connectives—those models are called *logical operations*.

Consider two sentences, *John loves Catty* and *Catty is loved by John*. Literally, they are distinct, but they express the same idea, have the same meaning. We can write down many other sentences carrying the same sense. Speaking about propositions, we keep in mind the meaning rather than the particular words the meaning is expressed in. Thus, despite many different words and meanings, we accept that there are only *two*[1] different *truth values*: disregarding its actual meaning, every proposition is either *true* or *false*, but not both. We reiterate this crucial property next.

A proposition *is a meaningful declarative sentence, having one and only one (so that not both simultaneously) of the two truth values—it is either true or false, but not both.*

We shall denote the truth values by the capital letters T (for Truth) and F (for Falsity), respectively. It is often more convenient to use digits 1 as a synonym for T and 0 as a synonym for F, respectively. If we consider a proposition as a whole thing without any inner structure, that is, if we are interested in the truth value only, we talk about a *simple proposition*. When we construct new propositions out of the given ones, using different connectives or their formal analogs (those will be discussed in a moment), these new propositions are said to be *compound propositions*.

Problem 45. *What is the difference between "Proposition" and "Preposition" in English?*

In arithmetic, people made calculations with individual numbers—1, 4, 23, etc., and learnt, for instance, that $2 + 3 = 3 + 2$. Then people realized that

---

[1] Three- or even infinitely-valued logics are beyond the scope of this book.

the same property—the commutative law of addition, is valid for any pair of natural numbers. At the next step, in algebra, people replaced individual numbers by symbols and wrote this law as a + b = b + a. We proceed now in the same way in the new algebra, which we are developing—the Algebra of Logic or propositional algebra. We shall denote the simple propositions by small letters p, q, r, etc., and the compound propositions by capitals A, B, C, and so forth. Sometimes, we use index notation as well: p_1, p_2, p_{123}, ..., A_1, ...

Problem 46. *Give two possible interpretations of the sentence: "You can buy a suit or a belt and shoes."*

Solution. We can set the parentheses as "You can buy (a suit or a belt) and shoes", or else as "You can buy a suit or (a belt and shoes)."

This example shows that the formulas may be ambiguous; thus, the language needs *separators*, that is, special symbols like *parentheses, brackets, curly brackets*, etc. Moreover, Exercise 2 on p. 30 implies that not every grammatically correct sentence is a proposition.

Hence, now we consider letters—symbols for propositions in the standard algebraic way—as elements of a certain alphabet, or as variables taking on values from a given set. In the usual algebra, a variable can take on, as a rule, an infinite number of different values—integers, rational numbers, and more. In the propositional algebra under construction, a variable can take on only two different values—T for *Truth*, and F for *False*. Instead of T, we often write 1 and instead of F, we can write 0. The writing $p \equiv T \equiv 1$ or $p \equiv F \equiv 0$ always means that the truth value T is assigned to the simple proposition p or that p has this truth value, and the same for F. The propositions can now be called the *logical functions*. We write $p \equiv q$ iff the propositions p and q have the same truth value, that is, they are either both *true* or both *false*.

Next, we study various logical operations with propositions in symbolic form. These formal analogs of connectives from regular human languages combine several propositions into a new compound proposition. In this process, we do not care about the actual meaning of the propositions involved. For instance, we can combine the proposition "2 + 2 = 4" with the proposition "Tomorrow I have a date with the witch." Unlike real-life speech, in the propositional algebra, such combinations are acceptable. In this algebra, we are interested in one question only—whether and in what way the truth value of the compound proposition depends on the truth values of its components and on the logical operations used in forming the final compound proposition.

Here, we have arrived at an important issue—what sequences of symbols, what *strings* are valid in this algebra? For example, we can write $x+ = 3$, but this string is probably, senseless. Every person, who would write a working program in any algorithmic language, knows this phenomenon—the strings must obey certain strict syntax; otherwise, the computer language program,[2] BASIC or C++, or Python, or any other does not understand you, and your code will be rejected. The rules in a textbook are not so precise and will be gradually developed in this section. We return to this issue at the end of the chapter.

The *strings* of symbols of various lengths, that is, pairs, triples, quadruples, etc., are also called *tuples*; for example, (a,b,c,d) is a 4-tuple or quadruple. The tuples are always *ordered*, that is, $(a,b,c,d) \neq (a,b,d,c)$ as long as $c \neq d$. In Problem 74 we listed all the eight 3-tuples of two symbols T and F. If a tuple must contain two or more identical symbols, we arrive at other combinatorial objects and discuss them later.

Given simple propositions p, q, ..., s, and using various logical operations, we can form different compound propositions. We shall write down the compound propositions using standard function notation: $A(p,q,..)$, $B(p_1,...,p_n)$, and so forth. Simple propositions p, q, ..., s, or $p_1, p_2, ..., p_n$ are called the *arguments*, or *independent variables*, or *indeterminates*; the index n, which can be 1, or 2, or 3, etc. means that the proposition B has n independent arguments. In particular, if $n = 1$, we call these propositions *unary*, that is, with a single argument; if $n = 2$, they are called *binary*, that is, with two arguments.

Definition 6. *Two propositions are said to be (logically) equivalent if they have the same set of arguments and for every tuple of the values of their arguments these propositions have the same truth value. The equivalence of propositions is denoted by the symbol $\equiv$, that is, writing $A \equiv B$ means that propositions A and B are logically equivalent.*

The same symbol indicates that a proposition takes on a truth value, either T or F, that is, $p \equiv F$ or $p \equiv 0$ means that p is a *false* statement.

Note that in mathematics, the symbol $\equiv$ means sometimes *identical equality* (or just *identity*)—for instance, we can write $x + x \equiv 2x$, if we want to emphasize that this equation is valid for all allowable values of x. In this text, $p \equiv q$ always stands for the (logical) equivalence of propositions p and q.

---

[2] Standard names like "C++" are exclusions.

Sometimes we consider propositions $A(p, q, \ldots, s)$ as logical expressions, but we mostly adopt the analytical point of view and consider these expressions as logical functions. Unlike the functions in calculus, the logical functions are mostly simple since the domain $\mathcal{Z}_2$ for every argument consists of only two values: $\mathcal{Z}_2 = \{T, F\}$, or equivalently, $\mathcal{Z}_2 = \{1, 0\}$, and all of these logical functions have the same finite range $\mathcal{Z}_2 = \mathcal{B} = \{1, 0\}$. Writing $A \equiv F$ or $A \equiv T$ always means that the simple or compound proposition A takes on the truth value F or T, respectively.

2.2 CONNECTIVES: TRUTH TABLES

It is convenient to represent propositions, or later on, Boolean functions by using the *truth tables*. For a proposition $A(p, q, \ldots, t)$ the table has a separate column for each argument p, q, ..., t. Since every cell can contain only one of exactly two values, F or T (or 0 and 1, which is equivalent), there are 2^n Boolean n – tuples, hence such a truth table, in addition to the header line, has exactly 2^n rows. The right-most column contains the truth values of proposition A, or the values of the Boolean function f. This column may be preceded by the columns for some parts—*sub-formulas* of A.

Problem 47. *Draw the truth table for the proposition* $A(p, q) \equiv p \vee (p \to q)$.

Solution. The proposition contains the disjunction $\vee$ and the conditional $\to$. These Boolean functions will be discussed in more detail later on in this section. To solve the example, we just mention that the conditional $p \to q$ is false iff its premise $p \equiv T$ and the conclusion $q \equiv F$, while the disjunction $p \vee q$ is false iff its both arguments are false, $p \equiv q \equiv F$.

Not counting the header, the two arguments, p and q generate $2^2 = 4$ rows of the table. The table contains the two columns for the arguments, an intermediate column for the sub-formula $p \to q$, and the resulting column for the truth values of the whole proposition A:

TABLE 2.1 The Truth Table for Proposition $A(p, q) \equiv p \vee (p \to q)$.

p	q	$p \to q$	$A(p, q) \equiv p \vee (p \to q)$
0	0	1	1
0	1	1	1
1	0	0	1
1	1	1	1

We see that the right-most column in Table 2.1 contains only 1; thus the proposition $A(p,q) \equiv p \vee (p \to q)$ (Table 2.1) is identically true.

When we develop some additional properties of the conditional $p \to q$, we will be able to prove many such properties analytically without constructing the truth tables. Now let us return to the tables.

Since $p \equiv 0$ or $p \equiv 1$, and for each of these two values, the proposition $f(p)$ can also take on any one of these two truth values, there are $2 \times 2 = 4$ unary propositions of one variable, and Table 2.2 shows all of them.

TABLE 2.2 Truth Tables for Four Unary Propositions (Boolean Functions).

p	$A(p) \equiv 0$	$A(p) \equiv 1$	$A(p) \equiv p$	$A(p) \equiv \neg p$
0	0	1	0	1
1	0	1	1	0

The first row (header) exhibits the argument p and the four functions: two constant propositions (identically false and identically true), the *coordinate function* $A(p) \equiv p$, and the proposition $A(p) \equiv \neg p$.

Definition 7. *This proposition $\neg p$ is called the negation of the given p.*

The negation is a *unary* operation, it has only one argument. Symbolically, if we denote the given proposition by p, then its negation is denoted as the upper bar $\bar{p}$, or as $\neg p$. These notations are synonyms; in different books, you can meet any of them and even some others. In this text, we use either the symbol $\neg$ or the upper bar. Clearly, if a proposition p has the truth value T, then its negation $\neg p$ has the opposite truth value $F \equiv \neg T$, and vice versa $T \equiv \neg F$. This last statement, which is actually the definition of the negation, can be written using a *truth table*, that is, the last column in Table 2.2. The only feature of the negation we mention is the *double negation* property $\bar{\bar{p}} \equiv p$, or

$$\neg(\neg p) \equiv p.$$

Sometimes we should distinguish between the negation of a statement and its *opposite statement*. Thus, the formal negation of the proposition

$$\text{"}P(X) \equiv \textit{Mr. X is a rich man"}$$

is *"Mr. X is not a rich man,"* whereas its opposite would probably sound like *"Mr. X is a poor man."*

Problem 48. *Is it correct that* $\neg T \equiv F$ *and* $\neg F \equiv T$ *are logically equivalent?*

Problem 49. *Simplify the propositions* $(\neg(\neg(\neg p)))$ *and* $(\neg(\neg(\neg(\neg p))))$.

Solution. We start with the interior part $\neg(p)$. In such cases, writing $\neg(p)$ or $\neg p$ is the same $\neg(p) \equiv \neg p$, and we will always write the latter expression, $\neg p$, since it contains less parentheses and usually is more readable. The original formula becomes $(\neg(\neg(\neg p)))$.

We see now that it is possible to safely remove the external parentheses $(\neg(\neg p)) \equiv \neg(\neg p)$ and write the formula as $\neg(\neg p)$. Certainly, this formula is not exactly the original one, but it is *equivalent* to the original, and using it in any solution or proof does not lead to any change in the results.

We see that the result depends upon the parity of the number of the pairs of the enclosing parentheses.

The reasoning above makes obvious the following *Principle of Substitution*.

Theorem 5. *Let a formula F depend upon the arguments* $p_1, p_2, \ldots, p_n$, *and a formula* F_1 *was derived from F by simultaneous substitutions of the formulas* $G_1, G_2, \ldots, G_n$ *instead of the arguments* $p_1, p_2, \ldots, p_n$, *respectively. If F is identically true, then* F_1 *is also identically true.*

This principle is closely connected with the following statement.

Theorem 6. *Let a formula F contain a sub-formula* F_1, *and the formula G be derived from F by substituting the formula* G_1 *instead of* F_1. *If* $F_1 \equiv G_1$, *then* $F \equiv G$.

Proof. This is obvious since if we imagine the truth tables of all the propositions involved, we see that F_1 and G_1 have the same last column of the truth tables, and so that the same is true for the truth tables of F and G.

In the formal exposition, to avoid any misunderstanding, we must consider strings like this: $\neq (p)$, $(\neq (p))$, etc.; however, these writings are more suitable for computers and are difficult to follow. In our less formal exposition, we will omit as many (pairs of) parentheses as possible, but without introducing any ambiguity. We study expressions in an alphabet consisting of several parts. Hereafter, small characters are reserved for simple propositions or for the arguments of Boolean functions, while capital letters are reserved for the functions or propositions. There is a very limited list of algebraic and logical operations, like $+, \times, \vee, \wedge$, etc. To avoid any ambiguity, we also have a small list of separators, namely, parentheses (,), brackets [,], curly braces {,}, etc.

If a formula contains too many separators, it is easy for computers but may be difficult to read for human beings. Therefore, after defining the logical operators, later on in this section, we describe the rules of *preference* or the *order of operations*, which make the formulas unique without using too many extra symbols. Thus, we study propositions as algebraic objects, subject to certain operations—negation, conjunction, etc., and without any discussion of their meaning. This way, propositions become *Boolean functions* (from now on we use this term interchangeably with logical *propositions*) $B(p, q, \ldots, s)$, where each variables $p, \ldots, s$ can independently take on the two values, T and F. If p takes the value T, we write this as $p \equiv T$, the second argument q can also be either T or F, etc. Thus, for two variables, p and q, we get two pairs of the argument values: (T, T) and (T, F). If $p \equiv F$, we get two more pairs of the argument values, (F, T) and (F, F). Together, we have four pairs of the values of arguments.

Remark 1. *We could write two-valued Boolean functions by following the "calculus way" as $\wedge(p, q)$, etc., but when a binary operation has a special symbol, as $2 + 6$, in particular, in logic, the connective is usually written between the arguments, as $p \wedge q, p \to q$, even though $\wedge(p, q)$ is also correct and has the same meaning.*

Thus, the domain of a two-valued proposition, which has two two-valued arguments, consists of $4 = 2^2$ points, or of $4 = 2^2$ ordered pairs or 2-tuples of the truth values. Here the base is equal to 2 because any argument can take on two different values, and the exponent is 2 because there are two simple arguments, p and q, of our proposition. Since in each of these four rows, a proposition can have one of the two values, true or false, there are $2^4 = 16$ non-equivalent binary propositions, all shown in Table 2.3. It contains two columns for two independent arguments p and q, and 16 columns for the 16 possible binary propositions with two independent arguments.

TABLE 2.3 Truth Tables for all the 16 Binary Propositions (Boolean Functions) with Two Independent Arguments.

p	q	$F_0 \equiv 0$	F_1	F_2	F_3	$F_4 \equiv p \wedge q$	$F_5 \equiv \neg p$	$F_6 \equiv \neg q$	F_7
0	0	0	1	0	0	0	1	1	1
0	1	0	0	1	0	0	1	0	0
1	0	0	0	0	1	0	0	1	0
1	1	0	0	0	0	1	0	0	1

p	q	F_8	$F_9 \equiv q$	$F_{10} \equiv p$	F_{11}	F_{12}	F_{13}	$F_{14} \equiv p \vee q$	$F_{15} \equiv 1$
0	0	0	0	0	1	1	1	0	1
0	1	1	1	0	1	1	0	1	1
1	0	1	0	1	1	0	1	1	1
1	1	0	1	1	0	1	1	1	1

Among these propositions, there are several familiar ones. Indeed, the very first and the very last propositions are identical logical constants, $F_0(p,q) \equiv 0$ and $F_{15}(p,q) \equiv 1$. We also recognize two negations $F_5 \equiv \neg p$ and $F_6 \equiv \neg q$, and two *coordinate* functions, $F_{10} \equiv p$ and $F_9 \equiv q$.

It is useful to mention that some functions actually depend only on one argument, for example, F_5 actually depends only on p. We say in that case that q is a fictive argument or non-essential argument of f_5. In particular, all the arguments of F_0 and F_{15} are non-essential. Of course, the other arguments are called *essential*.

Now we study the remaining propositions. The function $F_1 \equiv \neg(p \vee q)$ is called *NOR* and $F_{11} \equiv \neg p \wedge q$ is called *NAND*; they often appear in the theory of switching circuits, see Chapter 14. Consider F_4, which is 1 at only one case, when both $p \equiv 1$ and $q \equiv 1$. It is called the *conjunction* of p and q (or *logical multiplication*) and is denoted as $p \wedge q$ or $p \, \& \, q$, or similar to multiplication and if that does not result in any misunderstanding, without any symbol at all, as $F_4 \equiv pq$. If we look carefully at F_1 and compare its truth table with that of F_4, we observe that $F_1(p,q) \equiv F_4(\neg p, \neg q) \equiv (\neg p) \wedge (\neg q)$.

Problem 50. *Prove that* $F_2(p,q) \equiv F_4(\neg p, q) \equiv (\neg p) \wedge (q)$ *and* $F_3(p,q) \equiv F_4(p, \neg q) \equiv (p) \wedge (\neg q)$.

Next, we compare the functions $F_4 \equiv p \wedge q$ and F_{14}. Unlike F_4, the latter column contains only one 0 and three 1. It is called the *disjunction* (or logical *addition*) and is denoted as $F_{14} \equiv p \vee q$. It can also be said that conjunction is the *minimum* of its arguments, while disjunction is the *maximum* of the arguments. That is valid for any finite family of arguments.

It is easy to verify, and we leave that as an exercise to the reader, the following properties of these two connectives; we often call them (logical) *identities*. For the time being, we have only one way to prove these identities—by definition, that is, by constructing the truth tables of the left and the right parts, which must have identical right-most columns.

Problem 51. *Prove that the conjunction and disjunction are commutative, associative, and are connected by two distributive laws:*

$p \wedge q \equiv q \wedge p$—Conjunction is commutative
$p \vee q \equiv q \vee p$—Disjunction is commutative
$p \wedge (q \wedge r) \equiv (p \wedge q) \wedge r$—Conjunction is associative
$p \vee (q \vee r) \equiv (p \vee q) \vee r$—Disjunction is associative.

Proof. For any formula, construct the truth tables of the left and of the right parts; they must be identical.

Therefore, these formulas can be written without parentheses, as $p \wedge (q \wedge r) \equiv p \wedge q \wedge r$ or $p \vee (q \vee r) \equiv p \vee q \vee r$. Let us reiterate that, unlike high-school algebra, there are two symmetric distributive laws in this Boolean algebra:

$$p \wedge (q \vee r) \equiv (p \wedge q) \vee (p \wedge r)$$
$$p \vee (q \wedge r) \equiv (p \vee q) \wedge (p \vee r).$$

Look next at the Boolean functions F_7 and F_8, having two ones and two zeros in the truth table. The former is true, that is, it has 1 when both p and q have the same truth values, $p \equiv q \equiv 0$ or $p \equiv q \equiv 1$. Thus, we can write $F_7(p, q)$ as $(p)(q) \vee (\neg p)(\neg q)$ and $F_8(p, q) \equiv (\neg p) \wedge (q) \vee (p) \wedge (\neg q)$. This Boolean function is called the *exclusive disjunction* or *binary addition* and is often denoted as $XOR(p, q)$ or $p \oplus q$—see Table 2.6.

The features of the conjunction and disjunction are intuitively clear. The next Boolean function, the conditional F_{12}, deserves certain comments. It is called the *conditional*, because it claims the truth of the conclusion only under the condition that the premise is true; if the conclusion is false, then the conditional is true independently upon the premise. Let us elaborate a bit more on that.

The conditional has two arguments, and it is a logical function of two variables—the *binary* logical operation. The arguments of a function are also called its operands; thus, the negation $\neg p$ has only one operand p. We consider now a formal analog of the connective *IF...*, *THEN...* . The formal operation is called a *conditional* or an *implication* and is denoted by $p \rightarrow q$ or $p \supset q$ with two operands p and q; the function $p \rightarrow (q \vee r)$ has three operands, etc. The first term in a conditional, p is called the *premise*, or *hypothesis*, or *antecedent*; the second one, q is the *conclusion* or *consequent*. The implication is often used in colloquial speech, for example, as "If you pass the test, you will be rewarded," but its formal counterpart has certain particularities.

First, in our colloquial speech, the hypothesis and the conclusion of a conditional usually have something in common. If you are not put in a special situation, for example, when an Instructor wants to present properties of the formal conditional, you have probably never heard something like this: *If you enjoy eating integer numbers, then the Moon is my satellite.* That is why the implication in our natural language is sometimes called the *material conditional.* In the formal setting, though, the hypothesis and the conclusion may have nothing in common; in the algebra of logic, we are interested in one point only—in what way the true value of the conditional depends on the truth values of its arguments, that is, of the hypothesis and conclusion. This dependence is given in its truth table, see Table 2.4.

Second, we should not mix the truth value of a conditional itself with the truth value of its conclusion. Discussing a conditional, we verify the correctness, the validity of our reasoning, which departs from the premise p and results in the conclusion q. Therefore, if the *premise* of the conditional $p \to q$ is true, $p \equiv T$, but the *conclusion* is false, $q \equiv F$, then we have to conclude that our reasoning leading from p to q is wrong. Whence, it is quite reasonable to define the true value of the conditional $p \to q$ in the case when $p \equiv T$ and $q \equiv F$ as F. On the other hand, if the hypothesis and the conclusion both are true, $p \equiv q \equiv T$, we should believe in this reasoning and determine the truth value of the conditional to be true also: $p \to q \equiv T$.

In contrast, if we have a false hypothesis, $p \equiv F$, the conditional is defined to be T independently upon the truth value of the conclusion. An essential reason for such a definition is that if our premises are wrong, we can prove whatever we want. It does not matter how to argue starting from false premises—any arguing is good!

Everything that we said above was just motivation. Table 2.4 gives *the definition* of a formal logical operation and nothing else. Thus, the conditional $p \to q$ is defined by the following truth table; since the proposition has two arguments, it must have $2^2 = 4$ rows.

TABLE 2.4 Truth Table for the Conditional $p \to q$.

p	q	$F_8 \equiv p \to q$
0	0	1
0	1	1
1	0	0
1	1	1

Another important reason justifying the definition is given by the following example.

Example 2. *Consider the following implication:*

If a natural number n is divisible by 4 evenly, that is, with the zero remainder, then n is divisible by 2 evenly.

For $n \geq 4$, the statement is obvious. When $n = 3$, both the premise and the conclusion are false, which does not violate our logic. When $n = 2$, we have here the conditional with a false premise and a true conclusion. Therefore, if we want this simple theorem (and many other similar assertions) to be valid for all the natural numbers, we must define the conditional with a false premise to be true:

$F \rightarrow q \equiv T$ independently upon q.

Given a conditional $p \rightarrow q$, we often use the related propositions, *Variations of the Conditional*—see Table 2.5, are called
the *converse* $q \rightarrow p$,
the *inverse* $\neg p \rightarrow \neg q$, and
the *contrapositive* $\neg q \rightarrow \neg p$.

Problem 52. *Prove that a conditional and its contrapositive are logically equivalent. The same is true for an inverse and a converse of a given implication.*

Proof. It is enough to construct the truth tables and compare them pairwise.

TABLE 2.5 Variations of the Conditional.

Given Conditional	$p \rightarrow q$
The Converse of the Given Conditional	$q \rightarrow p$
The Inverse of the Given Conditional	$\neg p \rightarrow \neg q$
The Contrapositive of the Given Conditional	$\neg q \rightarrow \neg p$

A conditional $p \rightarrow q$ is usually read as "IF p, THEN q"; in this case, it is said to be in the *normal form*. For example, "If it is raining, then the streets are wet." In the normal form, the word "IF" precedes the hypothesis p of the conditional. Several other forms of conditional are also used. Sometimes we omit the word "THEN": "If it is raining, the streets are wet." Also, we can switch the two parts of a conditional: "The streets are wet IF it is raining." Again, the word "IF" here precedes the hypothesis of the conditional, and the expression "q IF p" is equivalent to the conditional in the normal form "IF p, THEN q." The form "q whenever p" is also equivalent to the conditional in normal form "IF p, THEN q."

The form "p ONLY IF q" is equivalent to the normal form as well. It tells us that the condition p can occur ONLY IF the condition q occurs. Thus, if $p \equiv T$, then q must also be T. On the other hand, this form tells us nothing about the situation when $p \equiv F$ or $q \equiv F$. Therefore, the expression "p ONLY IF q" is equivalent to the conditional in the normal form "IF p, THEN q."

Problem 53. *Compose the truth tables and find the normal forms for the propositions "p unless q" and "q unless p."*

Solution. The normal forms are $(\neg q \to p) \equiv p$ unless q and $(\neg p \to q) \equiv q$ unless p.

Problem 54. *Rewrite the following English statements in the normal form "If p, then q":*

1. *You will remember our appointment only if I remind you.*

2. *To pass the class, it is sufficient to study hard.*

3. *To pass the class, it is necessary to study hard.*

4. *To pass the class, it is necessary and sufficient to study hard.*

5. *It is necessary to buy a ticket to get on the train.*

6. *You can borrow a laptop only if you pay your fines.*

Thus, in Table 2.3, we can write $F_{12} \equiv p \to q$. Further analyzing their properties, we can write $F_{13} \equiv q \to p$, $F_{11} \equiv p \to \neg q$, and $F_{14} \equiv \neg p \to q$. The column F_{14} is now described twice, which immediately results in the following important formula.

Problem 55. *Prove that $p \to q \equiv \neg p \vee q$.*

The disjunction $p \vee q$ is sometimes called the *Inclusive disjunction* because it is true when either $p \equiv T$, or $q \equiv T$, or both $p \equiv T$ and $q \equiv T$. The *exclusive disjunction* F_8 is true iff exactly one argument is true, either $p \equiv T$ or $q \equiv T$, but not both p and q. The *exclusive disjunction* is also denoted as $\oplus$ and is defined by the following truth table (Table 2.6). It is also called "binary addition" because when we add $1 + 1 = 10$ in the binary system, we drop the 1 and preserve only 0.

TABLE 2.6 Exclusive Disjunction also called Binary Addition $F_8(p,q) \equiv p \oplus q$.

p	q	$F_8(p,q) \equiv p \oplus q$
0	0	0
0	1	1
1	0	1
1	1	0

Hence, all the formulas can be written *without* the conditional, but it is often useful to have it in our formulas.

Problem 56. *Construct the truth table for* $((p \to q) \wedge (q \to r)) \to (p \to r)$.

Problem 57. *State this property in words.*

This formula, called *syllogism*, is *identically true*. In a sense, the *identically true* formulas, like syllogism, represent the real rules of our thinking. The identically true formula is also called *tautology*. There are a few more useful tautologies, in particular, *modus ponens*

$$(p \wedge (p \to q)) \to q$$

and *modus tollens*

$$\overline{q} \wedge (p \to q) \to \overline{p}.$$

Problem 58. *Prove that* modus ponens *and* modus tollens *are identically true.*

Problem 59. *Prove that the following propositions are identically true:*

$((p \to q) \wedge \neg q) \to \neg p$

$(p \to q) \Leftarrow (\neg q \to \neg p)$—*The law of counterposition*

$(\neg p \to q) \wedge (\neg p \to \neg q) \to p$—*The proof by contradiction (getting to absurdity)*

Problem 60. *Prove that the negation of tautology is a contradiction and vice versa.*

Problem 61. *Rephrase the given conditionals below in the normal form and identify the hypothesis and the conclusion in every example.*

a. *"You will go to Florida if you get at least 90/100 on the Final."*

b. *"No Florida if you get less than 90/100 on the Final."*

c. *"No Florida if you get less than 75/100 on the Final."*

Tautologies are especially important in our reasoning since they represent the laws of our thinking. The arguing, based on tautologies, is correct always, independently upon any specific topic. For example, we employ the syllogism, when instead of the direct proof of a certain statement, we first prove a lemma and then deduce this certain statement from the lemma.

Problem 62. *Prove the next tautologies.*

a. $(p \to q) \equiv ((\neg q) \to (\neg p))$—*Contraposition rule*

b. $(p \to q) \land (p \to \neg q) \to \neg p$—*Proof by contradiction.*

Problem 63. *Calculate the truth value of each of the given expressions.*

a. $(((p \to q) \to p) \to q) \to p)$

b. $(p \lor q) \to (q \land (\neg r \leftrightarrow p))$

Problem 64. *Prove the following identities.*

$$\begin{cases} (\neg p) \land (\neg q) \equiv \neg(p \lor q) \\ (\neg p) \lor (\neg q) \equiv \neg(p \land q) \end{cases} \text{—De Morgan laws}$$

$$\begin{cases} p \land (p \lor q) \equiv p \\ p \lor (p \land q) \equiv p \end{cases} \text{—The Absorption Laws}$$

$$\begin{cases} p \land (\neg p) \lor q \equiv p \land q \\ p \lor ((\neg p) \land q) \equiv p \lor q \end{cases} \text{—The Deletion (Crossing-Out) Laws.}$$

Problem 65. *Design two examples of identically true and two of identically false propositions.*

Problem 66. *In Problem 47, we used the truth table to prove that the formula $p \lor (p \to q)$ is identically true. Now we can easily prove analytically that this proposition is identically true. Indeed,*

$$p \lor (p \to q) \equiv p \lor (\neg p \lor q) \equiv (p \lor \neg p) \lor q \equiv 1 \lor q \equiv 1.$$

Exactly as the formulas in Problem 47, any *tautology* and moreover, any *implication* can be a logical base for certain proofs. Thus, the conditional (Exer. 40, b)) $(p \to q) \land (p \to \neg q) \to \neg p$ has been used and will be used many times in the future for *proofs by contradiction*. As one more example of the *proof techniques* generated by simple logical formulas, we consider the next conditional.

Problem 67. *Prove the following identity and generalize it from two to n propositions.*

$$(p_1 \vee p_2) \to q \equiv (p_1 \to q) \wedge (p_2 \to q).$$

The method of proof, based on this identity, is called *The Proof by (separating the) Cases*. For example,[3] we use the proof by cases if we have to prove a property of the natural numbers, and we prove that property separately for the odd numbers and separately the property for the even numbers.

It is clear now why the *decidability problem*, that is, deciding whether a proposition is identically true or identically false or neither, is so important in mathematics. We will discuss it in more detail in the last chapter of the course.

Consider one more example of an implication in our speech. If a kid hears from the father, "If you get at least 90% on the next Discrete Math test, we will go to the "Six Flags", usually both of them understand this statement not only in its precise sense, as an implication $p \to q$, but also as a promise in the *opposite* direction:

"If you get less than 90% at the test, you will get no trip."

So that, sometimes in colloquial speech, we use a conditional but assume both this conditional and its converse as well. In mathematics, such confusion is unacceptable since it leads to errors. For example, the statement "Every integer number, which is multiple of 4, is an even number" is true, but the converse claim "Each even number is a multiple of 4" is clearly wrong, as the example of the even number 2 shows.

The case when the converse of the conditional is equivalent to the conditional itself is important and deserves special consideration.

The *biconditional* is a binary operation denoted by $p \leftrightarrow q$ and defined by Table 2.7.

TABLE 2.7 The Truth Table for the Biconditional $p \leftrightarrow q$.

p	q	$p \leftrightarrow q$
0	0	1
0	1	0
1	0	0
1	1	1

---

[3] For more about this method see Chapter 5 below.

The definition is quite natural: truth is equivalent to truth, false is equivalent to false, but truth and false are not equivalent to each other.

Problem 68. *(1) Prove that $p \leftrightarrow q \equiv (p \to q) \land (q \to p)$, that is, the biconditional is the formal analog of the equivalence $\equiv$.*

(2) Prove that the biconditional is the negation of the binary addition and vice versa.

(3) Prove the following properties of the XOR function, or the binary addition $p \oplus q$.

For any z, $z \oplus 0 \equiv z$ and $z \oplus z \equiv 0$. These properties are valid if z is a Boolean vector as well.

Problem 69. *Is the biconditional "$(2^5 > 30)$ if and only if (the Moon is plane)" true or false?*

Thus, we can define the *subtraction* in the algebra $\mathcal{B}_2$ as the binary addition, and this operation is self-inverse. The same is valid if we apply the XOR operation to Boolean vectors termvise, that is, for any Boolean vectors $b = (b_1, \ldots, b_n)$ and $k = (k_1, \ldots, k_n)$,

$$b \oplus k \oplus k \equiv b.$$

Moreover, due to the property of the conjunction, $b \land b \equiv b$, where $\land$ is done termvise. In this sense, the exclusive disjunction and conjunction are *linear operations*. Later on, when we study *stream ciphers*, we will often see these properties.

Problem 70. *(1) List all 16 ordered quadruples of the two letters T and F.*

(2) Prove by induction that the domain of a proposition in n arguments consists of 2^n n-tuples of the letters T and F.

(3) Are the two propositions (a) and (b) in every line below equivalent or not to each other?

(a) $3(x + y) = 3x + 3y$ *(b) $3(x + y) = 4x + 3y$*

(a) $p \land (x \lor y) = (p \land x) \lor (p \land y)$ *(b) $p \land (x \lor y) = p \land x \lor y$*

(4) Is the proposition $x \to y \equiv y \lor \neg x$ a tautology?

Consider a valid conditional $p \to q$, and suppose that the premise p is also true. Due to the features of the conditional, we know for sure that in this case, the conclusion q is true as well. Because of that, condition p is called the *sufficient* condition for condition q.

On the other hand, if the conditional $p \to q \equiv T$, and the conclusion $q \equiv T$, the premise p *cannot* be false; it must be true. That is why q is called the *necessary* condition for p. Thus if the biconditional $p \leftrightarrow q$ is valid, it is said that p is a *necessary and sufficient condition* for q; in turn, q is also a *necessary and sufficient condition* for p. This statement can also be expressed as "*p if and only if q*", often abbreviated as "*iff*." For example, the Pythagorean theorem can be stated as

A plane triangle with sides a, b, c has 90° angle if and only if the sides satisfy

$$a^2 + b^2 = c^2.$$

Thus, these two conditions are both necessary and sufficient for each other. Pairs of the necessary and sufficient conditions, that is, the equivalent conditions, are so valuable in mathematics because they sometimes give a new description, a new characterization of an object or property. That is why many theorems in mathematics are stated as the necessary and sufficient conditions.

The negations of conjunction or disjunction are also important and have special names since they often appear in applications. The negation of the conjunction, $\overline{x \wedge y}$ is called *NAND* connective or *NAND* gate, the negation of the disjunction, $\overline{x \vee y}$ is called *NOR* function, or *NOR* gate; these gates often appear in circuitry.

Problem 71. *Build the truth tables of the NAND and NOR Boolean functions.*

Above we considered three important problems about the propositions.

1. Is there an algorithm, which uses finitely many identical transformations to decide for every proposition whether or not it is identically true or identically false?

2. Is there an algorithm, which uses finitely many identical transformations to decide whether or not two given propositions are equivalent?

3. Is it possible to realize any (Boolean) proposition as a superposition of the Boolean operations above?

Later on, when we study Boolean functions in more details, we will learn that all the three answers are affirmative. However, we want to remark here that these functions are relatively simple objects; their domains are finite. If the domains are infinite, which for instance, may be for the predicates, the answers are negative. But more regarding that at the end of this course.

2.2.1 Order of Operations

To end this section, we study the *Order of Logical Operations*. Similarly to familiar elementary algebra, the order of operations in algebra of propositions is governed by parentheses. A formula with many parentheses is cumbersome and is difficult to read. Because of that, the logical operations are defined to follow the sequel hierarchy.

The *strongest connective* is negation, meaning that in formulas like $\neg p \vee q$ or $p \rightarrow \neg q$, we must first evaluate the negations, and only after that any other operations. Thus, the truth table of the first example above is

TABLE 2.8 The Truth Table of the Disjunction $\neg p \vee q$.

p	$\neg p$	q	$\neg p \vee q$
0	1	0	1
0	1	1	1
1	0	0	0
1	0	1	1

If in a particular problem we have to apply the negation to the disjunction $p \vee q$, the parentheses are necessary, as in $\neg(p \vee q)$.

Similarly to the multiplication and addition in elementary algebra, conjunction ties stronger than a disjunction. Moreover, the conditional is stronger than the conjunction but weaker than biconditional. The identical operations are done from left to right. Thus, in the example

$$p \vee q \wedge p \rightarrow r \rightarrow \neg p \leftrightarrow q \vee r$$

we proceed as follows:

1. $\neg p$

2. $\neg p \leftrightarrow q$

3. $q \wedge p$

4. $p \vee q \wedge p$

5. $\neg p \leftrightarrow q \vee r$

6. $p \vee q \wedge p \rightarrow r$

7. $p \vee q \wedge p \rightarrow r \rightarrow \neg p \leftrightarrow q \vee r$.

We remark that steps (1), (2), and (3) can be performed simultaneously, as well as steps (4) and (5). To override the natural order of operations and to clarify problems, we must use parentheses.

Problem 72. *Compute truth tables of the Boolean functions.*

1. $(p \vee q \wedge r) \vee \neg q$

2. $T \vee q \leftrightarrow r$

The argument of any unary Boolean function $f(p)$ can take on two and only two values, either T or F, but not both; we mention that $2 = 2^1$. Functions with two arguments are defined over 4 pairs of the truth values; see Table 2.2 and the following. Thus the *domain* of a Boolean function with n arguments consists of all n-tuples with 0 or 1 components.

Problem 73. *Prove that any such domain consists of 2^n tuples.*

Problem 74. *Prove that the domain of a proposition with three arguments consists of the following 8 ordered triples*

$$(T,T,T),(T,F,T),(F,T,T),(F,F,T),(T,T,F),(T,F,F),(F,T,F),(F,F,F).$$

Problem 75. *Prove that the domain of a Boolean function of n arguments consists of 2^n tuples.*

Remark 2. *One can calculate directly, that $2^{(3^2)} = 512$, but $(2^3)^2 = 64$, so that the "super-powers" a^{b^c} depend on the order of operations, and in general, $a^{b^c} \neq (a^b)^c = a^{bc}$. To avoid too many parentheses, we follow the standard agreement that $a^{b^c} \equiv a^{(b^c)}$; for example, $2^{3^2} = 2^9 = 512$.*

Problem 76. *The range of any Boolean function contains just two points, T(rue) and F(alse). Prove that there are exactly 2^{2^n} Boolean functions in n unknowns. In particular, we saw in Table 2.2 that if $n = 2$, there are $2^{2^2} = 16$ Boolean functions of 2 variables.*

Problem 77. *Let $NZ(x)$ be the statement $x \neq 0$, $NI(x)$—a numerator is increased by the number x, $ND(x)$—a denominator is increased by the number x. Use these symbols to write symbolically the proposition "If both the numerator and the denominator are multiplied by the same non-zero number, the value of the fraction is not changed."*

Problem 78. *Introduce symbols for simple propositions involved and use them to rewrite the following statement "9 divides a positive integer iff the 9 divides the sum of its digits."*

Solution. Let Z stand for the set of the integers and $\Sigma(n)$ stand for the sum of the digits of n. The biconditional can be written as follows:

$$\forall n \in Z(n \div 9 \in Z \leftrightarrow \Sigma(n) \div 9 \in Z).$$

Problem 79. *Prove the following logical identities; unnecessary parentheses are omitted.*

$$\neg\neg p \equiv p$$
$$p \wedge q \equiv q \wedge p$$
$$p \vee q \equiv q \vee p$$
$$p \vee (q \vee r) \equiv (p \vee q) \vee r \equiv p \vee q \vee r$$
$$p \wedge (q \wedge r) \equiv (p \wedge q) \wedge r \equiv p \wedge q \wedge r$$

$p \wedge p \equiv p; p \vee p \equiv p$. We remind that an algebraic operation $a \circ b$ is called idempotent if $a \circ a = a$. Hence, algebraic addition and multiplication are not idempotent operations, while conjunction and disjunction are idempotents.

$p \vee T \equiv T, \; p \wedge T \equiv p, \; p \vee F \equiv p, \; p \wedge F \equiv F$

$p \wedge \overline{p} \equiv F$—The contradiction law

$p \vee \overline{p} \equiv T$—The law of the excluded middle

$\neg(p \wedge q) \equiv \neg p \vee \neg q$

$\neg(p \vee q) \equiv \neg p \wedge \neg q$—De Morgan (1806–1871) laws

$p \vee (q \wedge r) \equiv (p \vee q) \wedge (p \vee r)$

$p \wedge (q \vee r) \equiv (p \wedge q) \vee (p \wedge r)$ —Two symmetric distributive laws

$p \vee (q \wedge r) \equiv (p \vee q) \wedge (p \vee r)$

$p \wedge (p \vee q) \equiv p \vee (p \wedge q) \equiv p$—Two symmetric absorption laws

$p \vee (\neg p \wedge q) \equiv p \vee q$

$p \wedge (\neg p \vee q) \equiv p \wedge q$—Two symmetric laws of crossing-out

$p \rightarrow q \equiv \neg p \vee q$—The property of conditional (implication)

$p \Leftarrow q \equiv (p \rightarrow q) \wedge (q \rightarrow p)$—The property of biconditional.

Problem 80. *Represent the Boolean functions F_1–F_{16} as superpositions of the disjunction and negation; as superpositions of the conjunction and negation.*

Problem 81. *The Principal of the Longhorn Middle School has to compose a schedule of classes, consisting of English, History, Chemistry, and Bull Fighting. The English teacher has other engagements so that she must be either the first or the second. The History teacher wants to have either first or third periods. The Literature teacher cannot come early to the first class. The Cowboy has no preference. How to compose the schedule?*

A Boolean function $f(p_1, p_2, \ldots, p_n)$ of n variables can be represented by the truth table, similar to the tables above. The table contains n columns for the arguments and the resulting column for the function itself. In the perpendicular direction, it has a row for the headers and 2^n rows for the vectors—arguments, since there are exactly 2^n different Boolean vectors with n components, as in Table 2.9.

TABLE 2.9 The Table of the Boolean Function of n Arguments.

p_1	p_2	...	p_{n-1}	p_n	$f(p_1, p_2, \ldots, p_{n-1}, p_n)$
0	0	...	0	0	$f(0,0,\ldots,0,0)$
0	0	...	0	1	$f(0,0,\ldots,0,1)$
0	0	...	1	0	$f(0,0,\ldots,1,0)$
...	...	...	...	...	...
1	1	...	1	1	$f(1,1,\ldots,1,1)$

Problem 82. *Are the next formulas identically true, identically false, or neither?*

$$x \oplus (y \rightarrow z) \equiv (x \oplus y) \rightarrow (x \oplus z)$$

$$x \rightarrow (y \wedge z) \equiv (x \rightarrow y) \wedge (x \rightarrow z)$$

Problem 83. *Write the implication as a superposition of the negation and disjunction.*

2.3   EXERCISES

Exercise 2.1 *Construct the truth tables for the propositions* $x \oplus (y \rightarrow z) \equiv (x \oplus y) \rightarrow (x \oplus z)$, *and* $x \rightarrow (y \wedge z) \equiv (x \vee y) \wedge (x \vee z)$.

Exercise 2.2 *Which of the following are propositions? Simple propositions? Compound propositions? Determine the truth values of the propositions.*

(1) $2 \times 2 = 4$

(2) $2 \times 2 = 5$

(3) Long Live Discrete Mathematics!

(4) $x \times 3 = 7$

(5) Discrete Mathematics is Indiscreet

(6) Is it true that you are a lier?

(7) I am a lier only when I say truth

(8) $3 \leq 1 = 2$.

Solution. Formulas (1) and (8) are true, (2) is false.

Exercise 2.3 *What are the negations of the following propositions.*

(a) $2 + 2 = 4$;

(b) $2 + 2 \neq 4$;

(c) If a whole number is multiple of 4, then it is a multiple of 2;

(d) If a whole number is multiple of 2, then it is a multiple of 4;

(e) Our Earth is flat like a pancake, and Sun is cold.

Solution. (a) $2 + 2 \neq 4$; (b) $2 + 2 = 4$.

Exercise 2.4 *Determine whether the conditions (A) and (B) in each pair are necessary and sufficient, or only necessary, or only sufficient for one another:*

(1) (A) A natural number m is a multiple of 6 ... (B) Each digit of m is 6

(2) (A) $\dfrac{1}{x} > 1$ *...(B)* $x > 1$

(3) (A) $\sin x > 0$ *... (B)* $0 < x < \pi$.

Exercise 2.5 *Is the formula* $(((p \rightarrow q) \wedge b) \rightarrow a)$ *a tautology, a contradiction, or neither?*

Solution. Neither.

Exercise 2.6 *Find other examples of idempotent operations.*

Exercise 2.7 *Rewrite the next sentences in the standard implication form "If p, then q."*

(1) Eat those sweets if you care.

(2) Eat those sweets and you will be sorry.

(3) You leave or the robot comes to you.

(4) I will do if you will do.

(5) I will go unless you will go.

Solution. (1) If you care, eat those sweets. (4) If you will do, I will do.

CHAPTER 3

NAÏVE AND FORMAL (AXIOMATIC) SET THEORY

3.1 CLASSES AND SETS. AXIOMS AND THEOREMS

So far, we have discussed *propositions*, denoted by p, q, etc. However, many symbols or combinations of symbols are not propositions. For instance, the expression $3 + 3 = 3$ is a proposition, but $x + 3 = 3$ is not, since we cannot determine whether it is *True* or *False*, its truth value depends upon the numerical value of x, and it becomes a proposition if we specify the value of x; say, if $x = 3$, we get $3 + 3 = 3$, which is a false proposition, etc. In formal theories, this distinction is important, and we call strings like $x + 3 = 3$ *propositional forms*. Propositions can be represented not only by logical operations but also by other symbols; for example, $\lim_{x \to \pi/2} \dfrac{\sin x}{x} = 1$ is a (false) proposition. Propositions and propositional forms are together called *formulas*.

Similarly, symbol 3 is a *name* of a number, while the symbol n is a *name form*. The names and name forms together are called *terms*. Formulas and terms make a frame of any formal theory. We do not study here axiomatic theories, leaving that for the formal mathematical logic. Instead, in the next lecture, we study the informal set theory. But first, we want to say a few more words about the language of formal theories.

Any collection of things, objects, possessing specified properties is called a *class*.

Similarly to descriptions of propositions in the preceding chapter, this phrase cannot serve as a definition. Indeed, any mature mathematical theory contains four basic ingredients:

I. Primary Concepts   *II. Definitions*
III. Axioms (or Postulates)   *IV. Statements, Lemmas, Theorems, etc.*

In this hierarchy, classes and propositions are primary, undefined concepts; therefore, they go to part *I*. Starting from these notions, we can introduce new definitions, like logical operations of negation, conjunction, and so forth—see the preceding chapter. In this chapter, we also give several definitions, which use the *primary notion of a class*. Having been equipped with primary and defined notions, we introduce axioms of a theory.

Axioms are not proved; we accept them as primary statements, such that the theory is based on. After that, we prove theorems by using the logical means of arguing. In theorems, we investigate various logical corollaries of our assumptions (of the primary concepts, of definitions, and axioms) and possible applications of the theory. If we are satisfied, if we see that the theory has meaningful applications or its own intellectual value, we can continue its investigation and development. If, for any reason, we are not satisfied, we can change some or all of our basic assumptions and try to build another theory with another set of axioms.

Thus, a class is a primary concept, which is not given any definition. Any statement like "A class is a collection of items such that..." serves only to some clarification of the language. Different things, entities that form a class, are called its *elements* or members. For example, "Julius Caesar" is an element of the class of all people who ever lived on the Earth, but his horse is not an element of that collection. The relation between a class and its elements is denoted by the symbol $\in$. Writing $x \in A$ (read: x is an element of the class A, or x is in A, or x belongs to A) means that we treat A as a class, and this thing x is one of the elements of A. Negation of this relation is denoted by $\notin$ or $\overline{x}$; writing $x \notin A$ means that x does not belong to A, or x *is not* its element. However, too liberal use of classes happens to be *controversial*, leading to paradoxes (or, which is the same, antinomies) like Russell's paradox.[1]

We consider these issues in more detail later, and here we say just a few words. Let us consider the collection $\mathcal{A}$ of all classes, which do not belong to itself as an element. For example, the class Z of all integers $Z \notin \mathcal{A}$ because Z itself is not an integer. We can consider the class of all the sets and denote it as $\mathcal{M}$. Then $\mathcal{M} \notin \mathcal{A}$. The class $\mathcal{A}$ can be written down as

$$\mathcal{A} = \{x \mid x \notin x\},$$

---

[1] Bertrand Russell (1872-1970).

where x is a variable for classes. It is clear that for every class, $\mathcal{M}$ we can now write

$$M \in A \Leftrightarrow M \notin M. \tag{3.1}$$

But since here $\mathcal{M}$ can be an arbitrary class, we can take as $\mathcal{M}$ in (3.1) the class $\mathcal{A}$ itself, and (3.1) becomes

$$A \in A \Leftrightarrow A \notin A$$

which is an obvious contradiction. This paradox was discovered by Russell in 1902. We see that no set consisting of 'all sets that do not contain themselves' can exist. Some other paradoxes were discovered in set theory even before Russell's. Still in 1897, Georg Cantor (1845–1918) considered the paradox, based on the *set of all sets*, which is internally contradictory.

Problem 84. *Explain why the following situation is paradoxical. A village barber follows the rule: He shaves all those and only those villagers who do not shave themselves. Must he shave himself?*

All these antinomies are based on the phenomenon of *self-reference*, that is, when one tries to define something through itself, and are to some extent based on the ancient *Liar Paradox*: Someone says: "I'm lying." Is this person a liar?"

Problem 85. *Try to answer the question above.*

Essential efforts were made to remove the paradoxes and restore the reliability of mathematics as a foundation of sciences and of computer science. For instance, if class A contains too many elements (more precisely, if its cardinal number is greater than that of the set $\mathcal{N}$ of natural numbers), any attempt to choose an element of this class is not a trivial problem and must be based on certain sophisticated mathematical tools, like the *choice axiom* or its equivalents.

However, if A is a finite or countable set, the problem can be dealt with in an "elementary" way. These topics are out of the scope of our text. Currently, arithmetic, crucial parts of analysis and geometry are considered to be consistent and contain no contradiction.[2] In particular, the consistency was achieved by distinguishing *classes* as primary entities, and *sets* as secondary *defined objects*, and carefully removing any possibility of self-referencing. However, in this introductory textbook, classes do not appear, and hereafter we only use the term *set*. We always remain at the level of informal, non-axiomatic set theory since we use only finite sets and those parts of mathematics, which firmly established to have no contradictions.

---

[2] The reader can consult [16] for more advanced exposition of those topics.

Problem 86. *What number is denoted by the phrase: "The least integer, which is not nameable with fewer than nineteen syllables."*

We say more regarding these issues in Chapters 18 and 23.

3.1.1   Classical Set Theory

First of all, we describe the ways to represent the sets, to write them down on paper, or on the blackboard, or in computer memory. In the example above, we just gave a description of a class in words (All people who ever lived on the Earth). This way may be vague, and the other methods of presenting classes were developed as well. The simplest way is to write down all the elements of a set A, if we are able to do this. This way is called *list* or *roster notation*. For instance,

$$S_1 = \{1, 2, 3\}$$

$$S_2 = \{A, B, C, \ldots, X, Y, Z\}$$

$$\mathcal{N} = \{0, 1, 2, 3, \ldots\}.$$

Set S_1 contains three natural numbers, set S_2 contains 26 capital letters of the English alphabet; such sets are called *finite*, because they contain only finite number of elements. "Finite" means that we can count all the elements of such a set using the natural (or counting) numbers, starting from 1 or from 0 and reaching but not exceeding some natural number n. This n is called the number of elements or *cardinality* of the class.

For instance, in the first example, 1 is an element of the set S_1, 2 is another element, and 3 is also an element of the set S_1; what is more, S_1 contains no other element. In the second example, A is the 1st element of the set S_2, B is the 2nd element, C is the 3rd element, etc., Y is the 25th element, and Z is the 26th element. We used the writing 1st element, etc., only for convenience – an order of listing of the elements of a set does not matter. We can write the same set as $S_2 = \{B, A, C, \ldots X, Y, Z\}$ or in many other ways; we shall soon discuss this issue in more detail. Because it turns out sufficient to use only the first 26 counting numbers, S_2 is a finite set containing 26 elements. The ellipsis, ..., in this example shows that we just saved some space and did not write all the 26 elements of S_2 explicitly, even though in this example, we can do that.

The ellipsis, ..., in the third example shows that we can continue this pattern, but cannot end it; we are to continue the given pattern to infinity. This is an example of an *infinite set*. The important finite set is the *empty* or *null set* $\varnothing$, that is, the set that does not contain any element. If we imagine a set like a shell or a bag with its elements inside, the empty set is the empty bag.

Another representation of a set is by specifying a characteristic property P, which describes the *elements* of the set, and *only its elements*. Thus, the statement

$$A = \{x : P(x)\}$$

is read as "A is the set of all elements x such that x possesses the property P." Instead of colon, we can use a vertical bar as below,

$$A = \{x \mid P(x)\}. \tag{3.2}$$

This method is called the *set-builder* or *set-generation* notation.

For example, S_2 can be written as $S_2 = \{x \mid x$ is a Latin capital character$\}$ or $\mathcal{N} = \{x \mid x$ is a positive integer$\}$. It is important in these problems to mention *explicitly* the *domain* for the variable x, that is, the largest collection of available values for x. We can write it as, for example,

$$\mathcal{N} = \{x \in \mathcal{R} \mid x \text{ is a positive integer}\}.$$

This symbol emphasizes that, say, complex numbers or triangles as possible values for x, are now beyond our consideration.

Dealing with collections of numbers, we usually call such a collection a *set*. For example:

$$\mathcal{N} = \{x : x \text{ is a natural number}\}$$

represents the set of all *natural numbers* 0, 1, 2, 3, …; or

$$PR = \{x : x \text{ is the current President of the U.S.A.}\}.$$

If we specify the date, for instance, 03/03/1800 or 03/16/2019, this set contains exactly one element.

Problem 87. *What set do we have in this example if the date is given as 03/03/1600?*

Solution. This set is empty.

Along with the set of *natural numbers*

$$\mathcal{N} = \{0, 1, 2, \ldots\}$$

several other number sets have standard notation:

$\mathcal{W} = \{0,1,2,3,\dots\}$—the set of whole numbers

$\mathcal{Z} = \{\dots,-3,-2,-1,0,1,2,\dots\}$—the set of all (positive, zero, and negative) integers

$\mathcal{P}$—the set of prime numbers

$\mathcal{Q}$—the set of rational numbers

$\mathcal{R}$—the set of all real numbers (rational and irrational)

$\mathcal{C}$—the set of complex numbers.

However, there is a pitfall in defining the sets *intensionally*, that is, through the appropriate predicate, as by (3.2). This situation is discussed elsewhere. Now we can continue to study this elementary non-axiomatic set theory.

Two sets are distinguished only on the basis of the elements they contain; the order in which their elements are listed does not count. Thus, the set $S = \{1,2,3\}$ can be written as $S = \{3,1,2\}$, or as $S = \{2,1,3\}$, or in three other ways; all these orderings represent the same set $S = \{1,2,3\}$. Moreover, the writing $\{1,1,2,3\}$ represents the same set S because we treat the repeating symbol 1 as presenting the same element, the natural number 1; indeed, there exists only one natural number 1. It is possible to consider *multi-sets*, but we do not introduce the multi-sets of elements of X.

Definition 8. *A set A is called a subset of a set B, iff each element of A is an element of B, i.e., $\{\forall x : x \in A \Rightarrow x \in B\}$. We denote this relation between sets as $A \subset B$ or $B \supset A$, even if $A = B$, and do not use symbols $\subseteq$ and $\supseteq$.*

If a set A is *not* a subset of a set B, then the definition implies that there is at least one element $x \in A$, which does not belong to B; $x \notin B$. Sometimes the symbol $\subset$ is used to emphasize the strict inclusion: $A \subset B$ means that A is a subset of B, $A \subset B$ but $A \neq B$.

Example 3. Thus, $\{1,2\} \subset \{1,2,3\}$ but $\{1,2\} \not\supseteq \{1,2,3\}$.

Problem 88. Let be $U = \{1,2,3,4,5,6,7,8\}$, $A = \{2,4,6,8\}$, $\{B = \{1,2,3\}$, $C = \{3,4,5,6\}$.

(1) Are the statements $A \subset B$, $A \subset C$, $B \subset A$, true or false?

Solution. *All are false.*

(2) Let the universe be U. Find $A \times C$, $\bar{C}, \bar{U}$, $(A \cap B) \cup C$, $B \setminus C$

(3) Represent $C \subset U$ as a bit string.

Making use of the statement above, we can prove by contradiction that there exists *only one* empty set $\varnothing$. Indeed, if there are two different empty sets, $\varnothing_1$ and $\varnothing_2$, then one of them must contain an element, which does not belong to the other set. But neither $\varnothing_1$ nor $\varnothing_2$ contains any element. Clearly, $A \subset A$ is true for each set A, because every element of the set A on the left is definitely an element of the same set A on the right. It is not so obvious, but it can be proved by contradiction, precisely like in the preceding paragraph, that the empty set is a subset of any set: $\varnothing \subset A$. Otherwise, the empty set would contain an element, which does not belong to A. However, by defini-tion, the empty set does not contain any element at all.

The next two subsets of any set A, the empty set $\varnothing$ and the set A itself, are called *improper* subsets of A; all the other subsets are called its *proper* subsets.

Problem 89. *What is the set of subsets of 1) An empty set? 2) Of the one-element set $\{0\}$?*

Solution. 1) One-element set $\{\varnothing\}$. 2) Two-element set $\{\varnothing, \{0\}\}$.

Problem 90. *Is there a set that does not have a proper subset?*

Definition 9. *Two sets A and B are said to be equal iff every element x belonging to A, also belongs to B, and in turn, every element belonging to B belongs to A. In other words, two sets are equal if they consist of the same elements. In symbols, for every x, , and vice versa, . We denote this with the standard equality symbol: $A = B$. Thus, $A = B$ iff both $A \subset B$ and $B \subset A$.*

This definition looks like a trivial tautology. Nonetheless, it is fundamen-tal because many mathematical statements reduce to proving the identity of two sets, which have seemingly different definitions. We consider three examples.

Example 4. *Let Z_6 denote the set of all natural numbers that are divided (evenly) by 6, and Z_2 and Z_3 stand for the sets of all natural numbers that are divisible evenly by 2 and by 3, respectively. The criterion of divisibility of a natural number by 6 claims that $Z_6 = Z_2 \cap Z_3$, where $\cap$ stands for the inter-section of two sets, i.e., for the set of all their common elements.*

Example 5. *Let $F_0 = \{2\}$ and*

$$F = \{n \in N \mid \text{the equation } x^n + y^n = z^n\}$$

has a non-trivial natural-valued solution, $x, y, z \in N$. Here "non-trivial" means that not all x, y, z can be zeros simultaneously. The Great Fermat theorem, ascending to the middle of 17th century and proven only in 1995-1996, asserts that $F = F_0$, that is, for no integer $n > 2$ the equation above has *a solution in natural numbers x, y, z. On the other hand, if $n = 2$, there are infinitely many Pythagorean triples, e.g., 3, 4, 5; 5, 12, 13, and more.*

Problem 91. *Derive a formula for all the Pythagorean triples.*

Example 6. *Let $F_{>2} = \{x \in N \mid x > 2\}$, and let G be the set of all numbers that can be represented as the sum of three prime numbers. The hypothesis of Christian Goldbach, which has not been proved or disproved now for about three centuries, claims that these two sets are equal. For instance, $6 = 2 + 2 + 2$, $7 = 2 + 2 + 3$, $8 = 2 + 3 + 3$, but we do not know whether we can find such representation for every integer greater than 2.*

These three examples demonstrate three different situations. The statement of Example 4 can be proved easily. Fermat's hypothesis was confirmed just a couple of decades ago, while it is still unknown, whether the Goldbach hypothesis is valid.

The following proof of equation $Z_6 = Z_2 \cap Z_3$ demonstrates standard reasoning used to prove the equality of two sets.

Solution of Example 4. Let $x \in Z_6$. By the definition of this set, x is a multiple of 6, that is, $x = 6k$, where k is another integer number. Since $6 = 2 \times 3$, we can write $x = 6k = 2 \times (3k) = 2 \times k_1$, where $k_1 = 3k$ is also an integer. Hence, we proved that x is a multiple of 2, i.e., x is an even number. However, we can also write $x = 6k = 3 \times (2k) = 3 \times k_2$, where $k_2 = 2k$ is also an integer number. Hence, we proved that x is a multiple of 3. We proved that $x \in Z_2$ and also that $x \in Z_3$, therefore, x is in their intersection, $x \in Z_2 \cap Z_3$. Reading the proof from the beginning, it is clear that we proved that *every* element of Z_6 is an element of the intersection $Z_2 \cap Z_3$. By Definition 2, we proved the inclusion $Z_6 \subset Z_2 \cap Z_3$.

It is worth repeating that we assumed only the inclusion $x \in Z_6$ and did not assume *anything else* about a number x.

Two sets are equal iff each of them is a subset of the other set. Thus to finish the proof of Example 4, we must establish a *reverse* inclusion $Z_2 \cap Z_3 \subset Z_6$. Proceeding in a similar way, we fix an element $x \in Z_2 \cap Z_3$. By definition of the intersection, we conclude that $x \in Z_2$ meaning $x = 2l_1$, and $x \in Z_3$, meaning $x = 3l_3$. But this is the same number x; thus, $2l_1 = 3l_2$. Since the left-hand side of the latter equation is even, its right-hand side, $3l_2$ is also even. However, the numbers 2 and 3 are *mutually prime* or *coprime*; therefore, the number $3l_2$ can be even only if the integer l_2 is even, $l_2 = 2l_3$, where l_3 is also an integer. We proved that $x = 3l_2 = 3 \times 2l_3 = 6l_3$. So that, x is a multiple of 6, and we proved the inclusion $Z_2 \cap Z_3 \subset Z_6$.

Problem 92. *Derive a criterion of divisibility of a positive integer by* $15 = 3 \times 5$.

Solution. Since 3 and 5 are mutually prime, an integer is divisible by 15 iff it is divisible by both 3 and 5.

Problem 93. *The reasoning above looks simple, and indeed it is, but to make sure you mastered it, try to prove that "since* $8 = 2 \times 4$, *a positive natural is divisible by 8 if and only if it is even and divisible by 4." Is this "test" correct?*

When we solve a problem, we never consider all possible sets and their elements. We consider relevant elements and their collections, which have something to do with the problem at hand. Usually, we assume that all the sets under consideration are subsets of a certain set, and all the elements under consideration are the elements of this set. In the specified problem, nothing exists for us outside this special set. This set is called a *universal set* U, or the *universe of discourse*, or just the *universe*.

The universal set is not always stated explicitly, but its existence has always been assumed. Despite its "universal" title and unlike the (unique!) empty set $\varnothing$, the concept of a universal set is relative, and it depends on a problem. For instance, if we solve a problem about the divisibility of natural numbers, the universal set may be the set of all integers Z. In problems of one-variable mathematical analysis (calculus), the universe usually is the set of real numbers $\mathcal{R}$ or $\mathcal{R}^n$.

Definition 10. *Collection of all subsets of a set A is called the set of subsets of A or the Boolean of A, and is denoted by P(A) or by* 2^A.

The latter notation was chosen because if A is a finite set, then the number of elements in 2^A is equal to $2^{|A|}$, where $|A|$ is the *cardinal number* (the number of elements in) of A. This formula is proved below in Theorem 19. For example, if $A = \varnothing$, that is, $|A| = 0$, then $|2^\varnothing| = 2^0 = 1$—in this case, the Boolean contains only one element $\{\varnothing\}$. If $A = \{a\}$, that is, $|A| = 1$, then the Boolean of A, $2^A = \{\varnothing, A\}$, and $|2^1| = 2$ elements. If A contains two elements, $A = \{a, b\}$, that is, $|A| = 2$, then $2^A = \{\varnothing, \{a\}, \{b\}, A\}$; obviously, the Boolean contains $2^2 = 4$ elements, etc.

Problem 94. *Give a roster and word description (set-builder notation) of the following sets:*

a. $\{1, 2, 3, 4\}$

b. $\{1, 4, 9, 16, 25, 36\}$.

Which of the following sets is a subset of some other set in the list below?

c. $\{1, 2\}$

d. $\{\{1\}, 2\}$

e. $\{1, 1, 2\}$

f. $\{1, 2, 3, 4\}$

State the inclusions between the number sets $\mathcal{N}, \mathcal{W}, \mathcal{Z}, \mathcal{Q}, \mathcal{R},$ *and* $\mathcal{C}$.

Problem 95. *Which of the following inclusions are correct?*
$2 \in \{1,2,3\}$; $\{2\} \in \{1,2,3\}$; $2 \in \{1,2,3\}$; $\{2\} \in \{1,2,3\}$; $\{2\} = \{1,2,3\}$; $7 \subset \{\varnothing, 3, 7, \{7\}\}$; $\{7\} \subset \{\varnothing, 3, 7, \{7\}\}$.

Problem 96. *Prove that if* $A \subset B$ *and* $B \subset C$, *then* $A \subset C$. *This statement means that the inclusion is a transitive binary relation, see Chapter 4.*

Find three sets A, B, and C, such that $A \in B, B \in C,$ *but* $A \notin C$. *This statement means that unlike the inclusion of sets, the relation* $A \in B$ *is not a transitive operation.*

Problem 97. *List all the subsets, that is, find the Boolean of the sets* $\{1\}$; $\{1, 2\}$; $\{1, 2, 3\}$; $\{1, 2, 3, 4\}$.

Problem 98. Compare the number of elements and the number of subsets in the sets in Problem 95 above. What is the relationship between these two quantities?

Problem 99. If $A \subset D$, what is the relation between the Booleans 2^A and 2^D?

Problem 100. For a set X and any two-element set $Y = \{a, b\}$ put the Boolean 2^X and the power-set Y^X into one-to-one correspondence.

3.1.2 Operations on Sets: Set Diagrams

The operations on sets are defined through the Boolean operations over the propositions. We always assume that all the sets under consideration are subsets of a certain **universal set** $\mathcal{U}$.

Definition 11. *The complement of a given set* $A \subset \mathcal{U}$ *is the set of all the elements of* $\mathcal{U}$ *not belonging to A. Sometimes the complement is denoted as an upper bar* $\bar{A}$ *or as a superscript* A^c *(c stands for complement). It is obvious that* $\mathcal{U} \setminus A = A^c$, *thus, the set-difference* $B \setminus A$ *may be called the relative complement of A to B.*

Example 7. *If* $\mathcal{U} = \{1, 2, 3, 4, 8, 9\}$ *and* $A = \{1, 2, 3, 4\}$, *then* $\overline{A} = \{8, 9\}$. *If* $B = \{1, 3\}$, *then* $\overline{B} = \{2, 4, 8, 9\}$. *However, if* $C = \{1, 5\}$, *then* $\overline{C}$ *is undefined, since C is not a subset of* $\mathcal{U}$. *To treat these cases, we introduce more set-theory operations. In what follows, all the sets are supposed to be subsets of the same universal set* $\mathcal{U}$.

Definition 12. Given two sets A and B, the set-difference $A \setminus B$ is the set of all the elements of A not belonging to B. Sometimes it is denoted as the regular difference of numbers, $A \setminus B = A - B$.

Definition 13. *The intersection of two sets A and B is the set of all the elements of U belonging to both A and B; it is denoted as* $A \cap B$. *The union of two sets A and B is the set of all the elements of U belonging to at least one of A and B; it is denoted as* $A \cup B$. *Equivalently these operations can be defined as*

$$A \setminus B = \{x \mid x \in A \wedge x \notin B\}$$

$$A \cap B = \{x \mid x \in A \wedge x \in B\}$$

$$A \cup B = \{x \mid x \in A \vee x \in B\}.$$

For example, if $A = \{1, 2, 3, 4\}$, $B = \{1, 3, 5\}$ and $C = \{6\}$, then $A \cap B = \{1, 3\}$, $A \cup B = \{1, 2, 3, 4, 6\}$, $A \cap C = \varnothing$, $A \cup C = \{1, 2, 3, 4, 6\}$, $A \setminus B = \{2, 4\}$, $B \setminus A = \{5\}$, $A \setminus C = A$.

Problem 101. *Let* $\mathcal{N}$ *and* $\mathcal{Z}$ *be the sets of the natural numbers and of the integers, respectively. What are the elements of* $\mathcal{N} \cap \mathcal{Z}$, $\mathcal{N} \cup \mathcal{Z}$, $\mathcal{N} \setminus \mathcal{Z}$, $\mathcal{Z} \setminus \mathcal{N}$, $\mathcal{Z} \setminus \varnothing$, $\varnothing \setminus \mathcal{Z}$, $\varnothing \cap \mathcal{Z}$, $\varnothing \cup \mathcal{Z}$.

These examples give rise to several plausible hypotheses. In particular, that the union and intersection of sets are commutative operations. Surely, no number of examples can prove a claim, but in this case, this is correct and easily follows from the commutativity of the disjunction and conjunction, respectively. Also, we see that $A \setminus B \neq B \setminus A$. This example confirms that the set-difference is not a commutative operation. This last statement means that there exists at least one pair of sets A and B such that $A \setminus B \neq B \setminus A$ and we presented above two such sets. Such examples in similar situations are called *counter-examples* because they disprove a possible hypothesis.

Problem 102. *For what sets A and B, the set-difference is commutative, i.e.,* $A \setminus B = B \setminus A$?

In the following equations, A, B, etc., represent arbitrary sets within an universe $\mathcal{U}$.

$\overline{\overline{A}} = A$—Double-Complementation Law

$\mathcal{U} \cap A = A$, $\mathcal{U} \cup A = \mathcal{U}$—Identity and Domination laws (Properties of the empty set and of the universe)

$A \cup \overline{A} = \mathcal{U}$—Complementation law

$A \cap \overline{A} = \varnothing$—Exclusion law

$A \cup A = A \cap A = A$—Idempotency laws for the union and the intersection

$A \cup B = B \cup A$, $A \cap B = B \cap A$—Commutative laws

$(A \cup B) \cup C = A \cup (B \cup C) = A \cup B \cup C, (A \cap B) \cap C = A \cap (B \cap C) = A \cap B \cap C = A \cap B \cap C$—Associative laws for the union and the intersection. Due to these rules, we can write down many formulas without parentheses at all, like the right parts above. Moreover, we can now define the union and the intersection of any finite collection of sets by induction, for example, as follows: $A \cap B \cap C \cap D = A \cap (B \cap (C \cap D))$. We will write the unions and intersections of any number of terms without parentheses as well, and no ambiguity occurs. $A \cup (B \cap C) = (A \cup B) \cap (A \cup C)$, $A \cap (B \cup C) = (A \cap B) \cup (A \cap C)$—Distributive laws for the union and the intersection. Unlike our familiar high-school algebra, in this algebra, which is a particular case of *Boolean Algebra*, the union and intersection operations are entirely symmetric; therefore, there are *two* symmetric distributive laws. These formulas can be used in two ways: from left-to-right when we want to "multiply" the outer set by the set-polynomial in parentheses and open these parentheses; or from right-to-left, when we want to factor out the set-common factor. Note also that we can prove by induction that the parentheses on the right-hand sides of these formulas can contain any finite number of terms; for example, $A \cup (B \cap C \cap D) = (A \cup B \cap (A \cup C) \cap (A \cup D)$. That is, to "multiply" such set-polynomials, we can proceed in a well-known way and combine every term from the first parentheses with every term from the second parentheses, as, for example,

$$(A \cap B \cap C) \cup (D \cap E) = (A \cup D) \cap (A \cup E) \cap (B \cup D) \cap (B \cup E) \cap (C \cup D) \cap (C \cup E).$$

$$A \cup (A^c \cap B) = A \cup B$$

$A \cap (A^c \cup B) = A \cap B$—These two formulas are called crossing-out rules because their right-hand sides are derived from their left-hand sides by crossing out the middle parts.

$$A \cup (A \cap B) = A$$

$A \cap (A \cup B) = A$—The Absorption Laws are called so because the set A on the left *absorbs* a combination of itself any B.

$$\overline{A \cup B} = \overline{A} \cap \overline{B}$$

$\overline{A \cap B} = \overline{A} \cup \overline{B}$—De Morgan (1806-1871) laws, two important formulas, showing that one can rewrite any equation without either union or intersection.

Problem 103. *Prove set-theory identities.*

$$A \setminus (B \cup C) = (A \setminus B) \cup (A \setminus C)$$
$$A\Delta(A\Delta B) = B$$

Problem 104. *Prove that $A \subset B$ iff $A \setminus B = \varnothing$.*

Remark 3. *In computer science, sets are often called types, for instance, the type of integers, real type, etc. It is important to remember that a type in computer science always is a finite set. Thus, integer type in any computer language contains not all integers $z \in \mathcal{Z}$ but only the integers in a specific range; the latter depends on the compiler, machine implementation, etc. Algorithmic languages of higher levels contain special tools, which are similar to the roster or set-builder notation and provide the user with the ability to create new types necessary for solving the problem in hand.*

Problem 105. *Solve the next equations for a set X, where $A, B, C \subset \mathcal{U}$ are fixed sets—coefficients or parameters of an equation and X is a variable for subsets of a fixed universe $\mathcal{U}$.*

$$A \cap X = B$$
$$A \cup X = B$$
$$A \setminus X = B$$
$$X \setminus A = B$$
$$A \cap X \cup B \equiv C$$

Problem 106. *Solve the systems of simultaneous equations.*

$$\begin{cases} A \cap X = B \\ A \cup X = C \end{cases} \quad \begin{cases} A \setminus X = B \\ X \setminus A = C \end{cases}$$

To visualize various relations between sets, people often use drawings, called sometimes the *Euler (1707-1783) —Venn (1834-1923) diagrams*. In such drawings, the universal set is usually represented as a rectangle, or an oval, a circle, any ambient curve that contains ovals representing sets in this universe, see, for example, Figure 3.1. The diagrams may be used to clarify a problem, to give a tip, or even to suggest the answer. Most of the equations above become obvious when we look at the corresponding diagram.

However, we have to emphasize that no picture is accepted in lieu of proof because there may be situations, which we can not even imagine while drawing the picture. An illuminative example of such a situation is a continuous function that does not have a derivative. That means that at no point its graph has a tangent. It is difficult not only to draw but even to imagine such a curve. Nevertheless, the existence of such a function can be proved by appropriate analytical reasoning. That is why all the time, we use drawings in mathematics, in particular, the Euler-Venn diagrams. We look at them trying to find clues to our problems, and then we have to produce an analytical—logical proof of the statement under discussion. Situations, when our intuition misleads us, will be discussed in the following section.

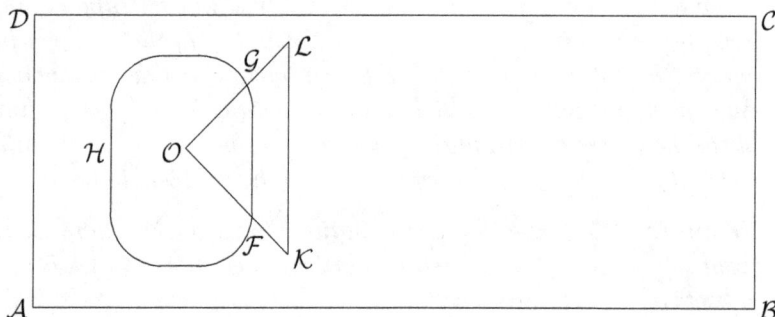

FIGURE 3.1 In this Euler-Venn diagram, the universe is the interior of a quadrilateral $ABCD$. The interior of the triangle $\triangle OFG$ is the *intersection* of the oval and the big triangle $\triangle OKL$. The interior of the figure $HGLKFH$ together with the big triangle $\triangle OKL$ in the union of these two sets.

Problem 107. *Draw Euler-Venn diagrams for the formulas above.*

Problem 108. *Prove that any integer can be written as 3k + 5l with some integers k and l.*

Proof. Depending upon the divisibility by 3, every integer can be written as $n = 3p$, or $n = 3p + 1$, or $n = 3p + 2$. In the first case, we set $k = p$, $l = 0$. In the second case, $n = 3p + 1 = 3(p - 3) + 2 \times 5$, thus we set $k = p - 3, l = 2$. In the third case, $n = 3p + 2 = 3(p - 1) + 5$, and we set $k = p - 1, l = 1$.

Problem 109. *Is it possible to represent every integer as a linear combination of 3 and 9 with integer coefficients? As a linear combination of 3 and 4 with integer coefficients?*

Descriptions of sets given above are good for human beings, but computers do not necessarily understand them. To make representations of sets understandable to computers, we must *digitize* the descriptions. One possible way to do that is as follows. First of all, we fix the *universe*; all the sets under consideration belong to. Therefore, every set in the universe is described as a collection of *binary strings*. For instance, if $U = a, b, c, d$, and we consider a *set* $A_{1,3} = a, c$, the latter is described as the string $\{1, 0, 1, 0\}$. There is a one-to-one correspondence between set-theory operations over the sets and termwise Boolean operations with the strings. For instance, we have $a, c \cup c, d = a, c, d$. These three sets are represented by the following three strings, $\{1,0,1,0\}$, $\{0,0,1,1\}$, and $\{1,0,1,1\}$; the termwise disjunction of the two strings is the third one: $\{1,0,1,0\} \vee \{0,0,1,1\} = \{1,0,1,1\}$.

3.2   EXERCISES

Exercise 3.1. *Prove that 2 and 3, and also 31 and 37 are mutually prime numbers. Are 14 and 49 mutually prime?*

Solution. It is enough to check the prime numbers not exceeding 7.

Exercise 3.2. *Prove that a natural number is divisible by 10 iff it is even and is a multiple by 5.*

Solution. Verify that 2 and 5 are mutually prime.

Exercise 3.3. *Prove that any integer can be written as a linear combination 3k + 7l with some integer coefficients k and l.*

CHAPTER 4

GROUPS, RINGS, AND FIELDS

4.1 GROUPS, RINGS, AND FIELDS

Up to now, we have used the commonly accepted current mathematical language, that is, the language of sets and maps, but without too much formalization. In this chapter, we consider certain structures, which Crypto appears in many problems. These structures sometimes look quite different but have very similar properties. Hence, this underlying formal structure deserves to be studied on its own. One does not have to repeat that study again and again, and it is enough to apply the general results in any particular application. As examples of structures, we consider the *groups*, *rings*, and *fields*. The reader will notice, of course, that at every step, we *add* one more operation to the previous structure. Thus, a field is a ring but not vice versa. Rings, in general, have no division; hence, they cannot be fields in general. A ring is a group, but not vice versa, etc.

Definition 14. *An additive group is a set G, whose elements can be added; that is, for every element $a, b \in G$, their sum is defined, and is an element of the group, and is associative, $(a + b) + c = a + (b + c)$. The group has the unique neutral element 0 and for any element $a \in G$, there exists the unique opposite element $-a \in G$, such that $a + (-a) = 0$. If the group operation is commutative, $a + b = b + a$ if it is an additive group or $a \cdot b = b \cdot a$ if it is multiplicative, the group is called commutative or Abelian.*

For example, the set of all integers is a commutative group with respect to the addition, and its neutral element is 0. Moreover, with respect to the multiplication, it is a different group, and the neutral element now is 1.

Problem 110. *Verify these statements.*

Problem 111. *Verify that the set of all the non-singular square matrices of order n with real elements is a non-commutative multiplicative group with respect to the matrix multiplication. What is its neutral element?*

Solution. The verification is straightforward. The neutral element is the square matrix of nth order with 1s on the main diagonal, and all the other elements are 0.

The existence of the inverse elements in multiplicative groups some-times serves as a substitution to division, which in a group is, in general, undefined. The following properties allow one to solve equations in groups.

Problem 112. *Let g,h,k, $g \neq 0$ be elements of a group G with a group multiplication $*$. Prove that if $g * h = g * k$, then $h = k$, and $h * g = k * g$ implies $h = k$.*

Solution. Indeed, since g^{-1} exists in the group, we can write $g^{-1} * (g * h) = g^{-1} * (g * k)$; thus, after simple identical transformations using group axioms, we conclude that $h = k$. The second equation is dealt with in the same way.

Problem 113. *Let G be a group with the identity element ε. Prove that if $x^2 = x$ for some $x \in G$, then $x = \varepsilon$. What about equations $x^3 = x^2$ and $x^3 = x$?*

Problem 114. *Consider the set of all the rotations of the three-dimensional Euclidean space $\mathbb{R}^3$, and define a consecutive rotation as their multiplication. Is it a group? Is it an Abelian group?*

Solution. It is a non-commutative group.

An additive group with a multiplication, subject to some additional properties, makes a ring.

Definition 15. *A ring is a non-empty set $K \neq \varnothing$ equipped with the two binary operations, denoted by the traditional symbols $+$ for the addition and $\cdot$ or $\times$ for the multiplication, mapping the Cartesian product $K \times K$ to itself, $+: K \times K \to K$ and $\cdot: K \times K \to K$, and satisfying the following axioms for any elements $a,b,c \in K$:*

$$K_1 : a + b = b + a$$
$$K_2 : a + (b + c) = (a + b) + c$$
$$K_3 : \forall a \forall b \exists! c (a + c = b)$$
$$K_4 : a \cdot (b \cdot c) = (a \cdot b) \cdot c$$
$$K_5 : a \cdot (b + c) = a \cdot b + a \cdot c$$
$$K_6 : (a + b) \cdot c = a \cdot c + b \cdot c.$$

Axiom K_3 guarantees that the subtraction in a ring is always possible, and the difference c is unique due to the definition of the symbol $\exists!$. The two distributive rules are needed since, in general, the ring multiplication *is not commutative*. For example, the standard "row by column" multiplication in the ring of square matrices of size $n \geq 2$ is not commutative. If this does not lead to misunderstanding, the symbol of multiplication is often omitted.

To simplify our writing, in all the axioms, where the quantifiers are not written down explicitly, it is assumed that all the "free" variables are within the scope of a certain universal quantifier. For example, axiom K_1 is an abbreviation of the following writing, $K_1 : (\forall a)(\forall b)(a + b = b + a)$.

The ring addition, unlike multiplication, is always commutative.

Definition 16. *A ring is called commutative, if its multiplication is commutative, that is, $a \cdot b = b \cdot a$ for all a and b.*

If $b = a$ in the axiom K_3, then one gets $\forall a \exists! c(a + c = a)$. The element c can be denoted as 0_a, that is, $a + 0_a = a$, and is naturally called the neutral element of the ring, corresponding to the element a. Quite similarly, one can write $b + 0_b = b$. Adding these two equalities, we get $(a + 0_a) + b = a + (b + 0_b)$, or $a + b + 0_a = a + b$, thus, $0_a = 0_b + b$ for any $b \in K$. Since by changing b, we can get as $a + b$ any element $c \in K$ (due to what axiom?), we conclude that all the neutral elements in a ring are equal to 0_a, that is, they are the same. This unique element is called the *neutral element* or *the zero* of the ring K; we denote it just as 0.

Problem 115. *Prove that in a ring, $\forall a(a \cdot 0 = 0 \cdot a = 0)$. Moreover, a ring can have no more than one unit, that is, an element e such that $\forall a \in K(a \cdot e = e \cdot a = a)$; a ring with such an element e is called unital.*

Problem 116. *Prove that the set of all even integers with the standard operations of addition and multiplication is a ring without a unit.*

Example 8. *Let $\mathcal{A}$ be any set, and $P(\mathcal{A})$ be its Boolean, that is, the set of all its subsets. Introduce on the Boolean $P(\mathcal{A})$ the operations addition and multiplication as $A + B = (A \cup B) \setminus (A \cap B)$ and $A \cdot B = A \cap B$. Prove that the Boolean of any set with these operations is a ring.*

If we look at the various mathematical structures, we observe that we have considered structures with more and more involved algebraic operations; for example, in rings, one can multiply elements, but division, in general, is undefined. Thus, in the previous problem, both 6 and 2 are even integers, but $6 \div 2$, though an integer, is *not even*. Now we introduce algebraic structures with *division*.

Definition 17. *An unital commutative ring, such that every non-zero element in it has a multiplicative inverse, is called a field.*

Example 9. *The set* $\mathbb{B}_2 = \{0,1\}$ *is a ring with the operations defined as addition and multiplication modulo 2, that is,*

TABLE 4.1 Addition and Multiplication in the Algebra $\mathbb{B}_2$.

p	q	$p + q$	$p \cdot q$
0	0	0	0
0	1	1	0
1	0	1	0
1	1	0	1

Problem 117. *Prove that* $\mathbb{B}_2 = \{0,1\}$ *is a field with just two elements. What are the inverses,* $(0)^{-1}$ *and* $(1)^{-1}$?

Problem 118. *Given two sets A and B, the set* $A \Delta B = (A \cup B) \setminus (A \cap B)$ *is called the symmetric difference of A and B. Prove that* $A \Delta B = (A \setminus B) \cup (B \setminus A)$.

Typical examples of fields are number fields, the field of real numbers $\mathbb{R}$, and the field of complex numbers $\mathbb{C}$. The former is a *one-dimensional vector space*, that is, every real number x can be written as $x = a_x \cdot x_0$, where x_0 is a fixed real number and $a = a_x$ is a real coefficient. The field of complex numbers $\mathbb{C}$ is a *two-dimensional space*, that is, every complex number z can be written as $z = a_z \cdot x + \beta_z \cdot y$, and the latter cannot be reduced to one real number. It must be noticed here, that while going from real to complex numbers, we not only grossly enlarged the supply of elements, but also we lost something. Namely, the field of real numbers is *ordered*, that is, given any two real numbers x and y, either $x < y$, or $y < x$, or $x = y$, however, two complex numbers are not comparable, in general. This leads to the following definition.

Definition 18. *An ordered field is a field with a binary relation of the strict order* <, *such that for any three elements a,b,c, either* $a < b$, *or* $a = b$, *or* $b < a$. *Moreover, the field addition and multiplication are monotone, in the sense that* $a < b$ *implies* $a + c < b + c$ *and* $(0 < a) \wedge (b < c)$ *imply* $a \cdot b < a \cdot c$.

Thus, the set of real numbers $\mathbb{R}$ with the standard operations of addition, multiplication, and with the relation of strict order is an ordered field, while the field of complex numbers $\mathbb{C}$ is not ordered.

Problem 119. *Let X be any set. Prove that a power-set has the ring structure, if one defines the operations on functions in B^X pointwise, as in B.*

Moreover, let K be a ring. For a natural number, $n \geq 2$ consider the Cartesian product $K^n = \underbrace{K \times K \times \cdots \times K}_{n\text{times}}$. The elements of the product are the n-strings of the elements of K. The ring operations on these strings are defined coordinate-wise, that is, $(x_1, \ldots, x_n) + (y_1, \ldots, y_n) = (x_1 + y_1, \ldots, x_n + y_n)$ and $(x_1, \ldots, x_n) \cdot (y_1, \ldots, y_n) = (x_1 \cdot y_1, \ldots, x_n \cdot y_n)$.

Problem 120. *Prove that these operations introduce a ring structure on K^n. What are the neutral elements 0 and 1 in K^n?*

In particular, the set B^n of n–tuples of zeros and ones with coordinate-wise operations is a ring.

4.2 MATRICES AND DETERMINANTS

We will also have several occasions to use matrices, even though they are usually studied more in Linear Algebra than in Discrete Mathematics. A $k \times l$ matrix is a rectangular table of k rows by l columns, whose elements are numbers or some other items, for example, functions; an $a_{x,y}$ stands for the item at the intersection of the xth row and yth column; the first element x always stands for the rows, while the second for the columns; for example, the 2×3 matrix

$$A = \begin{pmatrix} 0 & 1 & -2 \\ \pi & a+6 & 0 \end{pmatrix}.$$

Thus, each row and each column of a matrix is a vector, either a row-vector or a column-vector. We discuss here only some basic features of the matrices, needed in this course. The matrix A in the example above is $a_{i,j}, i = 1, 2; j = 1, 2, 3$, in particular, $a_{1,2} = 1$ and $a_{2,1} = \pi$.

Problem 121. *What are the other elements $a_{x,y}$ in the matrix A above?*

When we introduce any new entities in mathematics, we are, first of all, interested in what we can do with them; thus, first, we consider arithmetic operations with the matrices. Their addition/subtraction is done element-wise; therefore, all the familiar addition/subtraction properties of numbers, like the *commutativity, associativity*, existence of *neutral elements* are preserved. To multiply a matrix by a number (a scalar) b, we also do that component-wise, as

$$-3A = \begin{pmatrix} 0 & -3 & 6 \\ -3\pi & -3a - 18 & 0 \end{pmatrix}.$$

Matrix transposition is "flipping" a matrix with respect to its main diagonal and is denoted with the asterisk $*$ or with ^{T}, for example, with A as above,

$$A^* = A^T = \begin{pmatrix} 0 & \pi \\ 1 & a+6 \\ -2 & 0 \end{pmatrix}.$$

Matrix multiplication is not always possible. Consider two matrices, A of size $k \times l$ and B of size $p \times q$. It turns out, that the only reasonable way to define the product $A \times B$ is as follows. If $l \neq p$, the product is undefined. If $l = p$, the product $A \times B$ is defined and consists of the dot-products of row-vectors of A by column-vectors of B. Denote the matrix product $A \times B = (c_{i,j})$; every element $c_{i,j}$ is the dot or scalar product of the *ith* row of the matrix A by the $j-th$ column of the matrix B, that is, $c_{i,j} = a_{i,1} \cdot b_{1,j} + \cdots + a_{i,l} \cdot b_{l,j}$. For example, if

$$C = \begin{pmatrix} 0 & 1 \\ 1 & 6 \\ -2 & 0 \end{pmatrix},$$

and A is as above, then

$$A \times C = \begin{pmatrix} 5 & 6 \\ a+6 & \pi + 6a + 36 \end{pmatrix}.$$

Problem 122. *Unlike the product $A \times C$, which is a 2×2 matrix, the product $C \times A$ is a 3×3 matrix. Compute it, thus proving that the matrix multiplication is not commutative, even if both products do exist.*

Moreover, if both matrices A and B are square matrices, that is, $k = l$, and are of the same size, say, k, thus, both $A \times B$ and $B \times A$ exist and have the same size $k \times k$, they do not have to be equal. Matrix multiplication is not commutative.

Problem 123. *Find two 2×2 matrices such that both products $A \times B$ and $B \times A$ are defined and, therefore, have the same size 2×2, however, $A \times B \neq B \times A$.*

Inverse matrix also may not exist. Consider square matrices again and start with a 2×2 matrix $A = \begin{pmatrix} a & b \\ c & d \end{pmatrix}$. The quantity $a \cdot d - b \cdot c$ is called a *determinant* of second order; it is often written as

$$det(A) = \begin{vmatrix} a & b \\ c & d \end{vmatrix} = a \cdot d - b \cdot c.$$

If A is a number matrix, its determinant is a number. For example,

$$det(A \times C) = 5 \cdot (\pi + 6a + 36) - 6 \cdot (a + 6) = 5\pi + 24a + 144.$$

Determinants of higher orders will be defined *recursively*. The 2×2 case was just defined. It is natural to define the determinant of a number a, that is, a matrix of the first order, as the number itself, $det(a) = a$. The next case is about matrices of order 3. Given a 3×3 matrix

$$A = \begin{pmatrix} a_{1,1} & a_{1,2} & a_{1,3} \\ a_{2,1} & a_{2,2} & a_{2,3} \\ a_{3,1} & a_{3,2} & a_{3,3} \end{pmatrix}, \tag{4.1}$$

its Laplace expansion over the first row is an alternating sum

$$det(A) = a_{1,1} \begin{vmatrix} a_{2,2} & a_{2,3} \\ a_{3,2} & a_{3,3} \end{vmatrix} - a_{1,2} \begin{vmatrix} a_{2,1} & a_{2,3} \\ a_{3,1} & a_{3,3} \end{vmatrix} + a_{1,3} \begin{vmatrix} a_{2,1} & a_{2,2} \\ a_{3,1} & a_{3,2} \end{vmatrix}.$$

Every term of this sum is the product of an element $a_{1,j}$ of the first row by a 2×2 matrix multiplied by the + or – sign $(-1)^{1+j}$. These products, say for $j = 2$, are

$$(-1)^{1+j} \begin{vmatrix} a_{2,1} & a_{2,3} \\ a_{3,1} & a_{3,3} \end{vmatrix}.$$

They are called *co-factors* of the element $a_{1,j}$.

Problem 124. *Given the* 3×3 *matrix*

$$B = \begin{pmatrix} 1 & 2 & 3 \\ 0 & -2 & 2 \\ -3 & 3 & -2 \end{pmatrix},$$

find its co-factors and compute $det(B)$ *by making use of the Laplace expansion over the first row. Generalize the Laplace expansion over any row or column of the matrix, and compute* $det(B)$ *again, expanding it over every row,* $1 \le i \le 3$, *and every column,* $1 \le j \le 3$; *the result must be always* $det(B) = -32$.

Now one can define *recursively* the determinant of any square matrix. Indeed, since we know how to compute the determinants of the third order, we use the Laplace expansion to define the determinants of the fourth order, fifth, etc. The Laplace expansion itself has remained unproved; see any good book in Linear Algebra, for example, [34, p. 52].

There are certainly equivalent non-recursive definitions of the determinants. For example, we can notice that a determinant of order n is a polylinear form consisting of $n!$ products of its elements, whose signs alternate; each product contains n elements, one from each row and from each column. For instance, determinant (4.1) can be determined as the alternating sum of $3! = 6$ products

$$det(A) =$$

$$= a_{1,1}a_{2,2}a_{3,3} + a_{1,2}a_{2,3}a_{3,1} + a_{2,1}a_{3,2}a_{1,3} - a_{1,3}a_{2,2}a_{3,1} - a_{2,1}a_{1,2}a_{3,3} - a_{1,1}a_{2,3}a_{3,2}.$$

$$(4.2)$$

This assignment of signs of the triple products can be conveniently visualized by the Sarrus rule in Problem 125.

Problem 125. *In matrix (4.1), connect the triple of elements carrying the* + *sign in* $det(A)$ *with straight lines, and connect the triple elements carrying the* – *sign in (4.1) with dashed lines.*

Problem 126. *Extend the Sarrus rule for the determinants of the fourth order.*

A proof of the following important theorem can be found in many books in Linear Algebra, for example, in [34, p. 49].

Theorem 7. *If A and B are square matrices of the same size, then*

$$det(A \times B) = det(A) \cdot det(B).$$

Definition 19. *A square matrix with the vanishing determinants is called singular; otherwise, the matrix is called non-singular.*

Thus, matrix B in Problem 124 is non-singular, while the matrix $A = \begin{pmatrix} 1 & 2 \\ 3 & 6 \end{pmatrix}$ is singular.

Definition 20. *Let A be a square matrix. The matrix B is called inverse (to A) and is denoted as A^{-1}, iff $A \times A^{-1} = A^{-1} \times A = I$, where I is a square matrix of the same size as A. In this case, A is called the invertible matrix.*

Theorem 7 implies that if A is invertible, then A^{-1} is also invertible, and vice versa.

Problem 127. *Prove that if the matrix A is non-singular, then $det(A^{-1}) = (det(A))^{-1}$. The exponent on the left means the inverse matrix, while on the right, it stands for the reciprocal of a number $det(A)$.*

We will need the following definition.

Definition 21. *A square matrix $A = (a_{i,j})$ is called symmetric iff $a_{i,j} = a_{j,i}$ for all $1 \le i, j \le n$. For example, the matrix A in (4.1) is symmetric iff $a_{2,1} = a_{1,2}$, $a_{3,1} = a_{1,3}$, and $a_{2,3} = a_{3,2}$. The definition imposes no restriction on the diagonal elements $a_{i,i}$. On the other hand, the matrix B in Problem 41 is not symmetric.*

4.3   FINITE GROUPS

Groups were introduced above. Finite groups often occur in cryptography, and we go into some details here. A group G is finite if it consists of finitely many elements; their number is called the order $\rho = ord(G) = |G|$ of G. For example, the two-element set $\mathbb{B} = \{0, 1\}$ is a group with respect to the *binary addition*. Its *neutral element* is 0, and every element of $\mathbb{B}$ is *inverse* to itself. This two-element set has a rich algebraic structure, it is a ring and even a field. First of all, we discuss some examples.

Example 10. *Consider any finite set $A = \{a_1, a_2, \ldots, a_d\}$ of cardinality d, and all the permutations of its elements; we know that there are d! of these* *permutations. As a group operation ∗, we consider a superposition of two permutations, that is, $\rho_2 * \rho_1 = \rho_2(\rho_1)$, meaning that we first apply the permutation ρ_1 to A, and then ρ_2 to the image of the previous operation.*

Problem 128. *Prove that this is a non-commutative group, often denoted as $\mathbb{Z}_d$. Find its neutral element and the inverse elements for each permutation. This finite group of permutations is called the symmetric group S_d of order $ord(S_d) = d!$. It is useful to consider some special cases. For instance, write explicitly all the permutations of three and four elements.*

Let us consider a natural number d and the set $\mathbb{Z}_d = \{0,1,\ldots,d-1\}$ of natural numbers *modulo d*, see Chapter 9. This congruence relation is the equivalence relation on the symmetric group $\mathbb{Z}_d$, and its d equivalence classes are the arithmetic progressions of the integers having the same remainder after dividing by d. By $\mathbb{Z}_d^*$ we denote the set of natural numbers k, which are mutually prime with d, that is, $k \in \{0,1,2,\ldots,d-1\}$ and $gcd(k,d) = 1$. For example, $\mathbb{Z}_8 = \{0,1,2,3,4,5,6,7\}$, and $\mathbb{Z}_8^* = \{1,3,5,7\}$. The next claim will be proven in Chapter 8.

Theorem 8. *The set $\mathbb{Z}_d^*$ makes a commutative group, where the group operation is the multiplication modulo d, with the identity element 1.*

Problem 129. *Prove that $\mathbb{Z}_7^* = \{1,2,3,4,5,6\}$.*

Problem 130. *Prove that in general, if d is prime, then $\mathbb{Z}_d^* = \{1,2,\ldots,d-1\}$, thus, when d is prime, $\mathbb{Z}_d^*$ consists of d − 1 elements.*

Definition 22. *Let $\{G, *\}$ be a group with the group operation ∗, and an element $g \in G$ be such that $g^{*k} = \underbrace{g * g * \cdots * g}_{k\,times} = \beta$, where g repeats k times and β is the identity element of G. The smallest positive natural number k, such that g satisfies this equation in this definition, is called the order of the element g, denoted ord(g). The order of the identity element β is 1.*

Example 11. *Consider the congruence modulo 7, with $\mathbb{Z}_7 = \{0,1,2,3,4,5,6\}$ and $\mathbb{Z}_7^* = \{1,2,3,4,5,6\}$. Since $1^{*k} = 1$ for any k, ord(1) = 1. Next,*

$2^{*1} = 2$, $2^{*2} = 4$, and $2^{*3} = 8 \bmod (7) = 1$, $2^{*4} = 16 \bmod (7) = 2$, $2^{*5} = 32 \bmod (7) = 4$, and $2^{*6} = 64 \bmod (7) = 1$; the remainders repeat and $ord(2) = 3$. The powers of 3 are $3^{*1} = 3$, $3^{*2} = 9 \bmod 7 = 2$, $3^{*3} = \bmod 7 = 6$; $3^{*4} = \bmod 7 = 4$, $3^{*5} = \bmod 7 = 5$, $3^{*6} = \bmod 7 = 1$; thus, $ord(3) = 6$.

The next number is 4, thus, $4^{*1} = 4$, $4^{*2} \bmod 7 = 2$, $4^{*3} \bmod 7 = 1$; $ord(4) = 3$; $5^{*1} = 5$, $5^{*2} \bmod 7 = 4$, $5^{*3} \bmod 7 = 6$, $5^{*4} \bmod 7 = 2$, $5^{*5} \bmod 7 = 3$,

5^{*6} mod 7 = 1; $ord(5)$ = 6, and finally, 6^{*1} = 6, 6^{*2} = 1 mod 7; $ord(6)$ = 2. We see that different elements may have different orders, but all the different orders, in this example 1,2,3,6, are divisors of the order of the group, in this example, of the number $d - 1 = 7 - 1 = 6$.

That observation can be proved in general; see the following Theorem 9, but we omit a proof.

Theorem 9. *The order of any element of a finite group divides the order of the group* $|G|$.

Problem 131. *Let G be a finite group of order n with the unit element e. Prove that $a^n = e$ for every element $a \in G$.*

4.3.1 Cyclic Groups

These examples suggest that a periodical structure, which we observed in this particular case, should exist in general when we deal with the comparison of the integers. This periodicity is explored in more detail now. It is clear that an order of any element in a finite group cannot be bigger than the order of the group itself. Group elements having this largest possible order $ord(G) = |G|$ are called *primitive elements* or *generators* of the group. Thus, as we saw above, the group $\mathbb{Z}_7^*$ has two generators 3 and 5. since both these elements have the largest possible order 6 = 7 – 1. We also see that the powers of both these elements, 3 and 5, - of course, *modulo* 7, are 3,2,6,4,5,1 and 5,4,6,2,3,1, that is, each sequence is the entire group $\mathbb{Z}_7^*$; the elements with the maximal order are called *generators* of a group, and these groups are called *cyclic groups*.

Problem 132. *Are $\mathbb{Z}_{11}^*$ and $\mathbb{Z}_{13}^*$ cyclic groups? Compute the orders of elements of these groups and find, if any, their generators.*

Problem 133. *The additive group of all the integers is an infinite cyclic group.*

We omit a proof of the next statement, important in cryptography.

Theorem 10. *(1) If d is prime, then $\mathbb{Z}_d^*$ with a congruence as a group operation, is a commutative cyclic group.*

(2) If $|G|$ is prime, then all elements $a \in G$, $a \neq 1$, are primitive.

The next property says that all cyclic groups of a given order are, in a sense, the same.

(3) All the finite cyclic groups of the given order n are isomorphic to each other. All infinite cyclic groups are isomorphic to one another.

4.4   THE DISCRETE LOGARITHM PROBLEM

While developing the Affine Ciphers, we had to find the inverse elements of some group elements. It is easy if we work with real numbers, since x^{-1} exists for every real $x \neq 0$. However, in cryptography x is supposed to be integer, and its reciprocal must be also integer; hence the problem of finding the reciprocal may have no solution. Moreover, the exponent does not have to be -1. Again, solving an equation $a^x = b$, when $a > 0$ and b is a real number, straightforwardly leads to logarithms. Therefore, we have to extend that notion to a discrete setting.

Consider the finite cyclic group $\mathbb{Z}_p^*$ with prime p, its order is $p - 1$, and let $g \in \mathbb{Z}_p^*$ be a generator of this group. Let also another element be $h \in \mathbb{Z}_p^*$. *The Discrete Logarithm Problem* (DLP) requires finding the integer $x, 1 \leq x \leq p - 1$, such that $g^x = h \pmod{p}$. We denote the solution of this congruence, if it exists, as $x = \log_g h \pmod{p}$.

For example, computations in Example 11 tell that $5^{*4} \pmod{7} = 2$, therefore, we set $g = 5$, $h = 2$, and get $x = \log_5 2 \pmod{7} = 2$, which has nothing in common with $\ln 2 / \ln 5 \approx 0.43$. We can straightforwardly check that $5^{*4} = 625$ and $625 = 7 \times 89 + 2$.

This example shows why DLP is used in cryptography. We deal with a one-way function – see Def. 68 (p. 167). Given the value of the discrete logarithm, the verification is straightforward and fast. But the computations for finding this value currently, for really large parameters, are infeasible. For more about the DLP, the reader can consult, for example, [40] and the references therein.

4.5   EXERCISES

Exercise 4.1. *Solve the DLP for the values $g = 12$ and $h = 2, 3, 4$.*

Exercise 4.2 *Prove that $(g^{-1})^{-1} = g$ and $(gh)^{-1} = h^{-1} \cdot g^{-1}$.*

Exercise 4.3 *Prove that the set $B(0, 1)$ with the binary addition is an additive group, where every element is its own inverse.*

CHAPTER 5

PREDICATES AND QUANTIFIERS— ALGEBRAIC THEORY

5.1   PREDICATES

In Chapter 2, we studied propositions, that is, logical constants, which are either true or false statements, but not both. For instance, $5 = 4$ and $5 = 2 + 3$ are examples of a false and of a true proposition, respectively. However, there are many likewise sentences, which are not propositions. For instance, we cannot determine the truth value of a sentence $5 = x + 3$ if we do not know a value of x; therefore, this sentence is not a proposition. Since similar examples penetrate mathematics and sciences, it is necessary to include them in our theory. To give a definition, let us note that the linear equation $5 = x + 3$ is a *propositional function* with the argument x, whose domain may be the set of integers, and the range of this function is the two-element set *{true, false}*. Such maps are called *predicates*. We develop here a purely algebraic non-formal, non-axiomatic theory parallel and in a sense equivalent to the theory of relations in n indeterminates.

Definition 23. *A Boolean predicate in one unknown or one indeterminate x (unary predicate) is any function $P(x): D \to B$ with the domain D and the range $B = \{true, false\}$; this range is equivalent to the two-element set $B_2 = \{1, 0\}$.*

Example 12. *For example, let D be the set $\mathcal{R}$ of the real numbers and the predicate*

$$P(x): x^2 \text{ is positive.}$$

Here, $P(x)$ is true if $x \neq 0$ while $P(0)$ is false. This predicate takes on both true and false values, for example, $P(0) = 0$ and $P(3) = 1$. However, if we define the predicate as $Q(x) : x2$ is non-negative, then Q is an identically true predicate; it does not take on false values.

The predicates often describe not the properties of individual objects but relations among the objects.

Definition 24. *A function of several indeterminates*

$$P(x_1, x_2, \ldots, x_n) : x = (x_1, \ldots, x_n) \in D \to B_2$$

is called an *n*-ary Boolean predicate. A subset of the domain where the predicate is true is called its truth domain; the complement of the truth domain is called the falsity domain.

It is worth mentioning that if we fix an indeterminate in *n*-ary predicate, $n \geq 2$, it results not in a proposition yet; it is an $(n - 1)$-ary predicate. Propositions can naturally be considered as 0-ary predicates. It is also important that the collection of all the predicates with all domains is *not* a set; this collection leads immediately to the Russell paradox. So that the language of Boolean predicates is equivalent to the language of relations; sometimes it is more convenient to speak relations, sometimes—predicates.

Example 13. *Let the domain be $D = \{3, 4, 5, \ldots\}$; define a predicate $p(x) = 1$, if x is a positive number and $p(x) = 0$ otherwise. What is the truth domain of this predicate?*

Solution. Now, the falsity domain is empty, and the truth domain is the same as the entire domain D of the predicate. If, for the same propositional form, we choose another domain $D_1 = \{-1\}$, we get an identically false predicate. If we choose the domain as $D_2 = \{-1, 2, 3, 5\}$, the predicate is not identically false, nor identically true.

Example 14. *Let $D = B \times B$, and the binary predicate be given as*

$$P(x, y) = \begin{cases} 1, & x > y \\ 0, & x \leq y \end{cases}.$$

The truth set of P is an open half-plane below the bisectrix $y = x$. In particular, $P(1, 0) = 1$, while $P(1, y) \equiv 1 > y, \forall y \in \mathbf{R}$. Hence, this is not a proposition but a unary predicate.

In general, if we specify any group of k variables of an n–ary Boolean predicate, $1 \leq k \leq n$, we have an $(n - k)$-ary predicate. In particular, if $k = n$, we arrive at the proposition, that is, a logical constant.

Given two functions, we can perform several arithmetic operations with them. Given two predicates, we can do various Boolean (logical) operations. Let us start with the negation, which is a one-variable (*unary*) operation.

Definition 25. *The negation of the unary predicate P with the domain D is the predicate Q with the same domain D such that for every $x \in D$, $Q(x) \equiv \neg P(x)$. The same definition can be given for any n–ary predicate for negation with respect to any variable x_i, if we fix the values of all the other indeterminates but x_i. We usually write F = G for numerical functions, while for logical functions (predicates), we write $P \equiv Q$. This leads to no confusion.*

The binary operations are defined the same way. For instance, if

$$P(x, y) \equiv x \vee y \to y, \text{ then } \neg_x(P(x, y)) \equiv (\neg x) \vee y \to y.$$

Problem 134. *Give a formal definition of binary operations with the predicates. Describe precisely the domain of the predicate, say $P \wedge Q$, if P and Q have arguments with different domains.*

Solution. If the two indeterminates x and y with different domains appear in two or more predicates to be combined anyway, then the domain of these variables in any combination of the predicates is the intersection of the domains in the combined predicates.

5.2   QUANTIFIERS

However, there are two other operations with the predicates that have no similar operations with propositions—these are quantifiers, the existential quantifier and the universal quantifier. Consider the phrase "There exists an element x such that the statement $P(x)$ is valid." If $P(x)$ is identically false, then that statement is false; otherwise, it is true. Hence we have a map from one-variable predicates into the simplest Boolean algebra B_2, and we can straightforwardly translate it into the language of sets and maps.

Definition 26. *Consider a family of unary predicates {P} with the fixed domain D. (1) The map $\exists P$, denoted also as $\exists x(P(x))$, from that family to B_2, such that $\exists P \equiv 0$ if P is identically false and $\exists P \equiv 1$ otherwise, is called the existential quantifier (on that family).*

(2) The map $\forall P$, denoted also as $\forall x(P(x))$, from that family to B_2, such that $\forall P \equiv 0$ if P is not identically true and $\forall P \equiv 1$ if P is identically true, is called the universal quantifier (on that family).

For instance, the statement "There exists an integer divisor of 25" is true, the statement "There exists an integer divisor of 29" is false, and the statement "There are infinitely many prime numbers-twins" is not a proposition; as of *now* (May of 2021) we don't know whether it is true or false since the twin-primes hypothesis has not been proved or disproved yet. On the other hand, the statement "Every integer divides 25" is false, while the statement "The square of every real number is non-negative" is true.

Problem 135. *Give definitions of the existential quantifier and the universal quantifier for n-ary predicates. In the latter case, a quantifier maps any n-ary predicate to (n−1)-ary by "freezing" all the other variables except for the variable we apply the quantifier to. This variable is called the bounded variable, its domain is called the scope of the quantifier. All the other variables are called free in this calculation. It is worth repeating that bounding a free variable in an n-ary predicate results in an (n−1)-ary predicate.*

Problem 136. *Compute the truth values of the propositions*
(1) $\exists k(k \in \mathcal{Z})(k^2 = 36)$,   (2) $\forall k(k \in \mathcal{Z})(k^2 = 36)$.

Solution. (1) True. (2) False.

If P is an n-ary predicate and we want to apply the existential quantifier to one of its indeterminates, say x_i, we must fix all the other occurrences of the indeterminates, except for x_i, apply the quantifier to the resulting unary predicate, and then "unfreeze" the other variables.

Example 15. *Compute $\exists x(P(x,y))$, $x,y \in (R)$; the predicate P was defined in Example 14, p. 80.*

Solution. Now x is a bounded or quantified variable, while y is a free variable, and we can set y to be any real number. Thus, $\exists x(P(x,0)) \equiv 1$, moreover, we see that for every real number y_0, $\exists x(P(x,y_0)) \equiv 1$.

Example 16. *Let be $D = \mathbb{R} \times \mathbb{R}$, and a binary predicate is given as*

$$P(x,y) = \begin{cases} 1, & x^2 = y \\ 0 & x^2 \neq y \end{cases}.$$

Apply the existential quantifier to this predicate.

Solution. Now $\exists x P(x,1) \equiv 1$ and $\exists x P(x,-1) \equiv 0$.

Consider an (algebraic) predicate $P : D \to B_2$. The set

$$D_1 = \{x \in D : P(x) = 1 \in B_2\},$$

that is, the family of all the preimages of the element $1 \in B_2$, is called the truth domain of the predicate P. It is clear that $\varnothing \subset D_1 \subset D$. If P is identically true (Define this!), then it's $D_I = D$, and if P is identically false (Define this too!), then $D_I = \varnothing$.

Definition 27. *A predicate, which is not identically false, is called satisfiable.*

Problem 137. *Give examples of unary, binary, and ternary identically true, identically false, and satisfiable but not identically true predicates.*

5.2.1 Commutativity of Quantifiers and Boolean Operations

We know that $2 + 3 \times 4 = 14 \neq 2 \times 3 + 4 = 10$. Since the Boolean operations with predicates are defined through those operations with propositions, the Boolean operations also are not always commute with one another, for instance, $p \vee q \wedge r \not\equiv p \wedge q \vee r$, however, this is important, and we have to study how these operations "communicate" with each other. The definitions must provide for the *uniqueness* of the reading of every formula. To achieve that, we are to define the *order of operations*. Moreover, we often have to use *separators*—parentheses, etc. Since we work with logical objects, with the predicates and propositions, we often have to write that these objects are equivalent, that is, $p \equiv q$, etc. However, we treat them as algebraic objects; therefore, we sometimes write $p = q$ with the same meaning.

An important conclusion is that $p \vee q \wedge r \not\equiv p \wedge q \vee r$; and moreover, in general $\forall (P \vee Q) \not\equiv (\forall P) \vee (\forall Q)$. To derive valid formulas, first of all we consider a special case, when the domains of the predicates under consideration are *finite* sets. In the case of infinite families, the domains considered must be proper sets and not classes.

Example 17. *Consider a predicate $\{P(x) : x > 0, x \in Z\}$, that is, the property of the integers to be positive. The domain of the predicate is the set of integers Z and the truth domain is $\{1, 2, 3, \ldots\}$; thus this predicate is not identically true nor identically false, it is satisfiable.*

Example 18. *Next, we consider a predicate $\{P_9(x) : 0 \le x \le 9, x \in Z\}$, that is, property of the integers between 0 and 9 inclusive to be positive. The domain of the predicate is Z, and the truth domain is $\{1, 2, \ldots, 8, 9\}$; thus this predicate also is satisfiable but not identically true.*

The truth values of this predicate can be easily calculated in finitely many steps without any regard to the quantifiers since the predicate can be written as a finite conjunction

$$P_9 \equiv (0 > 0) \wedge (1 > 0) \wedge (2 > 0) \wedge \cdots \wedge (9 > 0),$$

and clearly, $\forall x(P_9(x)) \equiv False$, since $P(0) \equiv 0$; in other words, P_9 is not an identically true predicate.

If we try to do the same with the initial predicate P, we have to evaluate a conjunction, consisting of *infinitely many* terms. This calculation is not obvious and requires a separate definition of *infinite conjunctions*. To avoid that problem, we define the quantifiers as maps, which essentially means that we stay with the primary concept of the mappings. Now the following definitions are clear.

Example 19. *Consider again Example 17: $\exists x P(x,y)$; here y is a free variable and x is bounded, the scope of the existential quantifier is the set of real numbers. If we modify the example and set $\exists x P(x,y) \wedge (x^2 = 9)$, the first occurrence of x remains bounded, but the second its occurrence is free, since the scope of the existential quantifier ends before any Boolean operation. If we want to include the conjunction $\wedge$ in the scope of the quantifier, we can do, for example, as $\exists x(P(x,y) \wedge (x^2 = 9))$; now both occurrences of x are bounded.*

Problem 138. *For every variable in the examples above, determine whether it is free or bounded? Is it possible, and how, to make the free variables bounded, and vice versa, to free the bounded variables?*

We have already established De Morgan laws for propositions,

$$\neg(p \vee q) \equiv \neg p \wedge \neg q$$

and the dual formula

$$\neg(p \wedge q) \equiv \neg p \vee \neg q.$$

If the domain of the predicates is finite, then De Morgan laws for such predicates can be easily proved, for instance, by induction.

Problem 139. *For any propositions $p_1, \ldots, p_K$, prove the identity*

$$\neg\left(\bigvee_{k=2}^{K} p_k\right) \equiv \bigwedge_{k=2}^{K} \overline{p_k}$$

and the dual formula $\neg(\bigwedge_{k=2}^{K} p_k) \equiv \bigvee_{k=2}^{K} \overline{p_k}$. Analyze these identities if $K = 1$.

When the domains are infinite, though, De Morgan laws may look quite similar, but their proofs require new methods—compare the discussion of examples above.

Theorem 11. *Let P be a predicate with domain D. Then the following generalized De Morgan laws are valid,*

$$\overline{\forall x P(x)} \equiv \exists x \overline{P(x)}$$

and

$$\overline{\exists x P(x)} \equiv \forall x \overline{P(x)}.$$

Proof. We establish only the first equivalence. If P contains only the free variable x, then both sides are the propositions, and we have to prove the equivalence of these two propositions. If $\overline{\forall x P(x)} \equiv 1$, then $\forall x P(x) \equiv 0$, thus, the predicate P is not identically true, or $\neg P$ is not identically false. The latter, in turn, means that $\exists x(\neg P) \equiv 0$. All the other implications in the theorem are proved the same way.

Problem 140. *Rewrite the proposition* $\neg \forall y(\forall z \exists x T(x,y,z) \wedge \exists z \forall x U(x,y,z))$ *so that the quantifiers apply only to predicates.*

However, if a predicate P in Theorem 11 depends, besides x, upon other variable(s), we have to prove the equivalence of certain n-ary predicates. In this case, we have to fix all the other free variables and finish the proof exactly as before. The crucial observation is about those of predicates with a finite domain, which shows that the universal quantifier in this case is quite similar to the *conjunction*, while the existential quantifier is similar to the disjunction.

Theorem 12. *Let P and Q be the predicates with the same domain. Then*

$$\forall(P \wedge Q) \equiv \forall P \wedge \forall Q$$

and

$$\exists(P \vee Q) \equiv \exists P \vee \exists Q.$$

However, in general the universal quantifier does not commute with the disjunction, and the existential quantifier does not commute with the conjunction. But the following conditionals are identically true,

$$\forall P \vee \forall Q \rightarrow \forall(P \vee Q)$$

and

$$\exists(P \wedge Q) \rightarrow \exists P \wedge \exists Q.$$

Proof. We prove only the latter conditional, assuming that both P and Q are unary predicates since we know already how to deal in general case. A conditional can be false only if the premise is true, but the conclusion is false. So that, let us suppose that $\exists P \wedge \exists Q \equiv 0$, hence both $\exists x P(x) \equiv 0$ and $\exists x Q(x) \equiv 0$. These propositions mean that both unary predicates P and Q are identically false, so that their disjunction $P \wedge Q$ is also identically false. Hence $\exists(P \wedge Q) \equiv 0$ as well. An example $P(x) : x < 0$ and $Q(x) : x > 0$ shows that the inverse implication is not always true.

Problem 141. *Prove the other conclusions of the theorem.*

Problem 142. *The third and fourth statements of this theorem can be modified. Prove that if a is a proposition or more generally, the quantifiers are not applied to its variables, then*

$$\forall(a \vee Q) \equiv a \vee \forall Q$$

and

$$\exists(a \wedge Q) \equiv a \wedge \exists Q.$$

To conclude this topic, we study the commutativity of various quantifiers, which is important everywhere, from elementary mathematics to calculus (see, e.g., Example 17) and above, as well as in the sciences, and now even in political science.

Theorem 13. *For every predicate P with at least two variables,*

$$\forall x \forall y P(x, y...) \equiv \forall y \forall x P(x, y...)$$

$$\exists x \exists y P(x, y...) \equiv \exists y \exists x P(x, y...).$$

Moreover,

$$\left(\exists x \forall y P(x, y...) \to \forall y \exists x P(x, y...)\right) \equiv 1, \tag{5.1}$$

that is, this conditional is valid for any predicate P, even though there are predicates P such that the converse conditional $\forall y \exists x P(x, y...) \to \exists x \forall y P(x, y...)$ is false.

Proof. We prove that conditional (5.1) is identical, that is, for all the predicates, true. Again, in non-axiomatic language, the statement is obvious. Indeed, if there exists a certain x' such that for every y the conditional (5.1) is true, then of course, for each y such $x = x'$ does exist. The converse statement is not universally true since for different y these existing x may be different, and there is no unique x' common for all y. First of all, let us consider an illuminating example.

Example 20. *Let $f(x)$ be a real-valued continuous function defined on the set $D \subset \mathbb{R}$. The continuity of f at a point $x_0 \in D$ can be written as*

$$\forall x_0 (\in D) \forall \epsilon (>0) \exists \delta (=\delta(\epsilon, x_0) > 0) \forall x (\in D) (|x - x_0| < \delta \rightarrow |f(x) - f(x_0)| < \epsilon),$$

$$(5.2)$$

where parentheses indicate the scope of a corresponding quantifier. However, if f is uniformly *continuous on D, then*

$$\forall \epsilon (>0) \exists \delta (=\delta(\epsilon, x_0) > 0) \forall x_0 (\in D) \forall x (\in D) (|x - x_0| < \delta \rightarrow |f(x) - f(x_0)| < \epsilon)$$

$$(5.3)$$

and the only difference between (5.2) and (5.3) is the ordering of the two different *universal and existential quantifiers, $\forall x_0 \exists \delta$ and $\exists \delta \forall x_0$, since as we have already proved, the universal quantifiers can be written in either order. Thus, by Theorem 13, formula (5.2) immediately follows from formula (5.3). In other words, a well-known theorem of mathematical analysis about the continuity of any uniformly continuous function is an immediate corollary of the simple result of elementary mathematical logic. The converse conditional fails, in general.*

The other conclusions of the theorem can be proved exactly the same way, we omit their proofs. That concludes the proof of Theorem 13.

Let $P(x)$ be a unary predicate with the only free variable x. This means that one can fix the x to be any value from the domain D. For example, if $D = \mathbb{N}$, we can consider $P(1)$ or $P(77)$, etc., where now $P(77)\dots$ are propositions. Another way to get a proposition from a unary predicate is to bind the free indeterminate with a quantifier, universal or existential, as $\forall x(P(x))$ or $\exists x(P(x))$. The letter x here can be replaced with any other character without any change in the truth value of these propositions. It is, however, important to avoid the so-called *collision of variables*, that is, the same symbol **cannot** be used to denote different variables. We normally have enough symbols to avoid any collision.

Example 21. *Let P be any binary predicate, Q be any unary predicate, and $S \equiv \forall x(P(x,y) \wedge Q(x))$. Here one can replace x with any symbol with the only one restriction—it must be distinctive from y. For example, we can write $T \equiv \forall t(P(t,y) \wedge Q(t))$ and get $S(y) \equiv T(y)$. However, if we set*

$U \equiv \forall y(P(y,y) \wedge Q(y))$, then U is a proposition (a constant), and the conclusion $S \equiv \neg U$ is wrong.

Problem 143. *Is it possible to find the predicates P and Q, such that in the example above $S \equiv U$? Or $S \equiv \neg U$?*

To avoid the collision of variables, we assume hereafter, that no letter can denote both free and bounded variables.

If a certain variable appears in a formula several times, we talk about the first, second, third, etc., *occurrence* of the variable into the formula, counting from the left. For example,

$$A(y) \equiv \forall x(\underset{1}{x} \vee (\underset{1}{y} \wedge \underset{2}{x}) \to \exists y((\underset{2}{y}) \vee \underset{3}{x})).$$

Here every occurrence of each variable is bounded by the leftmost universal quantifier; however, the first occurrence of y is free, since it is within the scope of the x-quantifier. The second occurrence of y is bounded by the existential quantifier, and the third occurrence of x is bounded by the universal quantifier. All at all, only the first occurrence of y is free. $A(y)$ is a unary predicate; thus, $A(w)$ is another unary predicate with the free variable w, but

$$A(y) \equiv \forall x(\underset{1}{x} \vee (\underset{1}{y} \wedge \underset{2}{x}) \to \exists t(\underset{}{t} \vee \underset{3}{x}))$$

is the equivalent predicate $A(y)$ with the same free indeterminate y.

In other words, if no quantifier is applied to a predicate, then all the occurrences of all the indeterminates are free. When Boolean operations are applied to a predicate, then all occurrences of all the variables remain the same, that is, the free variables remain free and bounded variables remain bounded. Finally, if we apply a quantifier over the variable x to the predicate, then all free occurrences of x become bounded, and all the other occurrences of x remain the same as before.

While defining the free and bounded occurrences above, we assume that the quantifiers bound *stronger* than Boolean operations, that is, if there are no separators, one must first apply quantifiers, and only after those Boolean operations (negations, conjunctions, disjunctions, etc.) according to the order of Boolean operations.

It is also useful to analyze the following examples.

Problem 144. *Consider the binary predicate $P(x,y) \equiv \left(\dfrac{y}{x} \in \mathbf{N} \right)$ for $x, y \in \mathcal{N} \setminus \{0\}$, that is, y is a multiple of x, or, equivalently, x is a divisor of y for* *positive integers x and y. This binary predicate generates four unary predicates* $P_1(x) \equiv \exists y P(x,y),\ \ P_2(x) \equiv \forall y P(x,y),\ \ P_3(y) \equiv \exists x P(x,y),\ \ and\ \ P_4(y) \equiv \forall x P(x,y).$ *Binding the free variables in these four unary predicates, we generate* $4 \times 2 = 8$ *propositions. Compute the truth values of these propositions.*

Example 22. *A student of calculus has learned that every uniformly continuous function on a closed interval is continuous, but the converse statement is false, that is, there are continuous but not uniformly continuous functions. Why? To answer the question, write the definitions of both continuous and uniformly continuous functions as predicates with all the relevant quantifiers and compare the ordering of quantifiers. You will see that the answer is given by the third part of Theorem 13.*

The essence of Theorem 13 and the relevant examples, like in Problem 68, is the order of two different operations, in this case, the universal quantifier and the existential quantifier. That situation has occurred in mathematics and elsewhere many times before; see, for example, a detailed discussion of the relationship between the max and min operators in [38, Sect. 13.3].

Example 23. *Consider a predicate with three variables $x < z < y$. Choose the domains for the variables so that the proposition $\forall x \forall z \exists y (x < z < y)$ is a) true, b) false. Study the negation $\neg (x < z < y)$ and compare it with the first part of the example.*

Problem 145. *Let $P(m, n)$ be a proposition $m + n = m - n, m, n \in \mathcal{N}$. What is the truth value of each of the following propositions? Classify the following unary predicates.*

$$(\forall m)(\forall n)P(m,n); (\forall n)(\forall m)P(m,n); (\forall m)(\exists n)P(m,n); (\forall n)(\exists m)P(m,n);$$

$$(\exists m)(\exists n)P(m,n); (\exists n)(\exists n)P(m,n); (\exists m)(\forall n)P(m,n); (\forall m)(\forall n)P(m,n);$$

$$(\forall m)P(m,n); (\exists n)P(m,n).$$

Problem 146. *Let $P(x)$ be a sentence "x is a prime number."*

Compute $P(1);\ \neg \exists P(x);\ \neg \forall x \neg P(x);\ \exists x \exists y P(x+y);\ \neg \forall x \forall y P(x+y).$

Problem 147. *Let $P(x)$ be a unary predicate with the finite domain $D = \{x_1, x_2, \ldots, x_n\}$, that is, $P(x) = 0$ or $P(x) = 1$. Prove that, for the unary predicates, quantifiers can be defined as*

$$\forall x P(x) = \min\{P(x) : x \in D\}$$

and

$$\exists x P(x) = \max\{P(x) : x \in D\}.$$

They can be also defined as finite conjunctions or disjunctions, that is,

$$\forall x P(x) = P(x_1) \wedge P(x_2) \wedge \cdots \wedge P(x_n)$$

and

$$\exists x P(x) = P(x_1) \vee P(x_2) \vee \cdots \vee P(x_n).$$

The preceding examples show that the conditional converse to (5.1) is, in general, false. Therefore, it is of interest to study when this converse conditional

$$\forall y \exists x P(x, y) \to \exists x \forall y P(x, y)$$

is valid. Comparing with (5.1) and using the definition of the conditional, we see that the question is when is

$$\forall y \exists x P(x, y) \equiv \exists x \forall y P(x, y)$$

or when

$$\exists y \forall x \neg P(x, y) \vee \exists \forall y P(x, y) \equiv 1?$$

The answer to the last question is simple: There must exist a value y_0 such that $P(x, y_0) \equiv 0$ for all the values of x. Since the predicates under discussion can take on only two values, 0 or 1, we have no need to introduce here the notion of *saddle points*, etc.—cf. [38, Sect. 13.4].

For predicates with infinite domains, the quantifiers must be defined independently, and we leave this task to the reader.

To conclude this section, we collect together the major predicate equivalences, where 1 stands for the identically true predicate and 0 for the identically false one with an appropriate domain. Some symbols here can be omitted, for example, the second equivalence in (**10**) can be written as $P \vee PQ \equiv P$.

1. $\neg\neg P \equiv P$

2. $P \vee P \equiv P \quad P \wedge P \equiv P$

3. $P \vee 1 \equiv 1 \quad P \vee 0 \equiv P$

4. $P \wedge 0 \equiv 0 \quad P \wedge 1 \equiv P$

5. $P \vee \neg P \equiv 1 \quad P \wedge \neg P \equiv 0$

6. $P \vee Q \equiv Q \vee P \quad P \wedge Q \equiv Q \wedge P$

7. $P \vee (Q \vee R) \equiv (P \vee Q) \vee R \quad P \wedge (Q \wedge R) \equiv (P \wedge Q) \wedge R$

8. $P \vee (Q \wedge R) \equiv (P \vee Q) \wedge (P \vee R) \quad P \wedge (Q \vee R) \equiv (P \wedge Q) \vee (P \wedge R)$

9. $\neg(P \vee Q) \equiv (\neg P) \wedge (\neg Q) \quad \neg(P \wedge Q) \equiv (\neg P) \vee (\neg Q)$

10. $P \wedge (P \vee Q) \equiv P \quad P \vee (P \wedge Q) \equiv P.$

5.3 EXERCISES

Exercise 5 1. *Find the truth domains for all the predicates in this section.*

Exercise 5.2 *Prove the following properties of truth domains:* $D_1(\neg P) = D \setminus D_1; \ D_1(P \vee Q) = D_1(P) \cup D_1(Q); \ D_1(P \wedge Q) = D_1(P) \cap D_1(Q).$

Exercise 5.3. *Are the following formulas identically true with respect to the predicate variables P and Q, or identically false, or satisfiable?*

1. $\forall x(P(x) \vee Q(x)) \equiv \forall x P(x) \vee \forall x Q(x)$

2. $\exists x(P(x) \wedge Q(x)) \equiv \exists x P(x) \wedge \exists x Q(x)$

3. $\forall y \exists x P(x) \to \exists x \forall y P(x)$

4. $\forall x(Q(x) \to P(x)) \to (\forall x Q(x) \to \forall x P(x))$

5. $\forall x(Q(x) \to P(x)) \to (\exists x Q(x) \to \exists x P(x))$

6. $\exists x(Q(x) \to P(x)) \to (\forall x Q(x) \to \forall x P(x))$

7. $\exists x(Q(x) \to P(x)) \to (\exists x Q(x) \to \exists x P(x))$

8. $\forall x(Q(x) \to P(x)) \to (\exists x Q(x) \to \forall x P(x)).$

CHAPTER 6

BINARY RELATIONS AND RELATIONAL DATABASES

6.1   ORDERED TOTALITIES AND CARTESIAN PRODUCTS

Consider a two-element set {a, b} – the braces here indicate that this is a *set* without any ordering of its elements, thus, {a, b} = {b, a}, the latter is the same set, whether a = b or a ≠ b. However, in many problems, we have to distinguish *ordered totalities*. Thus, the car license plates A – 123 and A – 312 are two different plates. We emphasize that the sets are *unordered* totalities, denoted by (curly) braces, while the *ordered totalities* are denoted by *parentheses*. It is possible to *define* the ordered totalities through the maps, but we do not go into these details and treat ordered totalities as *primary* concepts. The major feature, distinguishing the *ordered pairs* from *non-ordered sets* is that (a, b) ≠ (b, a) as long as a ≠ b, while {a, b} = {b, a} always.

As another example, let a family have two sons, John and Peter. If we consider a set of these sons, then two writings, {*John, Peter*} and {*Peter, John*} represent the same set of children, {*John, Peter*} = {*Peter, John*}. On the other hand, if these two words represent the first name and the second name of an individual, then definitely Mr. John Peter Doe and Mr. Peter John Doe are two different persons. To work with ordered totalities, we need new concepts and definitions.

Likewise, we can define new objects, such that an *ordered triple* (a, b, c) with the first element a, the second element b, and the third element c, an *ordered quadruple* (a, b, c, d), ..., an ordered n-tuple $(a_1, a_2, a_3, ..., a_n)$, etc. For example, for any person, her first, second, and last names make an ordered triple, such as Mary Patty Doe. Social security numbers are ordered 9-tuples; we must regard 123-45-6789 and 213-45-6789 as distinct SS-numbers, even though the numerals in both numbers form the same set $\{1, 2, 3, 4, 5, 6, 7, 8, 9\}$. The same is true for telephone numbers: it is safe to call (718)900-0000, but it may cost you a lot of money to dial (900) 718-0000.

Problem 148. *Prove that for any ordered pair, there holds $(a, b) = (b, a)$ iff $a = b$.*

Problem 149. *Are the one-element set $\{John\}$ and the ordered pair with coinciding elements $(John, John)$, the same objects or different ones?*

Definition 28. *Given two sets X and Y, the collection of all ordered pairs (x, y), where $x \in X$ and $y \in Y$, is called the Cartesian[1] (or Direct) product of the two sets X and Y and is denoted as $X \times Y = \{(x,y) : x \in X, y \in Y\}$.*

Problem 150. *Let be $X = \{a, b\}$ and $Y = \{q\}$. Find $X \times Y$ and $Y \times X$.*

Solution. $X \times Y = \{(a, q), (b, q)\}$, while $Y \times X = \{(q, a), (q, b)\}$. If either $(a \neq q)$ or $(b \neq q)$, then $X \times Y \neq Y \times X$.

Thus, the Cartesian product $X \times Y$ *is not commutative*, unless $X = Y$.

However, there is an obvious one-to-one correspondence between these two Cartesian Products, so that they have the same cardinality.

Moreover, the Cartesian product is not associative, but there exist obvious one-to-one correspondences between the sets $X \times (Y \times Z)$, $(X \times Y) \times Z$ and $X \times Y \times Z$. Indeed, the elements of $X \times (Y \times Z)$ are ordered pairs whose *second* element is itself an ordered pair, the elements of $(X \times Y) \times Z$ are ordered pairs whose *first* element is an ordered pair, and $X \times Y \times Z$ consists of *ordered triples*.

Definition 29. *Given n sets $A_1, A_2, \ldots, A_n$, the set of all ordered n–tuples $(a_1, a_2, a_3, \ldots, a_n)$ with the elements $a_1 \in A_1, \ldots, a_n \in A_n$, is called the Cartesian product of these sets in this order and is denoted as $A_1 \times A_2 \times \ldots \times A_n$. In particular, the Cartesian product of two sets A_1, A_2 is the set of all the ordered pairs with the first element in A_1 and the second element in A_2; the Cartesian product of the three sets A_1, A_2, A_3 is the set of all the ordered triples (a_1, a_2, a_3) with the first element $a_1 \in A_1$, the second element $a_2 \in A_2$, and the third element $a_3 \in A_3$.*

---

[1] After French philosopher and mathematician René Descartes (1560–1601), whose Latin name was Cartesius.

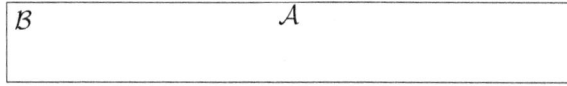

FIGURE 6.1 The Cartesian product of two sets A and B as the set of all the interior points of a rectangle with sides A and B. Whether the boundary of the rectangle belongs to A × B depends upon the boundaries of A and B belong to the sets or not.

It should also be mentioned that despite the term "product" traditionally used here, while forming a Cartesian product, we do not multiply anything but make a new set out of given ones and use the old term just to name it. However, such a choice of the old term usually has good reason. The term "product" is used here due to the feature that the cardinal number of the Cartesian product of two finite sets is equal to the arithmetic product of their cardinalities; we prove that later on.

Cartesian products of sets are sets; whence, one can perform all the set-theory operations with them. For example, if X and Y are sets, then

$$\{(a, X)\} \cap \{(a, Y)\} = \{(a, X \cap Y)\},$$

where (a, X), etc., stands for the set of all the ordered pairs (a, x) with x running through the set X. If we use this equation backward, that is, from right to left, it becomes one of the *distributive laws* for Cartesian products.

Problem 151. *When is a Cartesian product commutative? That is, for what sets A and B, A × B = B × A? Give necessary and sufficient conditions for this equation to be valid.*

The following statement will be used in the next chapter.

Lemma 3. *For any sets X_1, X_2, such that $X_1 \cap X_2 = \varnothing$, and any set Y, there is a one-to-one correspondence*

$$Y^{X_1 \cup X_2} \sim Y^{X_1} \times Y^{X_2},$$

where the symbol ~ denotes the one-to-one correspondence between the left- and right-hand sides of the formula above. If all the involved sets are *finite* sets, then both sides contain the same number of elements.

Proof. We have to construct a bijective map (*a one-to-one correspondence*) between the left- and right-hand sides of that formula. The set $Y^{X_1 \cup X_2}$ consists of all maps $f : X_1 \cup X_2 \rightarrow Y$. Together with every such a map, we consider two of its restrictions $F_1 = Y^{X_1}$ and $F_2 = Y^{X_2}$, define F as an ordered pair $F = (F_1, F_2)$, and prove that F is a bijection. To prove the injectivity of F, we consider two different maps $f, g \in Y^{X_1 \cup X_2}$. Since they have the same domain and the same codomain, we have to prove the existence of an element $x \in X_1 \cap X_2$, such that $f(x) \neq g(x)$. But that obviously follows since disjointness of the sets X_1 and X_2.

To prove that F is a surjective map, we choose a pair $(f_1, f_2) \in Y^{X_1} \times Y^{X_2}$, and define a map $f^0 : X_1 \cup X_2 \to Y$ as $f^0(x) = \begin{cases} f_1^0(x), & x \in X_1 \\ f_2^0(x), & x \in X_2 \end{cases}$. Thus, F is both injective and surjective, and therefore, bijective.

6.2   RELATIONS

Now we take up the main object of this section – relations. This is an example of overloading, that is, using known terms or concepts in a new sense. The term "relations" is used in mathematics in a sense, which is somehow similar to its everyday meaning. For instance, let a manager X have three employees A, B, and C. That means X has business relationships with them. Trying to describe this situation formally, we can consider three ordered pairs (X, A), (X, B), and (X, C). We have to use *ordered* pairs but not two-element sets because the pair (A, X) does not refer to the same relation since A is not a supervisor of X. If X has another employee D, then the pair (X, D) refers to this relation as well.

However, if a person E is not an employee of X, the pair (X, E) has nothing to do with this relation. Thus, to describe the "supervisor–employee" relation above, we should select the three *ordered pairs* (X, A), (X, B), (X, C) among all the *ordered pairs* (a, b), where a and b represent humans. Now, if H stands for the set of all people, then the ordered pair (a, b) belongs to the Cartesian product $H \times H$, and the three pairs (X, A), (X, B), and (X, C) make a subset of $H \times H$. Keeping in mind this model, we give the following definition.

Definition 30. *A binary relation ρ between elements of sets A and B taken in this order, or for short, from A to B, is an arbitrary subset of the Cartesian product $A \times B$. Thus, by definition, a binary relation is a subset $\rho \subset A \times B$.*

Example 24. *Let $X = Y = \{0, \pm 1, \pm 2, \ldots\}$ and the relation consists of the ordered pairs*

$$\rho = \{(1,0), (1,1), (1,-1)\}.$$

This binary relation contains only three ordered pairs.

Example 25. *Let A = R and B = {2, 3, 4}. Then the one-element set {(0, 2)} is a binary relation on A × B, while {(0, 0)} is not.*

Problem 152. *Similarly to Definition 30, define Ternary, Quaternary, and in general, n-ary relations. For example, the sphere*

$$\{(x,y,z) \in \mathcal{R}^3 : x^2 + y^2 + z^2 = 1\}$$

can be thought of as the graph of a ternary relation, representing the unit sphere in $\mathcal{R}^3$, centered at the origin. The ball $\{(x,y,z) \in \mathcal{R}^3 : x^2 + y^2 + z^2 < 1\}$ is the graph of another ternary relation, representing the open unit ball in $\mathcal{R}^3$, centered at the origin.

Definition 31. *Given binary relations $\rho \subset X \times Y$ and $\sigma \subset Y \times Z$, the relation $\zeta \subset X \times Z$ defined as*

$$\zeta = \left\{(x,z) \middle| \exists y \in Y, \text{ such that } (x,y) \in \rho \text{ and } (y,z) \in \sigma \right\}$$

is called the composition of ρ and σ and is denoted as $\zeta = \sigma \circ \rho$; it is a generalization of the compositions of functions. We usually write $\rho^2 = \rho \circ \rho$, and for any $n = 2, 3, \ldots$, define $\rho^n = \rho^{n-1} \circ \rho$.

Problem 153. *Prove that a binary relation ρ is transitive* (See Definition 38 below) *iff $\rho^n \subset \rho$ for every natural $n \geq 1$.*

Definition 32. *For a binary relation $\rho \subset X \times Y$, the relation*

$$\rho^{-1} = \{(y,x) \in Y \times X \mid (x,y) \in \rho\}$$

is called the inverse relation to ρ.

Problem 154. *Find, if any, the inverses to the binary relations in Examples 24 and 25, or prove that the inverse does not exist. In that case find a left- or a right-side inverse.*

Definition 33. *Let ρ be an n-ary relation on the Cartesian product*

$X = X_1 \times X_2 \times \ldots \times X_n$, that is, ρ is a subset of X. The set $proj_i(\rho) \subset X_i$ is called the *i*th projection of ρ iff every point $x_i^0 \in X_i$ is an ith component of at least one point $x \in \rho$.

For example, in Example 25, any real point belongs to the first projection of ρ, but only $2 \in proj_2, 3 \in proj_2, 4 \in Pr_2$, any projection of a unit sphere is the centered unit disc in the corresponding plane. For any map $f : X \to Y$, its first projection is the domain X and its second projection is the range $Rang(f)$.

Now it is possible to define the maps (functions) as special binary relations.

Definition 34. *A binary relation $\rho \subset X \times Y$ is called functional in x, iff its first projection $proj_x \rho = X$ and for every $x \in X$, there exists one and exactly one element $y \in Y$ such that the pair $(x, y) \in \rho$.*

Problem 155. *Define the injective, surjective, bijective maps as special classes of functional relations. Apply this definition to the functions in Problem 37.*

Remark 4. *This definition is sometimes useful, but we prefer hereafter to treat the maps as* primary concepts.

Problem 156. *Let A and B be two sets such that $|A| = m$ and $|B| = n$. Describe the set of binary relations $\rho \subset A \times B$ and the power-set B^A. Give examples and compare their cardinalities.*

Solution. There are 2^{mn} binary relations and n^m maps.

6.2.1 Special Classes of Relations

The n-ary relations are essentially used in relational databases and will be discussed elsewhere. Further in this chapter, we are primarily concerned with binary relations. The definition of relations above is too general to be used very often. Now, we introduce several classes of binary relations, which are used frequently in mathematics, computer science, and elsewhere. In what follows, we consider only binary relations given on Cartesian squares, that is, from a set to the set itself: $\rho : X \times X$. In such a situation, we shall say that the relation is defined *on* the set X, and instead of $(a, b) \in \rho$, we often write $(a \rho b)$.

Definition 35. *A binary relation ρ on a set X is called reflexive, iff it contains the diagonal, $\rho \subset X \times X$, and $\forall x \in X \,|\, (x, x) \in \rho$, that is, every element $x \in X$ is in this relation with itself.*

Example 26. *The relations $=$ ("equal") on the set of real numbers and $\geq$ ("bigger than or equal to") on the same set are reflexive, but the relation $>$ ("strictly bigger") on the same set is not, since the inequality $0 > 0$ is false: it is clear that the pair $(0, 0)$ does not belong to this relation.*

Definition 36. *A binary relation ρ on a set X is called symmetric or commutative, iff together with every pair (x, y), the relation contains the inverse pair (y, x).*

Example 27. *The relation "=" ("equal") on the set of real numbers is reflexive and symmetric, the relation $\geq$ ("bigger than or equal to") or $\leq$ is reflexive, but the relations $>$ ("strictly bigger") and $<$ on the same set are not.*

Definition 37. *A binary relation ρ on a set X is called antisymmetric, iff the two pairs (x, y) and (y, x) cannot belong to the relation, unless $x = y$.*

Example 28. *The relations $x \leq y$ and $x \geq y$ for $x, y \in \mathcal{R}$ are antisymmetric, the relations $x < y$ and $y < x$ are, due to the transitional property of the inequality and the observation, that the inequality $x < x$ (or $x > x$) has no real solution. The relation $x \neq y$ on any set is non-reflexive, symmetric, non-antisymmetric, and non-transitive.*

Problem 157. *Prove this statement, that is, prove that $\forall x (x < x) \equiv F$.*

Proof. The existence of an element x_0 such that $(x_o < x_0) \equiv T$ would contradict the trichotomy property of real numbers.

Definition 38. *A binary relation ρ on a set X is called transitive, iff it contains the pair (a, c) whenever it contains the two pairs (a, b) and (b, c). In other words, ρ is transitive iff for every three elements $a, b, c \in X$, if $(a \rho b)$ and $(b \rho c)$, then $(a \rho c)$.*

Problem 158. *Compute the square and the cube of these relations.*

Example 29. *Let $\mathcal{M}$ be any set, $X = 2^{\mathcal{M}}$ be its Boolean. Prove that the relation $\rho = \{(\mathcal{M}_1, \mathcal{M}_2) \subset \mathcal{M} \times \mathcal{M} \mid \mathcal{M}_1 \subset \mathcal{M}_2\}$ is reflexive, nonsymmetric, transitive, antisymmetric.*

Problem 159. *Let $\sim$ be the relation of similarity $T_1 \sim T_2$ on the set X of all plane triangles in the fixed plane. Prove that this relation is reflexive, symmetric, nonantisymmetric, and transitive.*

Problem 160. *Study the relations of being parallel, of being non-intersecting, and of being perpendicular on the sets of all lines in a line and of all planes in space.*

Problem 161. *Which of the relations in the examples above are transitive, and which are not?*

6.3   ORDERINGS AND POSETS

Definition 39. *A reflexive, transitive, and antisymmetric binary relation ρ on a set X is called a partial order on X or ordering of X. If the first projection of ρ is the entire set X, then ρ is called a linear ordering of X, and the set is called linearly ordered. The ordering is denoted as $\preccurlyeq$ or $\succcurlyeq$, If $x \preccurlyeq y$ or $x \succcurlyeq y$, and $x \neq y$, then these relations can be denoted as $\prec$ or $\succ$, respectively. If $x \succcurlyeq y$, then, the element y precedes the element x and x follows y. The relations $\prec$ and $\succ$ are called mutually dual, one is dual to another.*

We see that the relations ≤ or its inverse ≥ are linear orderings on the reals $\mathcal{R}$. Actually, the classical ordering of the real numbers served as a model for Definition 39 of ordering, exactly as the definition of abstract equivalence relation was modeled upon the equality relation on any set.

Problem 162. *Prove that if in any true statement about partially ordered sets replace the symbol $\preccurlyeq$ with $\succcurlyeq$ and vice-versa, then we again get a true statement.*

This is a *meta-theorem*, such as, for example, the *duality principle* in the propositional algebra or in Boolean algebra, see, for example, Chapter 13.

Problem 163. *Define the relations which are nonreflexive; nonsymmetric; nontransitive; non-antisymmetric. Do the classes of antisymmetric and non-symmetric relations coincide? Is there any connection between the classes of symmetric and antisymmetric relations?*

Problem 164. *Consider the next three properties on any non-empty set: reflexivity, symmetry, transitivity; for short, R, S, T. There are also their three negations, denoted as $\bar{R}, \bar{S}$, and $\bar{T}$. Prove that there are exactly $8 = 2^3$ combinations, $(R,S,T), (R,S,\bar{T}),\dots,(\bar{R},\bar{S},\bar{T})$, of these three symbols, containing either each symbol or its negation once. Prove that these three relations are independent in totality, that is, for each of these eight combinations, there is a binary relation with exactly this set of properties.*

Problem 165. *Prove that the three properties of reflexivity, transitivity, and antisymmetry are independent in totality, that is, provide eight examples of the relations with the appropriate set of properties:*
   A relation, which is reflexive, antisymmetric, and transitive
   A relation, which is nonreflexive, antisymmetric, and transitive
   A relation, which is reflexive, nonantisymmetric, and transitive

A relation, which is reflexive, antisymmetric, and nontransitive
A relation, which is nonreflexive, nonantisymmetric, and transitive
A relation, which is nonreflexive, antisymmetric, and nontransitive
A relation, which is reflexive, nonantisymmetric, and nontransitive
A relation, which is nonreflexive, nonantisymmetric, and nontransitive.

A few suitable examples have already been considered above. It is not necessary for an appropriate example to be very complicated. Often it is enough to consider a set containing very few elements and construct the table of the future relation with wanted properties.

Definition 40. A set equipped with a partial order relation is called a partially ordered set (a poset). The ordering is said to be partial because not every two elements are comparable. If any two elements of a poset are comparable, the relation of partial order is called a relation of linear order; and the corresponding set is called a linearly ordered set or a chain. If (X, ρ) is a poset and $(x, y) \in \rho$, the elements x and y are called comparable.

Consider the relation <, that is, "to be strictly less than." We are used to consider it as an ordering on the sets of reals or integers, etc., but it does not satisfy the reflexivity axiom. So that we are to introduce the following definition.

Definition 41. A binary relation ρ on a set X is called irreflexive, if it does not intersect with the diagonal of $X \times X$. In words, ρ is irreflexive, iff no element of X is in this relation with itself, or for every element $x \in X$, the ordered pair $(x, x) \notin \rho$.

The relation < (strictly less) is an example of an irreflexive relation. We emphasize that this property is not the negation of reflexivity.

Consider relations ≤ and ≥, both on the set of natural numbers. If we start with, for instance, the inequality $10 \leq 11$ and increase the second number by 1: $10 \leq 12$, $10 \leq 13$, $10 \leq 14$, etc., we will never finish – there is no largest natural number. However, if we decrease the first number in the inequality $10 \leq 11$, as $9 \leq 11$, $8 \leq 11$, $7 \leq 11$, etc., we very soon get to $0 \leq 11$, and if we work with natural numbers only, we cannot proceed down, since 0 is the smallest natural number. We see that in this regard the relations ≤ and ≥ have different properties, leading us to the following definitions.

Definition 42. Given a poset $(X, \preccurlyeq)$ and a subset $A \subset X$, an element $x \in A$ is called a minimal element of A iff it does not have a comparable element in

A, which is less than x. A minimal element $x \in A$ is called the least element of A iff it is comparable with every element of A. Thus, we have $x \preccurlyeq y, \forall y \in A$.

Definition 43. *Given a poset $(X, \succcurlyeq)$ and a subset $A \subset X$, an element $x \in A$ is called a maximal element of A iff it does not have a comparable element in A, which is bigger than x. A maximal element $x \in A$ is called the greatest element of A iff it is comparable with every element of A. Thus, we have $x \succcurlyeq y, \forall y \in A$.*

For example, a poset $(\mathcal{Z}, |)$, where $\mathcal{Z}$ denotes the set of all the integers and the vertical bar denotes the divisibility relation, contains neither the least, nor the greatest element; the poset $([0,1], |)$ contains both; and the poset $\mathcal{N}$ with the same relation of divisibility has the least, but does not have the greatest element.

Problem 166. *Prove that every greatest element is a maximal one, but the converse statement is false, that is, a maximal element does not have to be greatest. State and prove the corresponding statement for minimal and the least elements. Prove the uniqueness of the greatest and least elements.*

Problem 167. *Let $P(X)$ be the power set of a set X, and $X \subset Y$; we do not distinguish the symbols $\subset$ and $\subseteq$. Check that $(P(X), \subset)$ is a poset.*

We will remind a few auxiliary definitions, which are sometimes useful.

Definition 44. *A reflexive and transitive binary relation is called a preorder (or quasiorder) relation. An irreflexive, antisymmetric, and transitive binary relation is called a strict partial order relation (or strict partial ordering). For example, the irreflexive relation $<$ on the set $\mathcal{R}$ of all real numbers or any its subset is a strict partial order relation. Its transitivity is obvious, and it is an antisymmetric one, because a conditional with a false premise is true.*

To depict the posets, especially finite posets, it may be useful to use simple digraphs, called the *Hasse diagrams*. Of course, a poset can be drawn in many different ways. These diagrams are *directed graphs*, studied in detail in Chapter 17. The vertices of the graph are the elements of the poset, and an arrow goes from an element x to an element y iff $(X, \preccurlyeq)$.

Example 30. *Thus, Fig. 6.2 represents the Hasse diagram for the divisibility relation $x \,|\, y$ on the poset $\{1, 2, 3, 4, 5, 6\}$. Of course, we can accept the opposite definition, when the arrow shows another way; then the diagram will show the dual poset.*

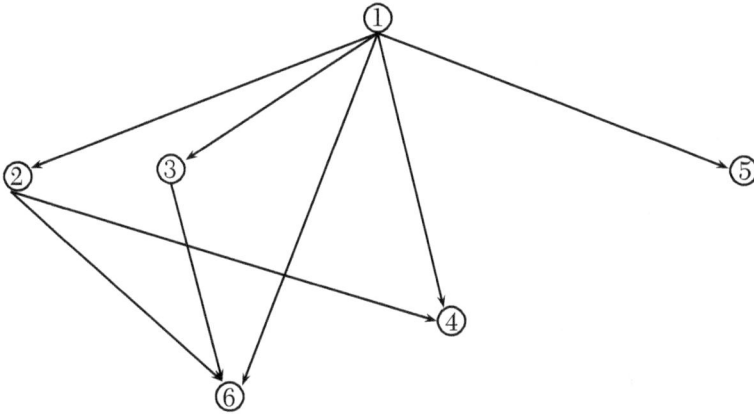

FIGURE 6.2 Hasse diagram for Example 30.

Problem 168. *Draw the Hasse diagram for the poset* $(\{2,3,4,5,8,9\},|)$, *where* $a \mid b$ *means that the integer a is a divisor of the integer b. In this poset, identify or state that it does not exist: a maximal element, a minimal element, the greatest element, the least element.*

An important property of the natural numbers is their *ordering*. Namely, given two natural numbers m and n, it is said that m is *less* than n or n is *bigger* than m, in symbols $m < n$, iff there exists a natural number $\kappa > 0$, such that the equation $m + \kappa = n$ holds good; equivalently, it can be written as $n > m$.

Problem 169. *Prove that this property of ordering is asymmetric, that is, the inequalities* $n > m$ *and* $n \leq m$ *are mutually exclusive.*

Proof. Indeed, $n > m$ means $n - m > 0$, then $0 > m - n$, and vice versa.

Problem 170. *Prove that this property is transitive, that is, the inequalities* $m < n$ *and* $n < q$ *imply* $m < q$.

Proof. The inequality $m < n$ means $0 < n - m$. Rewriting another inequality the same way and adding the inequalities, we conclude the proof.

Problem 171. *Trichotomy property. For any two natural numbers, one and only one of the three properties holds: either* $m < n$, *or* $n < m$, *or* $m = n$.

Thus, both sets of the integers and of the real numbers are posets. Moreover, the set of natural numbers is *well-ordered* in the following sense.

Definition 45 *A poset is called well-ordered, iff (1) its order relation is irreflexive, transitive, and possesses the trichotomy property and (2) every its non-empty subset has the least element.*

Problem 172. *Prove that the well-ordering property implies that any non-empty set of natural numbers. K has the unique least element, that is, a number $l \in K$, such that $l \leq q$ for any $q \in K$.*

Proof. This is just a part of the definition of well-ordering.

6.3.1 Equivalence Relations

Problem 173. *The properties of reflexivity, symmetry, and transitivity are independent in totality – also compare with Problem 165.*

To solve the problem, it is sufficient to show examples of the following eight relations:

A relation, which is reflexive, symmetric, and transitive.
A relation, which is nonreflexive, symmetric, and transitive.
A relation, which is reflexive, nonsymmetric, and transitive.
A relation, which is reflexive, symmetric, and nontransitive.
A relation, which is nonreflexive, nonsymmetric, and transitive.
A relation, which is nonreflexive, symmetric, and nontransitive.
A relation, which is reflexive, nonsymmetric, and nontransitive.
A relation, which is nonreflexive, nonsymmetric, and nontransitive.

One way to construct a relation with the necessary properties is by completing the table of the relation. We do not have to do something very complicated; let us start with a set with just three elements and see whether this is enough for our goal. So that, we consider a set $\{a, b, c\}$ and try to design the table of the reflexive, nonsymmetric, and transitive relation. To guarantee reflexivity, we have to include the diagonal in the relation. This is indicated by the + signs at the main diagonal of the table, indicating that the pairs $(a,a),(b,b),(c,c) \in \rho$. To assure the transitivity, we must have $(a,b) \in \rho \wedge (b,c) \in \rho \Rightarrow (a,c) \in \rho$, which is indicated as + signs at the corresponding cells. To have the nonsymmetry, we place the – signs below the main diagonal. The table below gives an example of a reflexive, nonsymmetric, and transitive relation ρ, we sought for.

ρ	a	b	c
a	+	+	+
b	–	+	+
c	–	–	+

Problem 174. *Study properties of the relation* $\subset$ *on the Boolean of the empty set* $\varnothing$.

Problem 175. *Is it possible to construct a reflexive, nonsymmetric, and non-transitive relation on a two-element set?*

Problem 176. *Is there a reflexive and simultaneously irreflexive relation?*

Problem 177. *Study the properties of the relation* $\neq$ *on any nonempty set.*
    Together with the orderings, the following class of binary relations often appears in mathematical practice.

Definition 46. *A reflexive, symmetric, and transitive relation is called an equivalence relation (or just an equivalence) and is denoted by the similarity (or tilde) sign* $\sim$. *If a binary relation* ρ *is an equivalence, and a pair* $(x,y) \in \rho$, *then we write* $x \sim y$ *and say that the elements* x *and* y *are equivalent (to one another).*
    Consider the equality relation $=$ on an arbitrary non-empty set X. Clearly, all the three axioms of the equivalence are valid. The reflexivity $x = x$ is valid for every $x \in X$. Next, if $x = y$, then evidently $y = x$ and if $x = y$ and $y = z$, then $x = z$. These properties of the equality sign were all known at the Euclidean times. Therefore, the equality relation on any set is the *equivalence*. Moreover, the three axioms of the equivalence relation are modeled from these properties of equality; we can say that any equivalence relation is a generalized equality.

Problem 178. *On the set of integers, consider the pairs* $k,l \in \mathcal{Z}$ *such that* $k - l = 3p$, *where* $p \neq 0$ *is any non-zero integer. We say that k and l are comparable or congruent modulo 3, and write this as* $k \equiv l \bmod 3$ *or* $k \equiv l(modulo\ 3)$. *In this example,* $5 \equiv 2 \bmod 3$ *and* $-4 \equiv -19 \bmod 3$, *but* $2 \not\equiv 3 \bmod 3$. *Prove that relation* $k \equiv l \bmod 3$ *is the equivalence relation.*

    Replacing here the number 3 with any non-zero integer $k \neq 0$, *we get the infinity of equivalence relations* $k \equiv l \bmod p$. *If* $p = 0$, *the relation becomes the equality relation* $k = l$ *on the integers* $\mathcal{Z}$. *Prove that statement.*

Problem 179. *Prove that the inverse relation to an equivalence is also the equivalence relation, and the inverse relation to a partial order is also a partial order, which coincides with the dual order.*

6.3.2   Equivalence Classes

Consider again the relation $k \equiv l \bmod 3$ on the integers, and all the integers equivalent to 1. These are 4, 7, 10, 13, ..., and in the opposite direction, –2, –5, –8, ... Therefore, together with 1, we get an *arithmetic progression* (arithmetic sequence) of integers, equivalent with 1 mod 3. We observe that this double infinite sequence: K_1 = –5, –2, 1, 4, 7, ... is non-empty, since it contains 1, and as we constructed it, its general term can be written down as $a_n = 1 + 3n$. So that, any two numbers in the sequence are *congruent* modulo 3. In particular, every term is equivalent to 3. Moreover, this sequence does not exhaust all the integers, for example, 2 is not congruent to 1, $2 \notin K_1$.

Therefore, we repeat the construction above, starting with 2, and deduce the set of integers K_2 = {... –7, –4, –1, 2, 5, 8, ...}. But the union $K_1 \cup K_2$ does not contain all the integers either, for example, it does not contain the 3. Thus, we generated another infinite series K_3 = {–6, –3, 0, 3, 6, 9, 12, ...}. We see that each of these three sets is non-empty, any two numbers in any one of them are congruent modulo 3, while two integers from any two of these three sets are not congruent modulo 3, so that, K_1, K_2, and K_3 are mutually disjoint. Moreover, the union of the three sets gives the whole set of the integers $\mathcal{Z}$.

The sets similar to K_1, K_2, K_3 can be built for any equivalence relation. They are called *equivalence classes* with respect to a given equivalence relation.

Let $\sim$ be an equivalence relation on a set X. Fix an arbitrary element $x_a \in X$ and choose all the elements $y \in X$, which are equivalent to x_a, that is, the pairs $(x_\alpha, y) \in \sim$, or $y \sim x_\alpha$. We denote the set of all the elements from X, which are equivalent to x_a, by X_a and call it an equivalence class generated by the element x_a. If occasionally all the other elements of X are not equivalent to x_a, then the class $X_a = \{x_a\}$, that is, it consists of one element only, but anyway, it is not empty.

On the other hand, it also might happen that all the elements of X are equivalent to a certain x_α; thus, they are equivalent to each other. In this case, there is only one equivalence class, which embraces the entire set X. If it is not the case, let us choose any element $\beta \in X \setminus X_a$, and exactly as before, build the equivalence class X_β, that is, we choose all the elements of X, or of $X \setminus X_a$, which is the same, that are equivalent to the element β; they form another equivalence class X_β. It is clear that these two equivalence classes are disjoint, and both are non-empty. If there is no more element left in X, then $X = X_a \cup X_\beta$, and we are done.

Otherwise, that is, if there is an element $\gamma \in X \setminus (X_a \cup X_\beta)$, we build the non-empty subset X_γ and continue as before until we exhaust the entire set X.

Definition 47. *Given a set X and an equivalence relation ~ on it, the totality of all the equivalence classes in the set X is called the factor-set of the equivalence relation ~ and is denoted by X/~.*

Problem 180. *In the example above regarding the congruence of the integers modulo 3, let us denote this relation by $\sigma(3)$. We constructed three equivalence classes K_1, K_2, K_3; hence, the factor-set consists of only three elements, $\mathbb{Z} / \sigma(3) = \{K_1, K_2, K_3\}$, which are non-empty, disjoint, and together exhaust the whole set of the integers $\mathbb{Z}$. Prove that claim.*

In Lemma 4 and Theorem 14, we prove that these properties are valid for every equivalence relation.

Lemma 4. *For an equivalence relation, any two equivalence classes either coincide or are disjoint.*

Proof. Given a set with an equivalence relation X/~, consider two equivalence classes X_a and X_β, and suppose that these two classes ere not disjoint. Hence, there is an element $x \in X$, such that $x \in X_\alpha \cap X_\beta$. Thus, $x \sim \alpha$ and $x \sim \beta$. It follows now from the axioms of the equivalence relation, that, $\alpha \sim \beta$, and finally that $X_a = X_\beta$.

Theorem 14. *The equivalence classes for an equivalence relation on a set X are nonempty pairwise disjoint sets, whose union is precisely the set X.*

Proof. An equivalence class is a nonempty set by construction. The second statement of the theorem was proved in Lemma 4. Now, since an equivalence relation is reflexive, every element of X generates an equivalence class, maybe containing this element only.

Remark 5. *In the case of infinite sets, the procedure of constructing the equivalence classes, described above, is not an obvious thing. To justify it completely, one needs the axiom of choice or any of its equivalent statements, like Zermelo theorem or Zorn lemma. For more details the reader can consult [16] or other advanced books about mathematical logic and set theory.*

The equivalence classes may contain many, even infinitely many elements. In many problems, it is more convenient to choose an element in every equivalence class, join these representatives in one set and work with that *system of representatives*. If in a problem, we do not distinguish between several different elements, we can join them in the equivalence class, and instead of considering the given set we can work with the factor-set or a system of representatives. The latter might often be simpler than the given set. Often, like in the example above, instead of a given infinite set, we arrive to a finite factor-set. Surely, it is easier to handle finite sets. Switching to a factor-set, we eliminate irrelevant information and simplify solving a problem.

For example, in the problem about the *modulo 3 congruence*, we can choose the three numbers, 0, 1, 2 as the representatives, and instead of the infinite set of the integers, we can work with the small set {0, 1, 2}, all of its features are obvious; moreover, the set contains the same information within our problem, as the set of integers.

Definition 48. *Consider a set X and a system of its non-empty disjoint subsets* $\{X_a\}$. *The system is said to be a partition of the set X, if the union of these sets is equal to the entire set X.*

Now, Theorem 14 can be stated in the equivalent form.

Theorem 15. *For any equivalence relation, the family of all its equivalence classes makes a partition of the basic set. Vice versa, any partition of a set X generates the equivalence relation on X.*

Proof. We have to prove only the second part of the statement. Denote the sets of the partition as

$X_a, X_b, \ldots,$ and define the relation ρ on X as follows: two elements $x, y \in X$ are in this relation, that is, $(x, y) \in \rho$ iff both x and y belong to the same set X_1 of the partition. Since the sets of any partition are nonempty and disjoint, we straightforwardly deduce that the relation ρ is reflexive, symmetric, and transitive.

Problem 181. *Let $\mathcal{Z}$ be the set of the integers, and the binary relation ~ on the Cartesian product of pairs (a, b) $\mathcal{Z} \times \mathcal{Z}$ is defined as $(a, b) \sim (c, d)$ iff $ad = bc$. Prove that this is an equivalence relation. What is the meaning of this equivalence relation on the integers?*

In many problems, especially if we want to represent a binary relation in the computer memory, it is useful to define the adjacency matrix of the binary relation.

Definition 49. *Let a binary relation ρ be defined on a set $\{a_1, a_2, \ldots, a_n\}$. An $n \times n$ matrix M is called the adjacency matrix of the relation ρ iff a matrix element $m_{ij} = 1$ iff $a_i \rho a_j$ is true; otherwise, the matrx element is 0.*

Problem 182. *Formulate the basic properties of binary relations (reflexivity, symmetry, etc.) in terms of the adjacency matrix of this relation. For example, a relation is reflexive iff the main diagonal of its adjacency matrix contains only ones, while all the other elements of the matrix are irrelevant.*

Problem 183. *Give a geometric interpretation of the properties of the order and of equivalence relations.*

Problem 184. *Let $(X, \preccurlyeq)$ be a partially ordered set. Prove that binary relation*
$$\rho = \left\{ (x,y) \mid (x \preccurlyeq y) \wedge (y \preccurlyeq x) \right\}$$ *is the equivalence relation and describe its factor-set.*

Example 31. *Consider a relation "To be a friend" on the set of all people. Suppose the statement "A friend of my friend is a friend of mine" is universally true. Study properties of this relation.*

Problem 185. *Which of the following binary relations are functional, that is, can be represented as a function, where the relations are*
$\rho = \{(x,y)\} \subset \mathcal{R} \times \mathcal{R}$; *(1)* $y = x^2$; *(2)* $x = y^2$; *(3)* $|x| = |y|$; *(4)* $2x + 3y = 4$; *(5)* $x \exp x = y$.

Problem 186. *For the functions above, find the composite functions. Find the domains, codomains, and ranges of the functions and their compositions.*

Problem 187. *For what pairs of functions from the previous problems do the inverse superpositions exist? Calculate, if any. Compare the domains, codomains, and ranges of the given functions and their compositions.*

Problem 188. *Let f and g be linear functions, that is, $f(x) = ax + b$ and $g(x) = cx + d$, where a, b, c, and d are constant (real-valued) coefficients. What are the composite functions $f \circ g$ and $g \circ f$? Under what conditions on the coefficients are they equal?*

Problem 189. *Let ρ be an equivalence relation on the set S. For the two elements $a, b \in S$, prove that the following three statements are equivalent.*

1. $a \rho b$

2. the equivalence classes, containing a and b, are equal

3. the intersection of the equivalence classes, containing a and b, is non-empty, and moreover, is equal to either of these equivalence classes.

Problem 190. *Consider a binary relation $\rho = \{(x,y)\,|\,x = y +1$ or $x = y -1\}$ on the set of the integers. Find its inverse ρ^{-1} and the square ρ^2. Describe the matrix and the graph of ρ. Is this an equivalence relation or a partial or linear order? If this is an equivalence relation, find the factor-set and a system of representatives.*

Problem 191. *Answer the same questions regarding binary relations given by the adjacency matrix M and the graph G (Fig. 6.3), where*

$$M = \begin{pmatrix} 0 & 0 & 0 & 1 \\ 1 & 1 & 1 & 0 \\ 1 & 1 & 1 & 0 \\ 1 & 1 & 1 & 0 \end{pmatrix}.$$

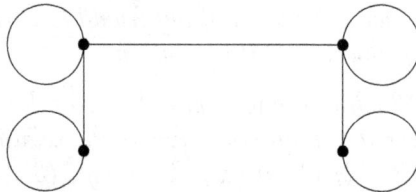

FIGURE 6.3 Graph G for Problem 191.

6.4 RELATION DATABASES

A *database* is exactly what it implicates – a collection of data. To store the data in and recover the data from a computer memory, the data must be appropriately arranged and formatted. Imagine a large library, where the books are randomly shelved – no reasonable usage of the library would be possible. There are various older and newer types of databases; however, we discuss here only one currently popular type called *relational databases*, because their construction essentially uses n-ary relations – see this chapter above. The ordered collections of items are often called *vectors*; in discrete mathematics and computer science, they are also called *strings* or *tuples*; more specifically, *n–tuples* if the tuple contains n components; these components are called *fields*.

We introduce some simple properties of *relational databases* and illustrate these properties by an example of the database **Student**. A database contains information on specific *objects*, also called *entities*, the *finite* set of entities is denoted hereafter as **ENT**; the entities in the small model example below are students enrolled at Data College. Every student now has enough experience with retrieving and using some information from databases, for example, finding her final grades and GPA score online. This information is a part of the entity's *record*. Each entity belonging to the database has its own record, which is an ordered *string (tuple)* of symbols.

This paragraph explains how to represent tuples in a computer memory. Let ω be the cardinality of the universal set $U = \{a_1, a_2, \ldots, a_\omega\}$ in a problem considered. Hence, no set in the current problem may have the cardinality bigger than ω, and it is enough to consider bit-strings of length ω. Let the string S_A represent the set A. If the set $A \subset U$ contains the element $a_l \in U$, that is, $a_l \in A$, then the *l*th element of the string S_A, representing A, is 1; otherwise, this element is 0. We call these bit-strings the *string-indicators* of subsets. For instance, if $\omega = 5$, $U = \{a_1, a_2, a_3, a_4, a_5\}$, and $A = \{a_2, a_4, a_5\} \subset U$, then $S_A = \{0, 1, 0, 1, 1\}$. The 1s in the latter vector occupy the second, fourth, and fifth positions, indicating, that the set A contains the elements with these indices.

Any object has various properties. In any specific case we are only interested in a certain set of these properties called *attributes* and denoted by $AT_1, AT_2, \ldots, AT_n$; the string

$$\mathcal{H} = (AT_1, AT_2, \ldots, AT_n)$$

is called a *heading*. Therefore, every record splits into *n fields* corresponding to attributes $AT_1, AT_2, \ldots, AT_n$, respectively. In our model example, we consider the attributes *Student's name, Student's ID, Student's Final Grade in Discrete Mathematics*, and *Student's GPA*, abbreviated to *SName, SID, SFGDiscrMath*, and *SGPA*; thus, in the example $n = 4$ and the heading is

$$\mathcal{H} = (SName, SID, SFGDiscrMath, SGPA).$$

For instance, the attribute AT_3 = Student's Final Grade in Discrete Mathematics (*SFGDiscrMath*).

The possible values of the attribute AT_i make a (finite) set called the *i*th *Domain*

$$D_i, i = 1, 2, \ldots, n,$$

of this attribute. Whence, a database consists of a basic set of entities **ENT**, the heading $\mathcal{H}$ of length n, of n domains corresponding to every attribute, and of n–tuples, called records, corresponding to every entity in **ENT**. An entry in ith field of each record is taken from the ith domain D_i.

Thus, a database can be thought of as a set of tuples from the Cartesian product

$$ENT \times \mathcal{H} \times D_1 \times D_2 \times \cdots \times D_n.$$

Since a relation is a subset of some Cartesian product, a database is an $(n + 1)$–ary relation, which explains the term "relational." Of course, all those parameters, including n and **ENT**, can depend on discrete time, but we do not consider this dependence explicitly here. Rather it may be more convenient to believe that the domains contain all possible, past and future, values of the attributes. A snapshot of a database can be conveniently represented by a table like that in the following example.

An important issue is the computer representation of sets, that is, how to represent sets not on paper but in computer memory. One convenient way for that is to use the *characteristic functions* χ of the sets and subsets, that is, the binary strings, marking unities 1 within a string. For instance, if we are given a subset $B = \{2, 3, 7, 9\}$ as a part of the set $A = \{1, 2, 3, 4, 5, 6, 7, 8, 9, 10\}$, then the characteristic function $\chi(B) = \{0,1,1,0,0,0,1,0,1,0\}$, where the 1s in the characteristic function flag the positions of the elements of the subset $B \subset A$ in the universal set A.

Problem 192. *For the same universal set $A = \{1, 2, 3, 4, 5, 6, 7, 8, 9, 10\}$, find the characteristic functions $\chi_C, \chi_D,$ and χ_E, where the subsets are $C = \varnothing, D = A,$ and $E = \{2\}$.*

The parentheses in the characteristic functions or in the *characteristic bit-strings* show that this is an *ordered* sequence; thus, the string $S_1 = (0, 1, 0, 1, 1)$ is different from the string $S = (1, 0, 0, 1, 1)$. We treat the symbols 0 and 1 here as the elements of the Boolean algebra $\mathbb{B} = \mathbb{B}^2$. The bit-strings are convenient to use because the Boolean operations on bit-strings are defined *termwise*. If $S = \{s_1, s_2, \ldots, s_k\}$ and $T = \{t_1, t_2, \ldots, t_k\}$ are two bit-strings of the *same length* k, then their disjunction is

$$S \vee T = \{s_1 \vee t_1, s_2 \vee t_2, \ldots, s_k \vee t_k\}$$

and likewise for conjunction and other Boolean operations. It is obvious that these operations on strings possess the same properties as the Boolean operations with propositions, that is, associativity, commutativity, etc.

Problem 193. *Find the negation, conjunction, disjunction, and conditional of the string*

$$S_1 = (0, 1, 0, 1, 1).$$

Let ω be the cardinality of the universal set $U = \{a_1, a_2, ..., a_\omega\}$ in a problem considered. Hence, no set in the current problem has the cardinality bigger than ω, and it is enough to consider bit-strings of length ω. Denote the string representing the set A as S_A. If the set $A \subset U$ contains the element $a_l \in U$, $a_l \in A$, then the lth element of the corresponding string S_A, representing A, is 1; otherwise, this element is 0. We call these representing bit-strings the *string-indicators* of subsets. For instance, if $\omega = 5$, $U = \{a_1, a_2, a_3, a_4, a_5\}$, and $A = \{a_2, a_4, a_5\} \subset U$, then $S_A = (0, 1, 0, 1, 1)$.

The following chart represents our model database.

Database STUDENT (Discrete Mathematics)

No.	Name	ID	FGDM	GPA
1	Al	1001	A-	3.75
2	Ben	1002	B	3.9
3	Cathy	1003	A-	3.9

Problem 194. *(1) How many different records can be stored in this database if the students have used 500 first names, 1000 last names, 201 possible values of GPA, and the school uses 19 different, including administrative grades and 6-digit integer numbers, from 000,001 to 999,999, as IDs?*

(2) The same problem in the case, if the school uses 9-digit SS-numbers as IDs?

To simplify the development of and work with databases, there were developed specialized programming languages, like DBASE or SQL - *Structured Query Language*. These languages allow to add/remove records to/from a database and include the standard set-theory operators, like the union, intersection, set-difference and Cartesian product. We describe here only two specific commands in SQL, *projection* and *join*.

Definition 50. *Given a database $\mathcal{D}(\textbf{ENT}, \mathcal{H}, D_1, ..., D_n)$, its projection on a sub-heading $\mathcal{H}_1 \subset \mathcal{H}$ of the heading $\mathcal{H}$ is a shortened database $\mathcal{D}_1(\textbf{ENT}, \mathcal{H}_1, ...)$, such that all the attributes in $\mathcal{H}_1$ have the same domains as in $\mathcal{H}$ and every record contains only the attributes (fields) from $\mathcal{H}_1$.*

For example, if $\mathcal{H}_1 = \{SName, SGPA\}$, then $n = 2$ and the table above abbreviates to

Database STUDENT(GPA)

No.	Name	GPA
1	Al	3.75
2	Ben	3.85
3	Cathy	3.9

Definition 51. *Given two databases*

$$\mathcal{D}_1(\mathbf{ENT}, \mathcal{H}_1, D_{1,1}, \ldots, D_{1,m})$$

and

$$\mathcal{D}_2(\mathbf{ENT}, \mathcal{H}_2, D_{2,1}, \ldots, D_{2,n})$$

*their join is the database with the same set of entities **ENT** and the heading*

$$\mathcal{H} = \mathcal{H}_1 \bigcup \mathcal{H}_2.$$

For example, consider two tables

Database STUDENT (Discrete Mathematics)

No.	Name	ID	FGDM	GPA
1	Al	1001	A-	3.75
2	Ben	1002	B	3.85
3	Cathy	1003	A-	3.9

and

Database STUDENT (Calculus)

No.	Name	ID	FGCalc	GPA
1	Al	1001	B	3.75
2	Ben	1002	B+	3.85
3	Cathy	1003	A-	3.9

The join of these databases is the table

Database STUDENT (Discrete Mathematics $\cup$ Calculus)

No.	Name	ID	FGDM	Calc	GPA
1	Al	1001	A-	B	3.75
2	Ben	1002	B	B+	3.85
3	Cathy	1003	A-	A-	3.9

To extract the information from the database, we look for an attribute that can distinguish every pair of their records: this attribute is called the *Primary Key*. If we study the small table above, we see that the GPA-column does not distinguish the second and third records in the sense that these records have the same GPA; thus, this attribute cannot be a primary key. Similarly, the SFGDM-attribute does not work as a primary key. However, the Name- and ID-attributes would do in this example, however, in many schools you can find several students with the same name.

For this reason, the schools assign special IDs, such that all the students, former, current and future, have different IDs. The Social Security Number *** – ** – **** would do, but usually it is preferable to avoid that sensitive information. However, there is an essential obstacle with the databases that will be studied in the next chapter. These discussions lead us naturally to the hashing functions and their applications, which, together with some applications in cryptology, will be also considered in the next chapter.

Problem 195. *Compute a sub-database containing the information about the students with the GPA of at least 3.8.*

6.5   EXERCISES

Exercise 6.1. *Compute the Cartesian product of A = {a, b} and P = {p, q, r}, and the product P × A.*

Solution. By the definition, the product $A \times P$ is

$$A \times P = \{(a,p),(a,q),(a,r),(b,p),(b,q),(b,r)\},$$

while $P \times A = \{(p, a), (q, a), (r, a), (p, b), (q, b), (r, b)\}$.

Exercise 6.2. *Let* $X = Y = \mathcal{N}$ *and the relation consists of the ordered pairs*

$$\rho = \{(1,1),(1,2),\ldots,(2,2),(2,4),(2,6),\ldots,(3,3),(3,6),\ldots,(4,4),(4,8),\ldots\}.$$

Prove that this relation can be expressed as

$$\rho = \{(m,n) \in \mathcal{N} \times \mathcal{N} \mid m \text{ is a divisor of } n\}.$$

Exercise 6.3. *Find the union, the intersection, and the complement of the relations* $<, >, =$ *on the set of all real numbers.*

Exercise 6.4. *Consider a relation of inclusion* $X \subset Y$ *on the totality of all sets in a given universe U. Clearly, this is a partial, but not linear ordering because not any two sets are comparable with respect to this relation. It might be that neither of them is a subset of another set. Analyze the case when the universe contains only one element.*

Exercise 6.5. *Consider a relation of divisibility on the set of non-zero natural numbers. It is denoted by a vertical bar, that is, if two positive integers k and l are in this relation,* $l = pk$ *with an integer p, it can be written as* $k \mid l$; *l is called a multiple or dividend of k, and p and k are divisors of l. Again, this is a partial, but not linear ordering because not any two integers are comparable with respect to this relation. For example,* $3 \mid 6$ *but* $6 \nmid 3$.

Exercise 6.6. *Find the negation, conjunction, disjunction, conditional of the strings* $S_1 = (1, 1, 1, 1, 1)$ *and* $S_2 = (0, 0, 0, 0, 0)$.

COMBINATORICS

7.1   FINITE AND INFINITE SETS: BASIC RULES

Consider two sets $\{1, 2, 3\}$ and $\{A, B, C\}$. They consist of different elements; thus, they are clearly *different*, but each of them contains *three* elements, everyone can count them. However, nobody can count all the integers because whatever number we reach, we can add 1 and continue... The power of mathematics is that we sometimes can find another procedure to accomplish the would be impossible task. Even if we cannot accomplish that recount, we can *compare* two sets and conclude that one set has more (or less) elements than another one.

Fundamental Principle of Combinatorics: If one can establish a one-to-one correspondence between a set X and another set Y, one says that X and Y have the same *cardinality*. In the case of the set of natural numbers $\mathbb{N} \subset Y$, the set Y is called *infinite*. All non-infinite sets are called *finite sets*. We do not study the infinite sets in this textbook. The sets, that we compare all the finite sets with, are the sets of the first n natural numbers, where n can be any natural number, $n = 0, 1, \cdots$; we call these sets *natural segments*, claim that for every n this set contains exactly n elements, and denote this set by $N_n = \{1, 2, \ldots, n-1, n\}$, $N_0 = \varnothing$. The number of elements of a set is called its *cardinality*. A natural segment N_n is a finite set, and its cardinality is n. For instance, the cardinality of the English alphabet (a,b,c,...x,y,z) is 26, the cardinality of the set of decimal digits (0,1,2,...,9) is 10, the cardinality of the bit set $\{0, 1\}$ is 2.

The set with *no* element is special, it is like an empty hull with nothing inside; it is called the *empty set* and is denoted as $\varnothing$.

Problem 196. *Compute* $\varnothing \cap \varnothing, \varnothing \cup \varnothing, X \cap \varnothing, X \cup \varnothing, X \setminus \varnothing, \varnothing \setminus X,$ *for any set* X

Problem 197. *Two school sections with 27 and 29 students, respectively, together with four teachers, go on a field trip. How many buses should the school book, if a bus can take 35 people and no person can stand when a bus is moving?*

Solution. Assuming that no student belongs to both classes, we have $27 + 29 = 56$ students. Adding four adults, the school needs two buses, since $56 + 4 = 60 < 2 \times 35 = 70$.

7.1.1 The Sum and Product Rules

This problem can be immediately extended as the general *Sum Rule*.

Lemma 5. *If k finite, pairwise disjoint sets $X_1, X_2, \ldots, X_k$ have the cardinalities $n_1, n_2, \ldots, n_k$, then the cardinality of the union is the sum of their cardinalities, that is,*

$$|X_1 \cup X_2 \cup \cdots \cup X_k| = n_1 + n_2 + \cdots + n_k. \tag{7.1}$$

Problem 198. *Prove that if A and B are finite sets and $A \subset B$, then $|A \cup B| = |B|, |A \cap B| = |A|, |B \setminus A| = |B| - |A|$.*

Problem 199. *Is Problem 198 correct, if A is not a subset of B? How to modify these formulas, if A is not a subset of B?*

Hint. Consider the set-differences $A \setminus B$ and $B \setminus A$.

To continue our study of the Sum Rule, let us consider another simple model problem.

Problem 200. *Every student in a section studies one and only one foreign language, 13 students take French, 12 students take German. How many students are there in the section?*

Solution. To answer the question, we have, obviously, to add $13 + 12 = 25$, but we want one more time to give a detailed interpretation of the problem in terms of the set theory. Let us consider the set X of all students in this section. The problem distinguishes two subsets of X, namely, the set X_F of the students studying French, $|X_F| = 13$, and the set X_G consisting of those taking German, $|X_G| = 12$. The sentence *"one and only one"* actually contains two opposite claims. It means that each student in this section belongs to *at least* *one* of these two subsets, that is, each student belongs to $X_F \cup X_G$. Moreover, it also means that no student belongs to both subsets simultaneously, that is, $X_F \cap X_G = \varnothing$. So that, in this example, we have to calculate the cardinality of a disjoint union of two finite sets, $|X| = 13 + 12 = 25$. In the same way, if every student in a group studies one and only one of the three foreign languages, then to find the total number of students in the group, we have to add up the cardinal numbers of the three subsets.

Problem 201. *There are 25 bags with apples of three different kinds, so every bag has apples of only one kind. Is it possible to find nine bags with apples of the same kind?*

Solution. Suppose for every kind we can find eight bags with the apples of this particular kind, 24 bags in total. Then the 25th bag will add the 9th bag to a certain kind.

Now we slightly change Problem 200.

Problem 202. *Let every student in a group study* at least one *of the two foreign languages, 13 students take French and 12 students take German. How many students are in the group?*

Solution. We have the same equation $X = X_F \cup X_G$; however, now some students may take both foreign languages; thus, the intersection does not have to be empty. Surely, if it is, then as before, $|X| = 13 + 12 = 25$. However, if the intersection $X_F \cap X_G \neq \varnothing$, that is, at least one student studies both foreign languages, the solution is different. The condition $X_F \cap X_G \neq \varnothing$ means that this intersection is not empty, and the solution must involve the information about this intersection.

The smallest possible size of a student section is 13, which means that 13 students study French and 12 *among them* in addition study German, that is, $|X| = 13$. But in general, the cardinality of the intersection is not known. This is an example of a combinatorial problem having several possible answers. Depending upon the cardinal number of the intersection, the group may contain any number of students, from 13 to 25, inclusive.

7.1.2 Unions, Cartesian Products, Booleans

Keeping in mind these examples, we are to learn how to calculate the cardinal numbers of unions, Cartesian products, and power sets.

Problem 203. *For a field trip, $n_c = 25$ first-graders brought sandwiches with cheese, and $n_s = 29$ students brought sandwiches with salami. How many students took part in the trip?*

Solution. The easiest way is to *add* $n_c + n_s = 25 + 29$, but that does not follow from the problem since the addition would mean that no student brought both kinds of sandwiches. This statement means that to solve the problem in a unique way, and we must include another piece of information; namely, one must know *how many students* brought both kinds of sandwiches. If this information is missing, the problem may have several answers. For example, if none has both things, then the sum $n_0 = n_c + n_s = 25 + 29 = 54$ is the answer. But if exactly one student has both kinds of sandwiches, then the sum n_0 counts this student twice. However, the school roster counts this student only once; thus, we must decrease the sum 54 by 1. Hence, the answer, in this case, is $25 + 29 - 1 = 53$. Next, if two students have both kinds of sandwiches, we must subtract 2 from 54, etc.

Some of the 25 cheese-loving students can bring another sandwich; then, the largest number to subtract is 25, and any number from 54 to 29 can be the answer.

The examples above illustrate the next important theorem, which will be used several times later on, in particular, when we will discuss Euler's totient function in the following chapter.

Theorem 16. The Principle of Inclusion-Exclusion. *Consider a set X, which is a union of n subsets, $X = X_1 \cup X_2 \cup \cdots \cup X_n$. Then*

$$| X | = | X_1 | + | X_2 | + \cdots + | X_n | - | X_1 \cap X_2 | - | X_1 \cap X_n | - \cdots - | X_{n-1} \cap X_n | +$$

$$+ | X_1 \cap X_2 \cap X_3 | + \cdots + (-1)^{n-1} | X_1 \cap X_2 \cdots \cap X_n |. \qquad (7.2)$$

Proof. We prove the theorem by the induction over n. As the basis of induction, we choose the case $n = 2$. In this case, the claim is

$$| X_1 \cup X_2 | = | X_1 | + | X_2 | - | X_1 \cap X_2 |. \qquad (7.3)$$

Problem 204. *Prove that $X_1 = (X_1 \setminus X_2) \cup (X_1 \cap X_2)$ where $(X_1 \setminus X_2) \cap (X_1 \cap X_2) = \varnothing$.*

Thus, the basis of induction follows straightforwardly from here and Lemma 2. To make the inductive step in (7.2), we write $X = X_1 \cup X_2 \cup \cdots \cup X_n \cup X_{n+1} = X_1 \cup X_2 \cup \cdots \cup X'$, where $X' = X_n \cup X_{n+1}$, apply (7.2) to the latter, use the distributive rule, and finally, twice employ the inductive assumption again.

The name of the principle stems from the procedure, since we alternatively include and exclude the elements of various subsets.

It is useful in many problems to re-state this principle in other terms [26, p. 18]. Consider a set $X, |X| = n$, such that its elements can have or have not q properties $P_i, 1 \le i \le q$. For every $x \in X$, we specify that either $P_i(x) = 1$ if the element x has the property P_i, or $P_i(x) = 0$ if x does not have that property; therefore, we assign q two-valued $0 - 1$ functions $P_i, 1 \le i \le q$, on the set X, and these properties are, indeed, mappings $P_i : X \to \{0,1\}$. Let also $X_i, 1 \le i \le q$, be the subset of X, whose elements possess the property P_i. Thus, these sets are the total preimages of 1, $X_i = P_i^{-1}(\{1\})$ and $n_i = |X_i|$ being their cardinalities. Similarly, let $X_{i,j}$ be the subset of elements of X, such that its elements possess both properties P_i and P_j, and $n_{i,j} = |X_{i,j}|$, etc. Denote also $n_0 = |X_0|$, where

$$X_0 = X \setminus (X_1 \cup X_2 \cup \cdots \cup X_n).$$

Now we can state an equivalent version of the Principle of Inclusion-Exclusion, which is also called the Eratosthenes[1] *Sieve Formula*.

Theorem 17. $n_0 = n - \sum_{i=1}^{q} n_i + \sum_{i_1 < i_2} n_{i_1, i_2} - \cdots$

$$+(-1)^s \sum_{i_1 < i_2 < \cdots < i_s} n_{i_1 i_2 \cdots i_s} + \cdots + (-1)^q n_{i_1, i_2, \dots, i_q}.$$

Proof *follows immediately if we notice that X_0 is disjoint with every set $X_i, 1 \le i \le q$.*

Problem 205. *Among a group of 21 students, 17 like dogs, and 13 favor cats. How many students like both dogs and cats? What if, among them, 3 do not like animals at all?*

Problem 206. *A license plate in the State of Duo consists of two symbols - an English letter and a digit from the set {0, 1, ..., 8, 9}. How many license plates can be issued by the Duo State?*

Solution. Before solving the problem, we must discuss several important issues. First of all, how do we treat ordering, that is, are the plates "A-3" and "3-A" identical or different? Actually, these two are *the different problems*.

---

[1] Eratosthenes of Cyrene 276 - c. 195/194 BC.

Mathematics cannot declare them different or identical; mathematics can only provide the tools for solving either of them.

Another issue is whether the big and small characters, that is, a and A, etc., are the same or different. Again, the latter is not a mathematical issue, and it just means that we have not 26 but 52 different symbols in the problem.

To be specific, we assume in this problem, that there are exactly 26 different symbols, suppose, 26 capital letters. Moreover, we suppose that a plate always starts with a letter, followed by a digit, like "A-5"; in a sense, the plates like 5 – A just do not exist. If in another problem, we declare the plates "A-5" and "5-A" *different*, that means that the order is important; thus, in the current problem, the order is not relevant.

One more issue, which is not evident in this problem but maybe important elsewhere, is the *repetition of the symbols*. In the current problem, we declare that no symbol can appear on the plate *twice*; even though, if in a different problem, one must resolve in advance, whether the plates "A-A" are allowed. In the latter case, we would talk about problems with *repetitions*.

Now we can formulate the problem precisely: How many are there two-symbol strings (x, y), if no symbol can repeat, the first symbol belongs to a set $X, x \in X, |X| = 26$, and the second symbol is from the set $Y, y \in Y, |Y| = 10$?

Let us consider a special case, when $|Y| = 1$, say, $Y = \{1\}$. Then all the plates can be depicted as $(A-1), (A-2), \ldots, (A-9), (A-0)$, thus, there are exactly 10 plates – as many as there are digits, which is the same as the cardinality of the set $|Y|$. One gets the same amount if A is replaced with B, or with C, … , or with Z. By the first Sum Rule, Lemma 2, there are $|X| \times |Y| = 26 \times 10 = 260$ license plates.

We are concerned here with various problems, in which we have to answer the question "How many?" or "In how many ways?" For example, a student has to take three exams. In how many ways can she end the entire examination session, if there are five different grades A, B, C, D, and F? Here we consider only finite sets, and it is obvious that there is only a finite set of possibilities in the problem (we calculate later that the number of different ways is $5^3 = 125$). We have already learned how to calculate the cardinal numbers of unions and of Cartesian products of finite sets. Now, considering several simple model problems, we interpret basic combinatorial constructs in the set-theory terms. We consider the classical permutations and combinations and introduce a few more combinatorial results such as the Dirichlet and Pigeonhole principles.

Definition 52. *The set of first n positive integers is called the natural segment or natural n–segment and is denoted as* $[n]$, *thus,* $[4] = \{1, 2, 3, 4\}$; *it is obvious that* $|[n]| = n$.

Definition 53. *Two sets have the same cardinality (the same number of elements) iff the sets can be put in a one-to-one correspondence one with another. If a set X can be put in a one-to-one correspondence (bijection) with a natural n-segment* $[n]$, *X called a finite set, containing n elements. Otherwise, a set is called infinite.*

Problem 207. *Find the cardinality of the English alphabet* $\{A, B, ..., Y, Z\}$.

Solution. Numerate the characters and establish a one-to-one correspondence between the English alphabet and the natural segment [26]. Indeed, there are 26 English letters. We do not distinguish here small and capital letters. It would be another problem, if we have to distinguish them.

Problem 208. *Prove that a set is finite iff it cannot be in a one-to-one correspondence with any of its proper subset. Another way, a set is infinite iff it can be put in a one-to-one correspondence with its proper subset.*

Problem 209. *Find a one-to-one correspondence between the sets of natural numbers and the set of positive natural numbers.*

Problem 210. *Prove that the set of integers and the set of even integers are infinite sets of the same cardinality. What about the set of prime numbers?*

Problem 211. *Establish a one-to-one correspondence between the set of integers and the set of all positive perfect squares.*

Problem 212. *Consider a power-set* Y^X *with* $|X| = k$ *and* $|Y| = n$; *find the cardinality* $|Y^X|$.

Now the following Product Rule is obvious.

Theorem 18. *Suppose that in a certain problem the solution set X can be put in a one-to-one correspondence with a Cartesian product*

$$Y = Y_1 \times Y_2 \times \cdots \times Y_k.$$

Then $|X| = |Y_1| \cdot |Y_2| \cdots |Y_k|$.

Problem 213. *A license plate in the State of New York consists of three English characters followed by 4 decimal digits, in this order. How many plates are there in NYS?*

A factor-set of any equivalence relation represents a *partition* of the basic set, that is, different equivalence classes are disjoint and together they exhaust this set completely. Therefore, the Sum Rule implies the following simple property of the factor-sets.

Proposition 1. *Consider the factor-set* $\{X_a, X_b, ..., X_p\}$ *of an equivalence relation on a set X. Suppose that there are p equivalence classes, and all of them have the same cardinality* $|X_a|$. *Then by Lemma 2 and Theorem 18,* $|X| = p \cdot |X_a|$.

All the model problems in this section were reduced to Cartesian products with the components of equal cardinality, which is not always the case.

Problem 214. *A motel has 12 guest rooms, among them 6 rooms, numbered* $R_{11} - R_{16}$, *for single guests, 4 rooms,* $R_{21} - R_{24}$, *with two beds, and 2 rooms,* $R_{31} - R_{32}$, *with three beds. How many guests can the motel accommodate at most?*

Solution. The Product Rule cannot be applied here. Therefore, we have to split the set of rooms into subsets R_1, R_2, R_3, such that R_1 contains six individual rooms, $|R_1| = 6$, four double rooms $|R_2| = 4$, and two rooms with three beds, $|R_3| = 2$. These subsets make a partition of the set of rooms, $R = R_1 \cup R_2 \cup R_3$, since $R_i \cap R_j = \varnothing$ for $1 < i, j < 3$, $i \neq j$. Therefore, the motel can accommodate at most $6 \cdot 1 + 4 \cdot 2 + 2 \cdot 3 = 20$ guests.

It is often convenient in problems like the above to employ the *tree of variants* or the *possibility tree*. The next drawing clearly shows this tree and how to draw it for Prob. 214.

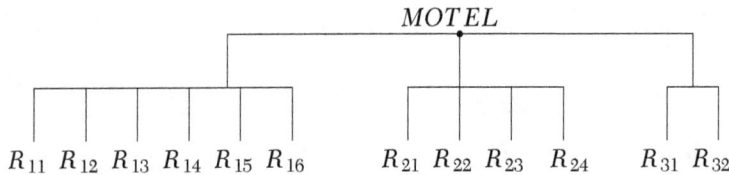

FIGURE 7.1 The tree of variants for problem 214.

7.1.3 Arrangements and Combinations

Next, we solve two more *model* problems.

Problem 215. *In how many ways can a group of 15 students elect their speaker and a recording secretary for the next discussion?*

Problem 216. *In how many ways can a group of 15 students elect two representatives for a campus assembly?*

Solution. Of course, we suppose that all people are different, and introduce a set $S = \{s_1, s_2, \ldots, s_{14}, s_{15}\}, |S| = 15$. We record the outcomes of elections as a two-element string of names, like (*Ann, Ryan*). Let us compare the strings, say, (*Ann, Bob*) and (*Bob, Ann*); are these two strings represent the same result or two different votings?

The right answer is that these are two *different problems*, two different mathematical models, namely, whether or not the order is important or not. What model to choose has nothing to do with *mathematics*, it depends upon the situation, while mathematics must provide methods for solving each of these problems.

In general, different persons perform different tasks differently. Hence in the first problem above, the answer depends ordering, that is, in this problem, we must treat the two strings above as different. If however, ordering is *irrelevant* like, probably, in our second problem, that is a problem without ordering. Let us reiterate that all the elements of any set are different.

To consider ordered totalities, like in Problem 215, we imagine a ticket with two boxes, marked by L (for "Leader") and R (for "Recording Secretary"); the names of students being elected to the corresponding office must be written after the voting in the appropriate box. Let us introduced the set $B = \{L, R\}, |B| = 2$. After this round is over, the names of the elected students are to be inserted into these boxes; say, the report B = $\{s_3, s_{14}\}$ indicates that the student s_3 was elected to be the leader and s_{14} the secretary. Therefore, we establish a *map* $V : B \rightarrow S$, whose domain is the two-element set B, and co-domain is the set of students S.

It is worth reiterating that given an office, we have chosen a student for the office, not vice-versa, that is, we establish a map $V : B \rightarrow S$, with the domain is B and the co-domain is S.

Therefore, in the problem, we immediately recognize the power-set S^B – see Definition 20, whose cardinality will be computed in Lecture 8. But we can intuitively conclude that there are 15 ways to find a student for the first box, and after that, only 14 ways to fill out the second box. The Product Rule would tell us to multiply these two numbers; thus, the answer in Problem 215 is $15 \cdot 14 = 210$.

Now we solve Problem 216. Unlike the previous problem, here we stipulate that both students – representatives have equal rights and responsibilities, so that both ballots $\{s_3, s_{14}\}$ and $\{s_{14}, s_3\}$ represent the same outcome; this is just a subset of the set S. Therefore, the answer to Problem 216 is only a half of the previous amount, it is $\dfrac{15 \cdot 14}{2} = 105$.

Definition 54. *A k-element subset of an n–element set X is called a combination of elements (without repetitions) of X; the number of k–combinations of elements of n-set is denoted by C(n, k). Since sets are non-ordered families of the elements, combinations are non-ordered.*

Definition 55. *A k-element ordered totality of an n–element set X is called an arrangement or permutation of elements of X (without repetitions); the number of k-arrangements of the elements of n–set is denoted by A(n, k). We also write P(n) = A(n, n).*

Since the permutations $P(n)$ often occur almost everywhere, we prove independently the formula for $P(n)$.

Proposition 2. *For any $n \geq 0$, $P(n) = n!$.*

Proof. If $k = 0$, we set $P(0) = 0! = 1$ by definition. Now we do mathematical induction over n. $P(n)$ stands for the number of permutations of n different symbols $\{x_1, x_2, \ldots, x_n\}$; if $n = 1$, there is, clearly, only one such an arrangement $\{x_1\}, |\{x_1\}| = 1$. Since $1! = 1$, we have $P(n) = n!$ when $n = 1$. Assume now that $P(k) = k!$ for all $k \leq n$, and compute $P(n+1)$. Pick any x' among the $X' = \{x_1, x_2, \ldots, x_n, x_{n+1}\}, |X'| = n+1$ elements, and arrange $X' \setminus x', |X' \setminus x'| = n$, in any order; by induction, there are $P(n) = n!$ such arrangements. What we have to do now, is to insert x' within this arrangement. There are $n - 1$ available spaces between these n x_i elements, and in addition two spaces before the left-most and after the right-most elements, in total, $n - 1 + 2 = n + 1$ positions. Thus, for $n + 1$ elements, there are $P(n+1) = n! \times (n+1) = (n+1)!$ permutations, and the Induction Principle implies Proposition 2.

Let us mention that if ordering is immaterial, then n items can be arranged in the unique way, since in this case, all the arrangements are equivalent. Next we consider the number of combinations $C(n, k)$ of n different elements taken k at a time.

Theorem 19. *Let X be any set of n different elements, and an integer $0 \leq k \leq n$. Then the number of k–combinations of these elements is $C(n,k) = \dfrac{n!}{k!(n-k)!}$.*

Proof. We do induction over integer n. This is a new problem, since it involves two integer variables or parameters, n and k, where the latter is subject to the restriction $0 \leq k \leq n$. The latter means that actually for every n we must prove $n + 1$ equations, for $k = 0, k = 1, \ldots, k = n$. If $k = 0$, then for any natural n, $C(n, 0) = 1$, since there is the only way to choose the empty set, which is a (unique) subset of any set. At the same time for any n, $\dfrac{n!}{0!(n-0)!} = 1$.

Now, let n be any positive integer, X be any n–element set, and $0 < k \le n$. By induction, the theorem is assumed to be valid for X, and we have to prove the statement for any $(n+1)$–element set $X' = X \cup x', x' \notin X$, and $k \le n+1$. If $k = n+1$, we are within Proposition 3, hence $C(n+1, n+1) = P(n+1) = (n+1)!$, and all is done. Therefore, we only have to consider the main case, when there are $n + 1$ entities, and $k < n + 1$. Fix any k–element subset $A \subset X', |A| = k$. There are two disjoint cases, either $x' \in A$ or $x' \notin A$.

In the first case, $|A \setminus \{x'\}| = k - 1$, hence by induction there are $C(n, k - 1)$ such subsets. In the second case, $A \subset X$ and $|A| = k$, thus again by the Inductive Assumption, there are $C(n, k)$ such subsets. The total is

$$C(n, k-1) + C(n, k) = \frac{n!}{(k-1)!(n-k+1)!} + \frac{n!}{k!(n-k)!} =$$

$$= \frac{(n+1)!}{k!(n+1-k)!} = C(n+1, k). \tag{7.4}$$

Theorem 20. *For any sets $X, |X| = k$ and $Y, |Y| = n$, with the integers n and $1 \le k \le n$, it holds*

$$A(n, k) = |Y^X| = k!C(n, k) = \frac{n!}{(n-k)!}.$$

Proof. Let a map $f \in Y^X$, its domain is $X, |X| = k$. If $k > 1$ and we permute these k elements, we get another map (consider an example!), but this is the same k–element subset of X; hence, these $k!$ subsets of X, that is, k different maps generate the same subset of X. Hence these subsets are equivalent, and Proposition 3 implies the statement.

The cardinality of the Boolean of X, that is, the set of all the subsets of X, denoted by 2^X, follows straightforwardly.

Theorem 21. *If $|X| < \infty$, then $|2^X| = 2^{|X|}$.*

Proof. Instead of an inductive proof, which is also simple, here we demonstrate another common technique of solving some combinatorial problems. Let us establish a one-to-one correspondence between the Boolean 2^X and the set D^X, where the set $D = \{0, 1\}$ is a two-element set, thus implying the equality of the cardinalities of these two sets. Indeed, if $|X| = n$, we represent every subset of X as n–string, where an element of the string $s_i = 1$ iff the element of X, x_i is in the string and $s_i = 0$ iff the element $x_i \in X$ is not in the string. Since there are 2^n such binary strings, the theorem follows.

The power-wet Y^X contains all the mappings from X to Y. We are often interested in the set of *injective maps* $Inj(Y^X)$ and in the set of all *surjective maps* $Sur(Y^X)$. Consider an injective map $f : X \to Y$. Since f is injective, it cannot 'glue' two preimages, thus independently upon the existence of any other maps, $|X| \leq |Y|$. Suppose, $|X| = k$ and $|Y| = n$. To realize an injective map $F : X \to Y$, we must choose an k–element subset in Y and order its elements. Therefore, to fix an injective map with these parameters is exactly the same as fix a k–arrangement of an n–element set. The number of those arrangements was computed in Theorem 20, and we immediately derive the following claim.

Theorem 22. *The number of injective maps* $|Inj(Y^X)| = \dfrac{n!}{(n-k)!}$, *where* $|X| = k$ *and* $|Y| = n$.

Exactly the same way, the bijective mapping $X \to Y$ exist iff $|Y| = |X|$, and we arrive at the next statement.

Corollary 1. *Bijective mappings from X to Y exist iff $|X| = |Y| = n$, and their number is $n!$.*

Maybe a bit surprisingly, but unlike the previous cases, no simple formula for the number of surjective mappings is known. Calculations, in this case, require the Inclusion-Exclusion Principle, and we leave the proofs to the reader.

Problem 217. *If $|X| = k \geq |Y| = n$, then*

$$|Sur(Y^X)| = n^k - C(n,1)(n-1)^k + \cdots + (-1)^{n-1}C(n,n-1).$$

In particular, $n! = \sum_{k=0}^{n}(-1)^k C(n,k)(n-k)^n$.

Problem 218. *How many positive factors does number 26 have?*

Problem 219. *In how many ways is it possible to place n different balls into m different boxes, so that no box is empty?*

Solution. The solution is obviously given by the number $|Sur(Y^X)|$ with $|X| = n \geq |Y| = m$.

However, if the boxes are equivalent (indistinguishable), then any permutation of the boxes without any change in their content is the same placement of different balls; thus, there are

$$S_2(n,m) = \frac{1}{m!} \left| Sur(Y^X) \right|$$

ways to place n different balls into m identical boxes without empty boxes.

7.1.4 Stirling Numbers

The numbers $S_2(n, m)$, occurring in many mathematical and not only mathematical problems, are called the Stirling numbers of the second kind.

Problem 220. *Prove that the number of ways to partition an n–set into m subsets is equal to Stirling number of the second kind $S_{n,m}$.*

Problem 221. *Prove that the subsets of size 3 and the subsets of size 9 from the set $\{1, 2, 3, \ldots, 11, 12\}$ have the same cardinality.*

Problem 222. *How many passwords of length 7, containing only 26 capital letters and 10 digits, can be made?*

It is obvious that if there are eight guests at a party, then at least two of them have a birthday at the same day of week. A slightly more general claim is given in the next lemma.

Lemma 6. The Generalized Dirichlet Principle. *Consider a finite set A and k subsets $A_1, A_2, \ldots, A_k$, such that every element of A belongs to at least t sets A_i. Prove that the average cardinality of the sets A_i is at least $t\,|A|/k$.*

Proof. The proof is a nice example of double-counting of the elements of A. If $t = 1$, one gets from here the basic Dirichlet Principle or the *Pigeonhole Principle,* which says that if you try to put n species into $n + 1$ cages, then at least one cage must be occupied twice.

The Dirichlet Principle has endless applications. For example, let m and n be positive integers. Applying the long division algorithm to the fraction $r = \dfrac{m}{n}$, we see that there are only finitely many different remainders. Therefore, for a rational number, its decimal expansion is periodical decimal.

Problem 223. *There are 22 students in a group. A promoter said that all the students will be rewarded if there are 4 of them born at the same day of week. Will this group be rewarded?*

Problem 224. *Let M be a subset of the set $\{1, 2, \ldots, 50\}$ with $|M| = 26$. Prove that M contains at least two consecutive integers.*

Problem 30. *Let n > 0 be an integer. Prove that if we choose any n + 1 among the first 2n positive integers 1, 2, ..., 2n − 1, 2n, then among them there is a pair p, q such that $\frac{q}{p}$ is a power of 2. For example, if n = 3 and {2, 3, 5, 6} is the set of four chosen numbers, then $\frac{6}{3} = 2 = 2^1$.*

Problem 226. *Introduce a binary relation ρ on the set of positive natural numbers, where p ρ q iff $\frac{q}{p} = 2^m$, m ∈ ℤ. Prove that this is an equivalence relation, and two positive integers are equivalent in this sense iff they have the same set of odd factors.*

The binomial coefficients $C(n,k) = \frac{n!}{k!(n-k)!}$ have been defined earlier. Consider a polynomial of degree n,

$$p(x) = p_n(x) = (1+x)^n.$$

We immediately see that

$$p_0(x) = 1, p_1(x) = 1 + x, p_2(x) = 1 + 2x + x^2, p_3(x) = 1 + 3x + 3x^2 + x^3, \qquad (7.5)$$

where the coefficients are $C(1, k)$, k = 0, 1, $C(2, k)$, k = 0, 1, 2, $C(3, k)$, k = 0, 1, 2, 3. It is clear why $C(n, k)$ are often called the binomial coefficients; they are also denoted as $C(n,k) = \binom{n}{k}$.

We can easily conclude that

$$(1 + x)^n = \sum_{k=0}^{n} C(n,k)x^k \qquad (7.6)$$

and prove (7.6) by induction, using any of the formulas for $C(1, k)$, $C(2, k)$, or $C(3, k)$ as the basis of induction. To make an inductive step, we write (7.6) as

$$p_{n+1}(x) = (1+x)^n \cdot (1+x) = 1 + \sum_{k=1}^{n}[C(n,k) + C(n,k-1)]x^k + x^{n+1},$$

where the coefficients are now $C(n + 1, k)$ – see formula (7.6).

Formula (7.6) is often called the Binomial Theorem, or the Newton Binomial, even though if n is a natural number, it was known for a couple of millennia before Sir Isaak Newton, when it can be written as $(x + y)^n = \sum_{k=0}^{n} C(n,k)x^k y^{n-k}$. Newton has proved it for complex n under appropriate restrictions on x.

Problem 227. *What is the coefficient of $x^2 y^4$ in the expansion of the binomial $(2x - 3y)^6$?*

There are infinitely many beautiful formulas involving the binomial coefficients, see, for example, collections [43] or [28]. We mention here just a few of them. It is almost obvious that

$$C(n,k) = C(n, n - k),$$

and if we set $x = 1$ in (7.6), we get $\sum_{k=0}^{n} C(n,k) = 2^n$, while if $x = -1$, (7.6) gives $\sum_{k=0}^{n} (-1)^k C(n,k) = 0$.

Problem 228. *Prove that 1)* $kC(p,k) = pC(p - 1, k - 1)$. *Moreover, if p is prime and $1 \le k \le p - 1$, then p divides $C(p, k)$*

2) $\sum_{k=0}^{n} \dfrac{1}{k+1} C(n,k) = .2cm \dfrac{1}{n+1} (2^{n+1} - 1)$

3) $\sum_{k=1}^{n} \dfrac{(-1)^{k-1}}{k} C(n,k) = 1 + \dfrac{1}{2} + \cdots + \dfrac{1}{n}$

4) $\sum_{k=l}^{m} (-1)^k C(m,k)C(k,l) = (-1)^m \delta_{ml}$, *where δ_{ml} is the* Kronecker delta, *called sometimes Kronecker - Wejerstrass symbol, which is equal to 1 if the indices are equal, $m = l$, and is equal to 0 otherwise.*

Theorem 20 computes the numbers of arrangements, with and without repetitions, that is, the number of maps from a set X to a set Y. In both cases, we choose a subset of X. The difference is that, in general, we are allowed to use an element of Y as an image more than once and get arbitrary mappings, but if this is not allowed, we get only injunctive mappings. In the first case, we have general arrangements *with repetitions*, whose number is $|Y|^{|X|}$, while in the second case, we have arrangements *without repetitions*, whose number is $\dfrac{|Y|!}{(|Y| - |X|)!}$.

Problem 229. *A combination lock has 5 disks with 10 different symbols on each disk. Only one combination opens the lock. If it takes 12 seconds to change a combination, what is the maximum time necessary to open the lock at random?*

7.1.5 Combinations With Repetitions

Since there are arrangements with repetitions, there should be combinations and permutations with repetitions. Next, we study combinations with repetitions and start with a model problem.

Problem 230. *There are four different kinds of trucks in a garage, denote them A, B, C, D. In how many ways can the manager send seven trucks to do a job?*

Solution. The problem does not specify the kinds of trucks; hence, we can permute any two of them; thus, we discuss combinations. However, we can repeat these *indistinguishable* entities; therefore, in such a problem, we must consider *combinations with repetitions*. Since elements of a set cannot repeat, we must introduce these new objects, but first, we finish the solution of this model problem. Since the order of trucks is immaterial, let us suppose that in this column of seven trucks, A–trucks go fist, then B-trucks, then C-trucks, and then D-trucks; thus forgetting about the trucks, we have a string of seven symbols. To describe these strings of seven symbols, it is convenient to introduce three additional symbols-separators, we denote them as S. For example, the string *AASBSCCCSD* describes a column of the two A-trucks, one B-truck, three C-trucks, and one D-truck, while the string *ASSSDDDDDD* marks one A-truck and six D-trucks. The solution is now obvious, we have a string of $10 = 7 + 3$ symbols, and there are $C(10, 3) = 120$ such strings.

Definition 5 . *Given a k–partition of the set X, $X = X_1 \cup \cdots \cup X_k$, define the equivalence relation on the set of r–element subsets of X, such that subsets A and B are equivalent, $A \sim B$, iff these subsets have equal r–cardinality, $|A| = |B| = r$ and the following k intersections*

$$|A \cap X_j| = |B \cap X_j|, j = 1, 2, \ldots, k,$$

also have pairwise equal cardinalities. Thus, the set of all r–element subsets of X is split into equivalence classes, called the combinations with repetitions. Their number is denoted by $C_{rep}(k, r)$.

Repeating this discussion, we derive the following formula.

Theorem 23. *If $1 \le r \le \min_{1 \le i \le k} |X_I|$, then*

$$C_{rep}(k, r) = C(k + r - 1, r) = \frac{(k + r - 1)!}{(k - 1)! r!}. \tag{7.7}$$

Problem 231. *Give a detailed proof of Theorem 23.*

Remark 6. *It is assumed in these problems that there are at least seven items of each kind, which is expressed by the inequality $r \le \min_{1 \le i \le k} |X_I|$. One can easily modify this assumption.*

Problem 232. *Under the conditions of Theorem 23, if we suppose that*

$$1 \leq r_1 = |X_1| < r \leq \min_{2 \leq i \leq k} |X_i|,$$

then

$$C_{rep}(k, r) = C(k + r - 1, k - 1) - C(k + r - r_1 - 2, k - 1).$$

Problem 233. *How many solutions in the integers does the equation*

$$x_1 + x_2 + \ldots + x_n = k$$

have?

Problem 234. *Prove that $C(n, 1) + C(n, 2) = n^2$. Generalize this equation for the sums $C(n, j) + C(n, j + 1)$ and $C(n, 1) + C(n, 2) + C(n, 3)$.*

Problem 235. *For any integers $m \geq 1$ and $n \geq 0$, prove the equation*

$$mC(m + n, n) \sum_{k=0}^{n} \frac{(-1)^k}{k + m} C(n, k) = 1,$$

either by induction or by considering the integral $\int_0^{\infty} e^{-mx} (1 - e^{-x})^n \, dx$.

7.1.6 Permutations With Identified Elements

We know that the elements (the letters) of the word "person" can be permuted by 6! = 720 ways because all of these six characters are different. But what if the characters can repeat? It must be mentioned that we consider strings of characters without attributing any meaning to them; the words *mom* and *mmo* are equally good in this problem.

Problem 236. *In how many ways can you transpose the letters of the word mom? Of the word baobab?*

Solution. To reduce the problem to a standard one with all the different symbols, let us make the two characters m distinguishable as m and M. The word becomes M, o, m, the solution is immediate as there are 3! = 6 such words, and we can easily list all of them,

$$\begin{Bmatrix} (o, M, m) & (M, o, m) & (M, m, o) \\ (o, m, M) & (m, o, M) & (m, M, o) \end{Bmatrix}$$

Now, if we erase the subscripts, the two words in every column become indistinguishable. So that, the two rows above are equivalent in this problem;

hence, there are three different permutations of the letters of the word *mom*, defined by the position of the vowel: the *o* can be either at first, or at the second, or at the third place.

If we compare that word with "baobab," we observe that in the latter, there are *two repeating symbols*. If the symbols *a* are fixed at the two positions, and we transpose them staying at these positions, these two permutations must be "glued," identified, as in the word "mom." Quite independently, if we permute the three symbols '*b*' when they remain at the three fixed places, we can do that in 3! = 6 ways, and these six words must be identified, that is, declared *equivalent*. To count each of them once, we divide the total number of permutations over 3! = 6. It is clear now that the symbols of word "baobab" can be transposed by $\dfrac{6!}{2!3!} = 60$ ways.

The solution above is transparent and is straightforwardly generalized. Consider a set X and any its partition $X = X_1 \cup \cdots \cup X_k$. By definition, all the elements of X are different, and if $|X| = r$, there are $r!$ transpositions of the elements of X; denote their set as $T(X)$. Define two elements of $T(X)$ equivalent, iff one can be deduced from another by transposing only the elements within any subset X_k at the places occupied by the elements of this subset.

Problem 237. *Prove that this is an equivalence relation.*

Definition 57. *The equivalence classes of the above equivalence relation are called the permutations of elements of X with repeating elements from the given partition. The amount of these permutations with repetitions is denoted by $C(r, r_1, \ldots, r_k)$. It is clear that $r_1 + r_2 + \cdots + r_k = r$.*

The next theorem is proved by exactly repeating the solution of the previous problem.

Theorem 24. $C(r, r_1, \ldots, r_k) = \dfrac{r!}{r_1 \cdot r_2 \cdots r_k}.$

Problem 238. *In how many ways can the letters of words "Mississippi" be transposed?*

The numbers $C(r, r_1, \ldots, r_k)$ are called the *multinomial coefficients* due to the following *Multinomial theorem*.

Theorem 25. *For all real $t_1, \ldots, t_k$,*

$$(t_1 + t_2 + \cdots + t_k)^r = \sum C(r, r_1, \ldots, r_k) t_1^{r_1} \cdots t_k^{r_k},$$

where the sum is taken over all the sets of integers $r_i, 1 \le i \le k$, such that $r_1 + \cdots + r_k = r$.

Problem 239. *Establish a one-to-one correspondence between the set of two-element subsets of the natural segment $\{1, 2, \ldots, 20\}$ and the set of solutions with positive integer elements of the Diophantine equation $x + y + z = 21$. How many such solutions are there?*

Problem 240. *Prove Multinomial Theorem 25.*

Problem 241. *Use Theorem 25 to find the expansion of the polynomial $(x_1 + x_2 + x_3)^5$ over the powers of x_1, x_2, x_3. How many terms does the expansion have? Does it contain the power x_1^6? What is the coefficient of $x_2^3 x_3^2$ in this expansion?*

Binomial coefficients can be conveniently exhibited by making use of the so-called *Pascal's Triangle*[2]. Here every number, except for the unities at the boundary, Blaise Pascal is equal to the sum of its two upper neighbors, namely, two 1s in the second row add up to the 2 in the third row, etc.,

$$2 = 1 + 1, 3 = 1 + 2 = 2 + 1, \ldots$$

The graph below shows the upper seven rows of the infinite triangle.

$$
\begin{array}{ccccccccccccc}
&&&&&& 1 &&&&&& \\
&&&&& 1 && 1 &&&&& \\
&&&& 1 && 2 && 1 &&&& \\
&&& 1 && 3 && 3 && 1 &&& \\
&& 1 && 4 && 6 && 4 && 1 && \\
& 1 && 5 && 10 && 10 && 5 && 1 & \\
1 && 6 && 15 && 20 && 15 && 6 && 1 \\
\end{array}
$$

$$\cdots$$

One can directly observe that the consecutive coefficients at x^k of the polynomial (7.6) of degree n are exactly the entries of the n-th row in Pascal's triangle. The connection between the binomial coefficients and Pascal's triangle entries is clear due to the following simple property.

Problem 242. *Check the identity $C(n, k) = C(n - 1, k - 1) + C(n - 1, k)$ for any natural $0 \le k \le n$.*

This connection implies many other useful and/or beautiful properties of the entries. For example, the sum of entries of the n–th row is 2^n for $n \ge 0$, the elements of kth diagonal follow certain pattern, and so on.

The numbers $C(n, k)$ are called *binomial* coefficients because they, in particular, appear as the coefficients of the binomial expansion (25). The coefficients of expansions $(x_1 + x_2 + x_3)^k, \ldots, (x_1 + \cdots + x_n)^k$ are called *multinomial coefficients* and have many properties similar to those of binomial coefficients, see, for example, [17, Vol. 1].

Problem 243. *Construct Pascal's pyramids, simplexes, etc.*

7.2   EXERCISES

Exercise 7.1. *Every student in a section studies at most one foreign language, 13 students take French, 12 students take German, 5 students take Chinese, and 3 students do not take any foreign language. How many students are there in the section?*

Solution. To answer the question, we have to add $13 + 12 + 5 + 3 = 33$.

Exercise 7.2. *There are 26 bags with apples of three different kinds, so as every bag has the apples of only one kind. Is it possible to find 9 bags with the apples of the same kind?*

Exercise 7.3. *Every student in a group studies at least one of the two foreign languages, 8 students take French and 9 students take German. How many students are in the group?*

Exercise 7.4. *The members of three political clubs in the Pluralistic City have hats of Blue, Green, and Purple colors. Each club lists one-half of the whole population. One-third of the population belong to each pair of the clubs. One-fourth of them belong to all the three clubs. How many people belong to at least one club? To exactly one club? To at least two? How many people hate any politics and belong to none?*

Exercise 7.5. *Among one hundred first-graders, 85 like "Chevrolet" models, 75 like "Dodge," and 65 like "Ford." What is the smallest possible amount of the kids who like all the three makes?*

Exercise 7.6. *Among the students at Illiter campus, 60% of the students read the magazine A, 50% the magazine B, 50% the magazine C, 30% both the magazines A and B, 20% the magazines A and C, 40% the magazines B and C, and 10% the magazines A, B, and C. How many students read 1) No magazine? 2) Exactly two magazines? 3) At least two magazines?*

Exercise 7.7. *Compute the binomial coefficients $C(10, 5)$, $C(31, 29)$.*

Exercise 7.8. *Car license plates in a small town Fewcarville contain only two characters, an English letter and a digit. How many different license plates can be issued in the town if they can use all 26 letters and 10 digits?*

Exercise 7.9. *There are two different roads from a town T_1 to a town T_2, and three different ways from T_2 to T_3. In how many ways can we arrange a trip from T_1 to T_3 with the transfer at T_2?*

Exercise 7.10. *For every natural $n > 0$, prove $\sum_{k=1}^{n} \dfrac{1}{2k^2 - 1} = \dfrac{n+1}{2n+3}$.*

Exercise 7.11. *A license plate contains either 1, or 2, or 3 English letters and 4 digits. How many are there such license plates?*

Exercise 7.12. *Three tokens, one with 6 faces, one with 8, and one with 10 faces are flipped simultaneously. In how many different ways can they fall?*

Exercise 7.13. *In how many ways can the same three tokens fall down, if we know that at least two of them fall on the face marked by 1?*

Exercise 7.14. *In how many ways can 12 identical balls be placed into 7 distinguishable boxes?*

Solution. Put the balls in one line, since they are identical, there is only one way to do that. We must split the balls into 7 parts, for that we need 6 partitions, hence, we arrive at the permutations with repetitions. The answer is $C(12 + 7 - 1, 7 - 1) = C(18, 6)$.

Exercise 7.15. *Given a set of eleven positive natural numbers. Prove that at least two of them must have the same right-most digit.*

Exercise 7.16. *How many integers must be picked up to insure that at least one of them is odd? At least one of them is even? At least one of them is a multiple of 9?*

Exercise 7.17. *Solve the following problem and prove that it is equivalent to the preceding problem about the trucks. A college cafeteria has pastries of four kinds. In how many ways can a student buy a seven pastries?*

Exercise 7.18. *A bus with 17 passengers has to make 9 stops. Each passenger can exit at every stop. In how many ways can the passengers leaves the bus?*

Exercise 7.19. *A Liberal Arts College runs three mathematical courses, seven sections of The Mathematics History, five sections of Mathematics in*

Arts, and two sections of Elementary Combinatorics. These classes have, respectively, 21, 17, and 10 seats. All 14 sections are taught by different instructors. There are $7 \cdot 21 + 5 \cdot 17$ and $2 \cdot 10 = 252$ students who satisfy the prerequisites. In how many ways can the students register for the classes?

Exercise 7.20. *How many paths are there in the Hasse diagram, Fig. 6.2, from vertex 1 to vertex 6?*

Solution. There are three paths.

Exercise 7.21. *Prove that the number $(p!)! \cdot (p!)^{-(p-1)!}$ is an integer for every natural p.*

Solution. That is a multinomial coefficient.

CHAPTER 8

ELEMENTS OF NUMBER THEORY

8.1   DIVISIBILITY AND FTA

Number theory is an ancient part of mathematics. The natural numbers were studied for several thousand years, and they have become quite modern now. In this book, we study only the sets of the integers and their closest relatives like the prime numbers, which are certainly discrete sets; thus, we undoubtedly remain within the discrete mathematics. The number theory is a fascinating theory and generates many open problems of any level, from elementary through advanced. *Elementary* here does not mean to be simple. By an elementary problem, we understand one, which can be stated but *not necessarily solved* by making use of the language and tools of high-school mathematics. In this chapter and the following, we study only a few number-theoretical topics relevant to cryptography.

8.1.1   Divisibility, the FTA, and the Euclidean Algorithm

If three integers $b, d \neq 0$, and q are connected as $q \times d = b$, then this equation can be rewritten as

$q = b \div d$; the number b is called the *dividend*, q and d are *divisors* of b, q is also called the *quotient*.

For example, $39 = 3 \times 13$; hence, 3 and 13 are divisors of 39 and the latter is the dividend (the product) of both 3 and 13.

We accept without proof *The Fundamental Theorem of Arithmetic* (*FTA*).

Theorem 26 (FTA). *Every integer can be written uniquely, up to the order of factors, as the product of prime factors.*

Remark 7. *There are many number systems where the factoring is not unique.*

The integers 9 and 15 have a common factor, which is 3; 1 is not counted as a common factor, that is why the integers 3 and 5 are called *mutually prime* or *relatively prime*; we can also say that their *greatest common divisor, abbreviated as "gcd,"* is 1. The *gcd* of any two integers a and b is written as $gcd(a,b)$. In general, the FTA implies that the *gcd* of several integers is equal to the product of all their common prime factors raised to the least common multiplicity. For instance, since $315 = 3^2 \cdot 5 \cdot 7$ and $375 = 3 \cdot 5^3$, the *gcd* $(315; 375) = 3 \cdot 5 = 15$, and the $gcd(1575; 375) = 3 \cdot 5^2 = 75$.

8.1.2 Euclidean Algorithm

Such procedures are called *algorithms*, and we discuss them in more detail later on. The *factoring algorithm* above is simple but not practical, since it initially requires to *factorize* the integers, which is currently(!) computationally infeasible. However, the following *Euclidean algorithm*, which is known for at least a couple of millennia, is efficient. The algorithm consists of several consecutive integer divisions with remainders. Let be given two positive natural numbers a and b. The $gcd(a, b)$, clearly, does not depend upon their ordering; hence, we can always assume that $a \geq b$; and even $a > b$, since if $a = b$, then $gcd(a, b) = a = b$.

To find the $gcd(a, b)$, assuming that $a > b$, we start by dividing the smaller number b into the larger a with a remainder $r \geq 0$. If $r = 0$, then b is a divisor of a and, obviously, is the $gcd(a, b)$. Hence, let us assume $r > 0$. It must be less than the smaller b of the two given integers; otherwise, we would continue the division. Now we again have two positive integers, the given smaller integer b and the first remainder r, which must be smaller than b. At the second step, and at each of the steps to follow, we divide the bigger number in this pair by the smaller. Since the sequence of the positive integer remainders is decreasing, after finitely many steps, we have to get the zero remainder. The last non-zero remainder is the $gcd(a, b)$ we sought for.

In the example above, $a = 375$, $b = 315$, and $r = r_1 = 60$. Therefore, $315 = 5 \cdot 60 + 15$, thus $r_2 = 15$. At the next step, $60 = 4 \cdot 15 + 0$, hence, the next remainder is zero, the last non-zero remainder is 15, and this is the $gcd(375; 315)$.

A generic step of the algorithm, that is, the division with remainder $a = k \cdot b + r$, can be written as

$$gcd(r, r_0) = gcd(\mathrm{r} - r_0, r_0),$$

where all the parameters are natural numbers, $k \geq 1$ is the first intermediate quotient, and the consecutive remainders satisfy $0 \leq r_0 < r < b$. It is obvious from the equation that if d is any common divisor of both a and b, then d must also be a factor of r, which together with the monotone decrease of the remainders, mentioned above, proves the correctness of the Euclidean Algorithm. *Correctness* of an algorithm means that the algorithm works. In this case, the correctness of the Euclidean algorithm means that for any two positive integers a and b, after finitely many steps, the algorithm returns the value of $gcd(a, b)$.

Tracing the algorithm back, we can restate the last equation as the (extended) Euclidean Algorithm or the Bésout lemma.

Theorem 27. *The $gcd(a, b)$ of any two integers a and b can be represented as their linear combination*

$$gcd(a, b) = s \cdot a + t \cdot b$$

with some integer, not necessarily positive coefficients s, t. In particular, if the given natural numbers are mutually prime, *their gcd is equal to 1, which 1 can be written as a linear combination of any two mutually prime integers.*

For example, $3 = gcd(15, 9) = 2 \cdot 9 - 1 \cdot 15$ and $1 = gcd(15, 8) = 2 \cdot 8 - 1 \cdot 15$.

Problem 244. *Represent the $gcd(375; 315)$ as a linear combination of these numbers.*

Solution. We have already found the $gcd(375; 315) = 15$. Solving now those equations back for the $gcd(375; 315) = 15$, we compute

$$15 = 315 - 5\times 60 = 315 - 5(375 - 1 \times 315) = 6\times 315 - 5 \times 375,$$

deriving the coefficients 6 and –5 representing the $gcd(375; 315)$ as the linear combination of the given integers 375 and 315. The procedure, obviously, works for arbitrary two integers a and b.

Problem 245. *Let a and b be two natural even numbers; prove that $gcd(a, b) = 2gcd(a / 2, b / 2)$. If a is an even natural number, and b is an odd natural, then prove that $gcd(a, b) = gcd(a / 2, b)$.*

Problem 246. *Analyze the algorithm of finding the gcd, if the integers have opposite signs.*

Definition 58. *The least common multiple $\operatorname{lcm}(m,n,\cdots,p)$ of the integers m, n, ..., p is the smallest among all the common multiples of these integers. For example, $\operatorname{lcm}(15, 99) = 495$, since both 15 and 99 are factors of 495, and no integer smaller than 495 is a multiple of both 15 and 99. The common multiple always exists; for example, the product of all these integers is their common multiple, but not necessarily their lcm. For example, $\operatorname{lcm}(6, 8) = 24$, while $6 \times 8 = 48$.*

Problem 247. *Find the $\operatorname{lcm}(15, 9)$.*

Solution. $15 = 3 \cdot 5, 9 = 3^2$, so that the $\operatorname{lcm}(15, 9) = 3^2 \cdot 9 = 45$.

Problem 248. *Prove that $\operatorname{lcm}(1; n) = n$ and $\gcd(1; n) = 1$ for any positive integer n.*

Problem 249. *Prove that $\operatorname{lcm}(m, n) \times \gcd(m, n) = m \times n$.*

Problem 250. *If we want to check whether a positive integer n is a prime, we do not have to try every integer up to n. Prove that if $k \cdot l = n$, then either $k \le \sqrt{n}$ or $l \le \sqrt{n}$.*

Solution. Indeed, if both $k > \sqrt{n}$ and $l > \sqrt{n}$, then $kl>n$. For instance, if we want to prove that 29 is a prime, it is enough to check only the primes 2, 3, and 5, since the next prime, 7 is too big, $7^2 = 49 > 29$.

8.1.3 Algorithms

This word has already been repeated several times. We have discussed the Euclidean Algorithm in some detail. More properties of the algorithms will be considered in the last chapter of this book. Now we consider a *mass* problem, meaning that it is not a singled-out question, even a very difficult one, but a typical representative of a class of similar problems. For instance, calculating the $\gcd(a, b)$ is a mass problem since there are infinitely many pairs of the integers; solving a quadratic equation is a mass problem, etc.

Above, we described how to find the *gcd* for any two integers. An algorithm is a precisely described sequence of operations that, after *finitely many steps*, either return the answer or claim that no answer exists. For example, if we execute the Euclidean Algorithm, that is, calculate the $\gcd(a, b)$ of the two integers a and b, the algorithm *must* return the $\gcd(a,b)$. We do not give a precise definition of an algorithm here but mention that this is a precisely described finite sequence of operations leading to a certain specific result.

Another important characteristic of algorithms is their *run time*. Consider again the Euclidean Algorithm. It is easily seen that if we divide $a \div b$, then either the remainder r or the difference $a - b$ is less than $a / 2$. Indeed, if both $a - b \geq a / 2$ and $r \geq a / 2$, thus, $a \geq 2b$, implying that $r \geq b$, which is impossible, for the remainder must be smaller than the divisor. So that, at every two consecutive iterations of the Euclidean Algorithm, the second remainder is less than a, $r_1 < a / 2$ or $2r_1 < a$. Iterating this inequality, we get $2^2 r_3 < a$, $2^3 r_5 < a$,..., $2^k r_{2k-1} < a$, thus,

$$k \leq C \times \log a,$$

where C is an absolute constant, that is, C does not depend upon k. Since this estimation involves logarithms, we say that Euclidean Algorithm has the *logarithmic run time*.

Problem 251. *Recover the detailed computations, involving the constant C above. How many arithmetic operations (addition/subtraction, multiplication, division) must be performed to compute gcd(5, 12), gcd(7, 21), gcd(13, 23)?*

Problem 252. *Find the quotient and the remainder when $789 \div 43$ and $789 \div (-43)$*

8.1.4 Modular or Clock Arithmetic: Linear Congruences

Up to now, we considered mostly division of the integers evenly, that is, with zero remainder. Now let us study the general situation. As an example, consider division by 5. If we divide the consecutive natural numbers 0, 1, 2, 3, ... by 5, we observe that the remainders make a periodical sequence 0, 1, 2, 3, 4, 0, 1, 2, 3, 4, 0, ... with the period of length 5, since the group of remainders {0, 1, 2, 3, 4} repeats endlessly. Thus, if in a certain problem regarding the integers we can only divide by 5, then in the problem there are just *five different* numbers, 0, 1, 2, 3, 4 – we cannot, for instance, distinguish the two numbers 1 and 6, or the 347 and –3. In Example 42, we considered the same phenomenon with regard to the divisibility over 3 and developed the equivalence relation σ (3) on the integers.

This binary relation has many important applications in mathematics, computer science, and elsewhere, including cryptography; now, we generalize it for any integer.

Definition 59. *Let $d \geq 2$ be a natural number. Two integers k and l are called congruent modulo d, iff their difference is a multiple of d, that is, the d is a factor of the difference $k - l$. This binary relation is denoted as σ (d), and we write the congruent numbers modulo d as $k \equiv l$ (mod d). Therefore, both k and l have the same remainders when we divide them by d. The* *smallest nonnegative remainder for x ÷ d is denoted as x* (mod *d*). *Thus,* 17 (mod 7) = 3, 17 (mod 17) = 0, *and* −17 (mod 7) = 4.

Problem 253. *It is obvious also that this relation is symmetric, that is, k ≡ l* (mod *d*) *iff l ≡ k* (mod *d*). *Moreover, it is also clear that k ≡ k* (mod *d*), *that is, the congruence relation is reflexive.*

Problem 254. *Compute* 3 · 4 (mod 17), 27 · 4 (mod 17), 27 · 43 (mod 17), −3 · 4 (mod 17).

Problem 255. *For the natural numbers a, b, n, is it true that:*

1. *If* $a \cdot b \equiv 0 \bmod n$, *then either* $a \equiv 0$ (mod *n*) *or* $b \equiv 0$ (mod *n*)?

2. *If* $a^2 \equiv a$ (mod *n*), *then either* $a \equiv 0$ (mod *n*) *or* $a \equiv 1$ (mod *n*)?

3. *If* $a \equiv b$ (mod *n*), *then* $2^a \equiv 2^b$ (mod *n*)?

Problem 256. *Solve the congruences for x.*

1. $8x \equiv 1$ (mod 13)

2. $8x \equiv 4$ (mod 13)

Solution. The equations with the integer coefficients are called Diophantine after Diophantus of Alexandria (200 - 284), tentatively. Consider the congruence $8x \equiv 1$ (mod 13), which can be written as $8x = 1 + 13k$, where the parameter k can be any integer, and replace it with an equivalent diophantine equation $8x - 13k = 1$ with two integer unknowns x and k. Since 8 and 13 are mutually prime, the Euclidean algorithm gives the solution, $x = 5$ and $k = -3$, or $1 = -3 \cdot 13 + 5 \cdot 8$. But 5 is only the smallest positive solution, and the general solution is given as $x \equiv 5$ (mod 1)3, or $x \in \{...-8, 5, 18, 31, 44, ...\}$.

This system of operations with the integers is called *modular* or *clock* arithmetic. The coefficients of congruence equations, due to their properties, can always be reduced modulo d to the elements of d–element set $\{0, 1, ..., d - 1\}$. For example, the congruence $82x^2 + 23x + 1 \equiv 10$ (mod 9) can be reduced to the congruence $x^2 + 5x + 10 \equiv 1$ (mod 9).

Problem 257. *Prove the statement above.*

Problem 258. *Let* $k \equiv l$ (mod *d*) *and* $l \equiv m$ (mod *d*). *Prove that* $k \equiv m$ (mod *d*), *that is, the congruence relation on the integers, is not only symmetric and reflexive (Problem 255) but also transitive; so that this is an equivalence*

relation. Its factor-set consists of d equivalence classes, and every class con-
sists of infinitely many integers congruent to one another; another way, each
class is an arithmetic progression with the difference of d, similar to …,
–d–1, –1, d–1, 2d–1, …. Each term of every progression can be written as

$$a_k = a_0 + k \cdot d, k = 0, \pm 1, \pm 2, \ldots \tag{8.1}$$

where a_0 is an integer.

Problem 256. *Is $5 \equiv 37 \pmod 6$? Is $12 \equiv 16 \pmod 4$?*

Now we define an algebraic structure of the factor-set $\mathcal{Z} / \sigma (d)$. This set is a commutative group of order d; moreover, it is a ring. The addition and multiplication of congruences are described in the next lemma.

Lemma 7. *For every positive integer d, if $k \equiv l \pmod d$ and $m \equiv n \pmod d$,*
then $k + m \equiv l + n \pmod d$, $k - m \equiv l - n \pmod d$, and $k \cdot m \equiv l \cdot n \pmod{d}$, thus, the congruence relations can be added/subtracted and multiplied termwise.

Proof. The statement follows immediately if we represent every term of each congruence by formula (8.1) and perform arithmetic operations of the Lemma 7.

Corollary 2. *Setting in this lemma $m = n = -l$, we get that $k \equiv l \pmod d$ iff*
$k - l \equiv 0 \pmod d$, that is, any additive term of a congruence can be trans-
posed from one side of the congruence to another by negating this term.

Problem 260. *Prove that $115 \equiv 21 \pmod{10}$. Hint: Is it true that $23 \equiv 7$*
$\pmod{10}$ and $5 \equiv 3 \pmod{10}$?

Problem 261. *Let $a > b$ be integers. Prove that $a(\mathrm{mod}\, b) < \dfrac{a}{2}$.*

Problem 262. *Prove that if $k \equiv l \pmod d$ then*

$$(k + l) \pmod d = (k \pmod d + l \pmod d) \pmod d$$
$$k \times l \pmod d = (k \pmod d) \times (l \pmod d) \pmod d$$

Proof. Again, it is enough to use representation (8.1).

Problem 263. *(1) Check the numerical identity*

$$x \cdot x_1 - y \cdot y_1 = x(x_1 - y_1) + (x - y)y_1.$$

(2) Prove that if $x \equiv y \pmod m$ and $x_1 \equiv y_1 \pmod m$, then

$$x \cdot x_1 \equiv y \cdot y_1 (\mathrm{mod}\, m).$$

Division of congruences is not always possible and requires certain additional assumptions.

Problem 264. *Let $d > 0$ be a divisor of m, p and q. Then a linear congruence $p \equiv q \pmod{m}$ is equivalent to the congruence $\dfrac{p}{d} \equiv \dfrac{q}{d} \left(\bmod \dfrac{m}{d}\right)$.*

Problem 265. *Prove that if p, q, r is a Pythagorean triple, that is, these are the integers such that $p^2 + q^2 = r^2$, then $pqr \equiv 0 \pmod{60}$. The smallest such an example is the so-called Egyptian triangle with $p = 3$, $q = 4$ and $r = 5$.*

Problem 266. *Let $x_1, x_2, \ldots, x_n$ be n arbitrary integers. Prove that*

$$10^{n-1} x_n + 10^{n-2} x_{n-1} + \cdots + 10 x_2 + x_1 \equiv x_n + x_{n-1} + \cdots + x_1 \pmod{9}.$$

We see that the factor-set $\mathcal{Z} / \sigma(d) \equiv \mathcal{Z}_t$ is a finite set consisting of d infinite arithmetic progressions, that is, sets $\mathcal{Z}_0, \mathcal{Z}_1, \ldots, \mathcal{Z}_{d-1}$. For instance, if $d = 2$, then $\mathcal{Z}_2$ consists of the two sets, the set of all even integers, and the set of all odd integers. We also mention that, given l and m, the equation

$$(k \pmod{d}) + (m \pmod{d}) = (l \pmod{d})$$

with respect to k has the unique solution $k \equiv l - m \pmod{d}$; moreover, 0 $\pmod{d}$ is a neutral element in the set of linear congruences modulo d on the integers.

In some problems, rather than use an initial infinite set of elements or a factor-set of sets, it is more convenient to employ a *system of representatives*, that is, another set, containing exactly one element from every equivalence class. A system of representatives can in some examples be a finite set. In the case of modular arithmetic, we can choose either the system $\{0, 1, \ldots, d - 1\}$ or the system $\{1, 2, \ldots, d - 1, d\}$ – these sets of representatives are equivalent, and our choice depends upon the convenience only. In what follows, we will use the system $\mathcal{D} = \{0, 1, \ldots, d - 1\}$.

In Lemma 7, the set $\mathcal{D}$ was equipped with two operations, analogous to the addition and multiplication of the integers. You can straightforwardly verify that these operations are *commutative*, *associative*, and have the *neutral elements* 0 and 1, respectively. Moreover, the multiplication is distributive over addition. Of course, here both operations must be considered modulo d, that is, in this algebra the neutral element 0 is the whole of the two-sided sequence $\{\ldots, -2d, -d, 0, d, \ldots\}$, and 1 is equivalent to the set $\{1 \pmod{d}\}$.

Lemma 8. *Prove, which is straightforward, these properties.*

What is not obvious, it is the existence of inverse elements, which is not always the case; we first consider an example.

Example 32. *Let $x \equiv 5 \pmod 3$, $k = \ldots, -4, -1, 2, 5, 8, \ldots$, and $m \equiv 2 \pmod 3$, that is, $m = \ldots, -4, -1, 2, 5, 8, \ldots$. Thus, $k - m$ is the sequence $k - m = \ldots, -6, -3, 0, 3, 6, 9, \ldots$, or $k - m \equiv 0 \pmod 3$. We see that $-m$ is the inverse to the addition $\pmod 3$, but $-m$ can be written in this algebra as $-m \equiv 3 - x \pmod 3$.*

Since all these computations are valid for any congruences and modules, we have actually proved the existence of the *inverse element* for addition modulo any natural number.

Problem 267. *Prove that if $a \in \mathcal{Z}_d$, $a \neq 0$, or just $a \in \mathcal{D}$, $a \neq 0$, then its additive inverse exists and is equal to $d - a$, that is, $a + d (d - a) = 0$. Moreover, in this algebra, $0 + 0 = 0$.*

Now the finite set of the integers modulo any natural number with respect to addition becomes a *commutative group*, where we can *add* the equivalency classes.

To learn how to multiply congruences, we start with a linear *equation*, say, $3x = 5$. To solve it, we must obviously divide by 3. Now, if we have to solve a congruence $3x \equiv 5 \pmod d$, we must develop an analog of the division in the algebra of congruence relations. However, division by 3 can be interpreted as multiplication by the *inverse element* of 3. Thus, we must introduce the *inverse elements* with respect to multiplication. We again start with an example.

Problem 268. *Solve a linear congruence $3x \equiv 5 \pmod 6$.*

Solution. By definition and property (8.1), the congruence is equivalent to the infinite system of equations $3x = 5 + 6k$, where $k \in \mathcal{Z}$ is an integer parameter. From here, $3(x - 2k) = 5$. However, the left-hand side of the latter equation is *always* a multiple of 3, while 3 does not divide 5. Therefore, our equation has no integer solution. In particular, we have proved that the inverse of 3 modulo 6 in this algebra does not exist.

Problem 269. *Does 4 have a multiplicative inverse in the ring $\mathcal{Z}_6$?*

We see that not every nonzero element of $\mathcal{D}$ has a multiplicative inverse, and we must develop techniques for determining whether this inverse does exist and for finding it.

Definition 60. *Let $k \in \mathcal{Z}_d$; here k is an equivalence class or its representative modulo d, that is, k is an integer in the range $0 \le k \le d - 1$. The number k is invertible in $\mathcal{Z}_d$ iff there exists an integer $k^{-1}, 1 \le k^{-1} \le d$, such that*

$$k \cdot k^{-1} \equiv 1 (\bmod\, d).$$

Problem 270. *Compute the addition and multiplication tables for the modulo 5 arithmetic, and the addition and multiplication tables for modulo 6 arithmetic. It is essential that 5 is a prime, while 6 is composite. For what factors the products exhibit any periodicity or comprise all the integers {0, 1, 2, 3, 4, 5}?*

Solution. By definition, we compute $3 + 4 = 7 = 1 \cdot 5 + 2 \equiv 2 \pmod 5$. Finding all the other sums and products modulo 5, we compute the following Tables 8.1 and 8.2.

We observe that the addition tables contain all the numbers {0, 1, 2, 3, 4} or {0,1,2,3,4,5}. Each row of the first multiplication table also contains a permutation of the entire set {0,1,2,3,4}. However, when we multiply (mod 6) and the second factor is 2 or 4, the product contains only three numbers, 0, 2, 4. This observation is clearly connected with the invertibility.

TABLE 8.1 Addition-Multiplication Modulo 5 Tables for Problem 270.

+	0	1	2	3	4	×	0	1	2	3	4
0	0	1	2	3	4	0	0	0	0	0	0
1	1	2	3	4	0	1	0	1	2	3	4
2	2	3	4	0	1	2	0	2	4	1	3
3	3	4	0	1	2	3	0	3	1	4	2
4	4	0	1	2	3	4	0	4	3	2	1

TABLE 8.2 Addition-Multiplication Modulo 6 Tables for Problem 270.

+	0	1	2	3	4	5	×	0	1	2	3	4	5
0	0	1	2	3	4	5	0	0	0	0	0	0	0
1	1	2	3	4	5	0	1	0	1	2	3	4	5
2	2	3	4	5	0	1	2	0	2	4	0	2	4
3	3	4	5	0	1	2	3	0	3	0	3	0	3
4	4	5	0	1	2	3	4	0	4	2	0	4	2
5	5	0	1	2	3	4	5	0	5	4	3	2	1

The definition clearly implies that both k and k^{-1} are not vanishing in $\mathcal{Z}_d$. Thus, the operation of multiplication in $\mathcal{Z}_d$ has the unity (the neutral element). The set $\mathcal{Z}_d$, equipped with the addition as above and with this multiplication, is a *commutative ring with the unity*.

Remark 8. *Rings are determined in Definition 15 in Chapter 4. For more detail, see any book in general algebra, for example, [27]. It is useful, however, to mention that in this context, we can work with the equivalence classes in $\mathcal{Z}_d$ almost as with the integers, except that we cannot divide them.*

Problem 271. *Prove that for any module p, $a = b^{-1}$ iff $b = a^{-1}$.*

Theorem 28. *Let k be an element (i.e., an equivalence class) in $\mathcal{Z}_d$. The following two conditions are equivalent:*

1. $gcd(k, d)=1$

2. *There exists a multiplicative inverse k^{-1} in the finite ring $\mathcal{Z}_d$.*

Proof. Due to the Extended Euclidean Algorithm, there are two integers p and q, such that $p \cdot k + q \cdot d = 1$. Since the equivalence class of $q \cdot d = 0$, one has the equivalence class of $p \cdot k + q \cdot d = p \cdot k = 1$; therefore, k is invertible in $\mathcal{Z}_d$. Therefore, $(1) \Rightarrow (2)$. To prove the converse implication, we remark that every phrase in the paragraph above is reversible, thus, $(2) \Rightarrow (1)$.

Corollary 3. *If d is a prime, then every integer $n, 1 \le n \le d - 1$, is mutually prime with d, thus, every such integer has a multiplicative inverse in $\mathcal{Z}_d$.*

Problem 272. *In $\mathcal{Z}_7$, compute the multiplicative inverses of its six non-zeroterms $n = 1$, 2, 3, 4, 5, 6.*

Solution. $1^{-1} = 1$; $2^{-1} = 4$; $3^{-1} = 5$; $4^{-1} = 2$; $5^{-1} = 3$; $6^{-1} = 6$. Pay attention that the inverses make a permutation of the initial set.

When we add more and more natural numbers, the result is getting bigger and bigger. However, if we do addition modulo an integer, the sum changes cyclically. Indeed, consider $q(k) = (1 + k)(mod\ 3)$. For $k = 0, 1, 2, \ldots$, $q(k) = 1, 2, 0, 1, 2, 0, \ldots$, thus, the value *loops*. If we choose a bigger module, say, 300, the point $q(k)$ will very neatly describe a circle. That is why arithmetic in $\mathcal{D}$ or $\mathcal{Z}_d$ is often called *clock arithmetic*.

Problem 273. *What time does a 24-hour clock (the military format) read 30 hours after it shows 18:00?*

Congruence Equations, or Congruence Relations. We can easily solve a *linear equation* $2x + 5 = 7$. We have even known how to solve systems of linear equations. Aiming at the applications, we now study how to solve *simultaneous linear congruence equations*, and we are equipped for that task now. It is easy to cancel a common factor of an equation, even though sometimes we can get an extra root. However, as we have seen, we cannot always cancel a common factor in a congruence. The conditions, when it can be done, were studied by Pierre de Fermat (1601-1665). To avoid confusion with the Great Fermat Theorem, this result is commonly called the *Fermat's Little Theorem*.

Theorem 29. *(1) For any integer n and prime p, the congruence $n^p \equiv n \pmod{p}$ is valid.*

(2) In particular. if p is a prime and n is not a multiple of p, then $n^{p-1} \equiv 1 \pmod{p}$

(3) If n and p are not mutually prime, then $n^{p-1} \not\equiv 1 \pmod{p}$.

Remark 9. *Thus, to cancel out the congruence in part (1) by n, we have to require an additional assumption that n is not a multiple of p.*

Proof. (1) The congruence in question can be rewritten as $n^p - n = q \cdot p$, where q is an integer. For the proof, first, we assume that $n > 0$ and apply the induction on n. If $n = 1$, then the congruence is obvious. If the congruence is valid for a certain n, that is, $n^p - n = q \cdot p$, we have

$$(n+1)^p - (n+1) = n^p + p \cdot n + p \cdot n^{p-1} / 2 + \cdots + p \cdot n + 1 - n - 1.$$

The latter can be written as $(n^p - n) + a \cdot p$. The expression in the parentheses is multiple of p by the inductive assumption, while due to properties of the binomial coefficients, one can see straightforwardly that the a is an integer. If n is negative, it is enough to consider the statement for $-n$.

(2) We have $n^p \equiv n \pmod{p}$, thus, $n \cdot (n^{p-1} - 1) \equiv 0 \pmod{p}$, but n is not divisible by p; hence $n^{p-1} - 1$ is a multiple of p.

(3) Is obvious due to parts (1) and (2).

Problem 274. *We proved that the condition (1) implies (2). Prove the converse implication, that is, that the condition (2) implies (1).*

Example 33. *Calculate $17^{1226} \pmod{13}$. It's unlikely that any human being will evaluate the huge power 17^{1224} to find the remainder after dividing it by 13; it takes time even for a good computer. However, if one learns the modular arithmetic and Fermat's little theorem, the problem is straightforward.*

Indeed, the integer 13 is prime and 1226 = 102 × 12 + 2, hence we can apply Theorem 29, and compute $17^{1226} = 17^{12 \times 102 + 2} = (17^{12})^{102} \times 17^2$. Since 12 = 13 − 1, then by part (2), $(17^{12})^{102} \times 17^2 \equiv 1^{102} \times 289 \pmod{13} \equiv 289 \pmod{13} \equiv 3$.

Problem 275. *Prove that $3^{4n} \equiv 1 \pmod{10}$ and then find the unity digit in the number 3^{2000}.*

Problem 276. *Solve the congruences $3x \equiv 4 \pmod{20}$, $6x \equiv 4 \pmod{10}$, $4x \equiv 3 \pmod 5$.*

Problem 277. *Prove that if p is prime, the congruence $x \equiv y \pmod p$ is an equivalence relation on the set of integers. How many equivalence classes are there? Describe them.*

Problem 278. *Let p be a prime. Prove that the congruence $x^2 \equiv 1 \pmod p$ has only two series of solutions, $x \equiv 1 \pmod p$ and $x \equiv -1 \pmod p$.*

Linear congruences resemble, to some extent, the familiar properties of linear equations. We are to develop an algorithm for solving the linear congruences.

Problem 279. *Find the general solution of an equation $ax = b$ for any real a and b and state the conditions on the coefficients a and b. For example, if $a = 0$ and $b \neq 0$, then the equation has no solution.*

A similar theory for solving the linear congruence $a \cdot x \equiv b \pmod m$ is done in the next lemma and illustrated with the following examples.

Lemma 9. *Given natural non-zero numbers a, b, and m, set $d = \gcd(a, m)$, $A = \dfrac{a}{d}$, and $M = \dfrac{m}{d}$. If b is not a multiple of d, then the congruence*

$$a \cdot x \equiv b \pmod m \tag{8.2}$$

has no solution. Otherwise, that is, if d divides b, then the congruence (8.2) has d solutions, which are all congruent modulo M to the unique solution of $Ax \equiv B \pmod M$.

Proof. Let $B = b/d$. By definition, d divides a. In the case that d does not divide b, the congruence (8.2) clearly cannot have a solutions. Now let d divides b. If the congruence (8.2) has a solution x_0, then

$ax_0 = b + mt$, where m is the module and t is an integer. Dividing the latter through over d, we get
$Ax_0 \equiv B \pmod M$, so that, M is a divisor of the difference $Ax_0 - B$, and in turn, the x_0 is a solution of $Ax \equiv B \pmod M$.

On the contrary, let x_1 be a solution of $Ax \equiv B \pmod{M}$. Then there exists an integer t_1 such that $Mt_1 + Ax_1 = B$. Hence, $dMt_1 + dAx_1 = dB$, or $mt_1 + ax_1 = b$. Therefore, m is a factor of $b - ax_1$, whence x_1 is a solution of (8.2). Hence, $Ax \equiv B \pmod{M}$ has the same integer solution. Let x_0 be the unique smallest positive solution of the latter equation. Since $M = m/d$, the integers $x_0, x_0 + M, \ldots, x_0 + (d-1)M$ are all the integer solutions of these equations, which are between 0 and m.

Example 34. *Solve the congruence* $3x \equiv 1 \pmod{2}$.

Solution. The $\gcd(3; 2) = 1$, thus this example has the unique solution, consisting of all the odd numbers.

Example 35. *Solve the congruence* $3x \equiv 1 \pmod{6}$.

Solution. Now the $\gcd(3; 6) = 3 \neq 1$, hence the congruence has no solution. Indeed, one must satisfy from here $3(x - 2k) = 1$, but 3 is not a factor of 1.

Example 36. *Solve the congruence* $3x \equiv 9 \pmod{6}$.

Solution. Here $m = 6$, $A = 1$, $M = 2$, $B = 3$, and the $\gcd(3; 6) = 3$ divides 9. The congruence simplifies to $x = 3 + 2k, k \in \mathcal{Z}$, thus the solutions of the congruence are all odd numbers, thus, there are three solutions 1, 3, 5 in the integer segment $(0, 6)$. On the other hand, the congruence $Ax \equiv B \pmod{M}$, which is now $x \equiv 3 \pmod{2}$, has the same odd solutions.

8.1.5 Totient Function

An important assumption in the Fermat Little Theorem is the primality of p. Departing from this condition, Euler introduced the *totient function*

$$\varphi : \mathcal{N} \to \mathcal{N}.$$

For a positive integer n, this function $\varphi(n)$ is equal to the number of natural numbers, not exceeding n and relatively prime with n. For example, $\varphi(12) = |\{1, 5, 7, 11\}| = 4$; and for every prime p, $\varphi(p) = p - 1$. The totient function has appeared in many problems of discrete mathematics and cryptography. Its explicit expression can be straightforwardly found by making use of the Inclusion-Exclusion Principle (Theorem 16, p. 116).

First we introduce some notation. List the prime numbers, $2 = p_1$, $3 = p_2$, $5 = p_3$, $\ldots$, $11 = p_5$, etc. For a positive integer a, let $a = p_1^{\chi_1} \cdot p_2^{\chi_2} \cdots p_k^{\chi_k}$ be its prime factorization, where the p_i are different prime numbers indexed as above, and $\chi_i \geq 0$ are their multiplicities in the a. Denote also the natural segment $\mathbf{A} = \mathbf{A}(a) = \{1, 2, \ldots, a\}$. In what follows, the a is fixed, and is omitted in the notations. A natural number p_i can be a factor of the a, or can be not. We say that the a has the property $P(i)$, iff p_i is a factor of a, $p_i \mid a$, that is, the a is a multiple of p_i. Finally, let $\mathbf{B}(i)$ be the set of numbers in $\mathbf{A}$, which are multiple of p_i.

Definition 61. *Given a natural number n, the number of natural numbers, including 1, which do not exceed n and are mutually prime with n, is called the Euler totient function $\varphi(n)$; $\varphi(1) = 1$ by definition.*

For example, $\varphi(2) = 1$, $\varphi(3) = 2$, $\varphi(4) = 2$, $\varphi(5) = 4$.

Problem 280. *Compute $\varphi(6), \varphi(7), \varphi(8), \varphi(9)$.*

Theorem 30. *Prove that for any natural n,*

$$\varphi(n) = n \prod_{k=1}^{n} \left(1 - \frac{1}{p'_k}\right),$$

where p'_k are all the prime factors of n without counting their multiplicity.

Proof follows immediately from Theorem 16 on p. 116, if we notice that $|P(p_i)| = \dfrac{n}{p_i}$. Thus, Theorem 6 gives

$$\varphi(n) = n - \sum_{1 \le i_1 \le n} \frac{n}{p_{i_1}} + \sum_{1 \le i_1 < i_2 \le n} \frac{n}{p_{i_1} p_{i_2}} + \cdots + (-1)^n \frac{n}{p_1 p_2 \cdots p_n},$$

which after small algebra results in

$$\varphi(n) = n \left(1 - \frac{1}{p'_1}\right)\left(1 - \frac{1}{p'_2}\right)\cdots\left(1 - \frac{1}{p'_n}\right).$$

We omit the proof of the next statement, which follows almost directly from the previous theorem.

Theorem 31. *For any positive $a \in \mathcal{N}$ and n mutually prime with a,*

$$a^{\phi(n)} \equiv 1 \pmod{n}.$$

If n is prime, we again get the Fermat Little Theorem. However, if n is composite, it does not have to be 1; for example, if $n = 12$, hence $\phi(12) = 4$, we get $a^4 \equiv 1 \mod 12$, which can be checked immediately. For example, $7^4 = 2401$ and $11^4 = 14641$.

Problem 281. *For an integer $n > 1$, the ring of congruence classes modulo n is a field iff n is a prime number.*

8.1.6 Pseudorandom Numbers

Out of the numerous applications of the linear congruences, we discuss now the generation of *pseudorandom* numbers. Many problems, and not only in mathematics, require to employ random numbers. Those are the sequences of integers, such that, roughly, one cannot predict any next number. Currently, we believe that physical quantum processes generate random numbers, but mathematical sequences are *pseudorandom*, that is, they generate *periodical* sequences with so large period, that one cannot observe their periodicity.

Example 37. *Consider the sequence of integers $\{x_0, x_1, x_2, \ldots\}$, satisfying the linear congruence $x_n = (7x_{n-1} + 5) \pmod 9$ with $x_0 = 5$; the number x_0 is called the seed of the generator. What sequence is generated by this congruence?*

Solution. It is easy to see that if $x_0 = 5, x_1 = 4, x_2 = 6$, then $\{5, 4, 6, 2, 1, 3, \ldots\}$ and next the sequence is cycling with the period $\{3, 8, 7, 1\}$. Hence, to design a good generator of pseudorandom numbers, we must find the parameters of the linear congruence so that the period is really big.

Problem 282. *Let k be a square of an integer. Prove that, modulo 8, k is congruent to either 0, or 1, or 4. What residues must be considered in the problem, if instead modulo 8, we consider any other module $k = 1, 2, 3, 4, 5, 6, 7$?*

In problems, the residues are often masked as the remainders, as in the next problem.

Problem 283. *Prove that among any 7 integers there are two, whose difference is a multiple of 6.*

Solution. With regard to dividing by 6, any 6 integers have at most six, not necessarily different remainders. Thus, among the 7 remainders, there are two equal. The difference of the two integers with equal remainders is divided by 6 with the zero remainder.

Instead of remainders, we can talk here about the congruences modulo 6. After solving a single linear equation, in elementary algebra we usually move to solving the systems of simultaneous equations. Before considering the systems of simultaneous linear congruences, we again consider an example.

Problem 284. *A group of kids have several candies. They decide to divide them in equal parts. First, they put them in groups of two candies, but then one candy was left unassigned. Then they divide the candies in groups of five, but now three candies were left unassigned. What was the smallest number of candies the kids had? Find the general expression for the number of candies in the problem.*

Solution. Let they have x candies. The problem claims that $x - 1 = 2k$ and $x - 3 = 5l$, where k and l are some integers. If $k = 1$, then $x = 3$, but now the second condition fails, since we want at least one group of 5. It also fails if we set $k = 2, 3, 4, 5$, and only for $k = 6$, that is, when $x = 13$, we can find an integer solution $l = 2$. Thus, the smallest solution is 13 and we derived the general solution by excluding x. Hence, $2k = 5l + 2$, or $x = 2 \cdot 5m + 3$.

Observe, that in the example we actually had to solve the system of two congruences $\begin{cases} x \equiv 1 \pmod{2} \\ x \equiv 3 \pmod{5} \end{cases}$. Moreover, the product of the modules $5 \cdot 2$ naturally appears in the solution. Such systems are studied in the next theorem, which is traditionally called the *Chinese Remainder Theorem* or CRT.

Theorem 32. *Let $d_1, d_2, \ldots, d_r$ be mutually prime natural numbers, and $a_1, \ldots, a_r$ be any integers. Then the system of linear congruences*

$$\begin{cases} x \equiv a_1 \pmod{d_1} \\ x \equiv a_2 \pmod{d_2} \\ \ldots \\ x \equiv a_r \pmod{d_r} \end{cases}$$

has the unique solution modulo the product $D = d_1 \cdot d_2 \cdots d_r$.

Proof. Set $D_j = D / d_j$, $j = 1, 2, \ldots, r$. It is clear that $gcd(D_j; d_j) = 1$ since the moduli above are mutually prime. Next, $D_i \equiv 0 \pmod{n_j}, i \neq j$. So that, the congruence class of D_i has a multiplicative inverse, let us denote its representative as k_i; hence, $D_i \cdot k_i \equiv 1 \pmod{d_i}, i = 1, 2, \ldots, r$. Therefore, the number $x = \sum_{i=1}^{r} D_i k_i a_i$ solves the given system of congruences. Moreover, each step in the above proof is uniquely determined, therefore, the solution x is unique modulo the product D.

Remark 10. *The condition that the moduli are pairwise prime, is sufficient but not necessary, as the next problem shows.*

Problem 285. *Prove that the system* $\begin{cases} x \equiv 1 \pmod 4 \\ x \equiv 3 \pmod 6 \end{cases}$, *has the general solution* $x \equiv 9 \pmod{12}$, *where 12 is not the product of the moduli 4 and 6, but rather their least common multiple. Pay attention that the moduli 4 and 6 are not mutually prime.*

Problem 286. *Prove that the system of two congruences*

$$\begin{cases} x \equiv a \pmod m \\ x \equiv b \pmod n \end{cases}$$

has a solution iff $a \equiv b \pmod{\gcd(m,n)}$. *In particular, if m and n are mutually prime, then the system has the unique solution* $x \equiv x_0 \pmod{mn}$, *where* x_0 *is any integer solution of the given system.*

Problem 287. *Solve the systems of congruences and find their smallest positive solutions.*

$$1) \begin{cases} 2x \equiv 5 \pmod 7 \\ 3x - 4 \equiv 2 \pmod 8 \end{cases} \quad 2) \begin{cases} 3x \equiv 5 \pmod 7 \\ 2x \equiv 6 \pmod 8 \end{cases} \quad 3) \begin{cases} x \equiv 1 \pmod 2 \\ x \equiv 3 \pmod 5 \end{cases}.$$

Problem 288. *Ann has several candies. She distributed them to her friends by 2, but one candy remains unassigned. Then she distributed them to the same group of friends by 3, and also one candy remained unassigned. And when she distributed the candies by 4, one was left unused. What is the smallest amount of the candies, she had?*

The CRT theorem is a powerful tool, but there may be other instruments, which sometimes are preferable.

Problem 289. *Find the remainder after dividing a positive integer over 6, if after dividing over 2 and over 3 the integer gives the remainders 1 and 2, respectively.*

Solution. It is often convenient to begin at the end. When an integer is divided over 6, there are 6 possible remainders, 0, 1, 2, 3, 4, 5. But the condition regarding divisibility by 2 tells that the number, we sought for, is odd, therefore, there are only three possible remainders, 1, 3, 5. If the remainder were 1 or 3, then, after simplification, we get that ± 1 is a multiple of 3, which is impossible. Thus, the remainder can be only 5, and indeed, the integers 5, 11, 17, ..., etc., satisfy the problem.

8.1.7 Cryptography Application: Sharing Secrets

The CRT has many applications in Number Theory, Cryptography, and beyond. One of these applications is belonging to A. Shamir threshold scheme, which allows to share a secret among several people without distributing it too wide. Without going into some technicalities, we follow here [9]. To introduce the matter, consider a small example. Three friends (or not so much...) F_1, F_2, F_3 left a piece of secret information in a bank safe with three locks. Motivations aside, they developed a special rule for opening the safe. According to the protocol, to open the safe, at least two out of three people (any two!) must be present, and these two people must learn a secret number S, unknown to each of these people individually. However, everyone of the three people was given an additional number a_i, and was told that to find the S, they must solve any of the three systems, each consisting of two linear congruences, namely,

$$\begin{cases} L_1(x) \equiv 0 \pmod{q} \\ L_2(x) \equiv 0 \pmod{q}, \end{cases} \quad \begin{cases} L_1(x) \equiv 0 \pmod{q} \\ L_3(x) \equiv 0 \pmod{q}, \end{cases} \quad \begin{cases} L_2(x) \equiv 0 \pmod{q} \\ L_3(x) \equiv 0 \pmod{q}. \end{cases}$$

Here $L_i(x) = S + a_i x, i = 1,2,3$, are the linear functions where S is unknown, the coefficient a_i is known only to the person $F_i, i = 1,2,3$, and q is a natural number bigger than S, which is, actually, irrelevant for the solution, as long as $q > S$.

To be more specific, introduce a small chart below,

F_1	F_2	F_3
(1,9)	(2,11)	(3,13)

The points $(i, L_i(i)), i = 1,2,3$, in the second row belong to the lines L_i, respectively; each ordinate $L_i(i)$ must be given to person F_i, $i = 1, 2, 3$, but S is still unknown.

Now, when any two friends, say, F_1 and F_2, meet at the vault and want to open it, they can compose and solve the system

$$\begin{cases} S + ax_1 \equiv y_1 \pmod{q} \\ S + ax_2 \equiv y_2 \pmod{q}, \end{cases}$$

where the points $(x_1, y_1) = (1, 9)$ and $(x_2, y_2) = (2, 11)$ are given in the chart. The smallest positive solution of the system is easily found to be $a = 2$ and $S = 7$.

Problem 290. *Make sure, that if we choose the equations $\{L_1, L_3\}$ and $\{L_2, L_3\}$, the answer is the same $S = 7$.*

The protocol described in the example, is called the (n, w)– threshold scheme of sharing a number S among w people or computers, such that any n of them can compute the S, while for any $n - 1$ it is impossible.

8.1.8 Affine Ciphers

The shift ciphers involve only one integer parameter. Let us consider more advanced ciphers, which are the substitution ciphers with two integer parameters, *dilation* χ and *shift* ω, but first of all we must fix the terminology. In the language of high school and the elementary algebra, *linear functions* are expressions

$$y(x) = kx + b, \tag{8.3}$$

where k is called the *slope* and b the *y–intercept*.

However, in calculus an operator $T(f)$ is called *linear* iff it is *homogeneous* and *additive*, that is, $T(\lambda f) = \lambda T(f)$, where λ is a scalar, and $T(f + g) = T(f) + T(g)$. Hence, a linear function (8.3) with $b \neq 0$ is not a linear operator, it is neither homogeneous nor additive. In what follows, functions (8.3) are called *affine functions*.

Now we return to the cryptography. A substitution cipher on the alphabet of 26 letters[1] is called an *affine cipher*, iff its encryption function is defined as a substitution map

$$f = f_{\chi,\omega} : \mathbb{A} \mapsto \mathbb{A}, f_{\chi,\omega}(n) = \chi n + \omega \quad (\text{mod } 26), \tag{8.4}$$

where both operations are done in modular arithmetic, that is, *modulo* 26, and the encryption procedure is evident.

To find the formula for the *decryption procedure*, we will employ properties of the linear congruences we have derived earlier in this section. To solve equation

$$m = \chi \cdot n + \omega \quad (\text{mod } 26) \tag{8.5}$$

for n, given a number m, one has to transpose the symbol ω to another side of the equation and multiply both sides of the equation by χ^{-1}. One gets the result *formally* as $m \equiv \chi^{-1}(m - \omega)(\text{mod } 26)$. However, we can proceed this way only if the element χ is invertible in the ring $\mathbb{Z}_{26}$. The condition for that

---

[1] This number is of no importance, we need it just to consider English-language examples.

is, as we have showed above, that the numbers χ and 26 are *mutually prime*, that is, iff $gcd(\chi, 26) = 1$.

Since $26 = 2 \cdot 13$, there are 12 integers between 1 and 26 inclusive, satisfying this condition; indeed, out of 26 numbers between 1 and 26, we must exclude 13 even numbers and the number 13 itself, thus, there are 12 possible values for χ:

$$\chi \in \{1, 3, 5, 7, 9, 11, 15, 17, 19, 21, 23, 25\}. \tag{8.6}$$

Problem 291. *Find* 11^{-1} *in the ring* $\mathbb{Z}_{26}$.

Solution. To compute the inverse of $\chi = 11$ in the ring $\mathbb{Z}_{26}$, we must find the integer r, such that $11 \cdot r^{-1} \equiv 1 \,(\mathrm{mod}\, 26)$. Among the integers in the set $\{1, ..., 26\}$ only the number $r = 19$ satisfies the condition; indeed, $11 \cdot 19 = 209 \equiv 1 \,(\mathrm{mod}\, 26)$. Thus, $11^{-1} \equiv 19 \,(\mathrm{mod}\, 26)$.

Problem 292. *Find the inverses* χ^{-1} *for every integer* $\chi \in \{1, 3, 5, 7, 9, 15, 17, 19, 21, 23, 25\}$. *The computations above show in particular, that since* $11^{-1} = 19 (\mathrm{mod}\, 26)$, *it must be* $19^{-1} = 11 (\mathrm{mod}\, 26)$.

The key for affine cipher is the ordered pair (χ, ω). It is several times larger than the key space for the shift cipher, but it is quite manageable for the modern personal computers, therefore, it is not safe from the cryptographic point of view. For example, if $\chi = 12$ and $\omega = 26$, then the cardinality of the key space is $12 \cdot 26 = 312$ keys. Later on, we consider essentially safer ciphers. however, the examples above show that these ciphers also are not absolutely safe. Any such a cipher can be deciphered if you have enough computer power. A natural question occurs now, whether does exist an "unbreakable cipher"? We answer this question at the end of Chapter 13.

Problem 293. *Solve the congruences.*

1. $2x + 1 \equiv 0 \,\mathrm{mod}\, 13$

2. $10x \equiv 3 \,\mathrm{mod}\, 49$

3. $x^2 \equiv -1 \,\mathrm{mod}\, 13$

4. $x^2 + 3x + 10 \equiv 0 \,\mathrm{mod}\, 19$

5. $15x \equiv 12 \,\mathrm{mod}\, 33$.

8.2 EXERCISES

Exercise 8.1. *Compute the gcd(72, 17) and find the integers s, t, such that gcd(72, 17) = 72s + 17n.*

Exercise 8.2. *Find the gcd's and represent them as linear combinations with integer coefficients: gcd*$(1; 67), gcd(1013; 765), gcd(199; -320).$

Exercise 8.3. *Find* $gcd(0; 12), gcd(1; 12), gcd(12; 12)$ *and* $gcd(0, n)$ *for any natural n.*

Exercise 8.4. *How many are there positive integers not exceeding 60, which are mutually prime with* 60?

Exercise 8.5. *Are the integers* 128 *and* 125 *mutually prime?*

Solution $128 = 2^7$, $125 = 5^3$, therefore, these integers have no common factor and are mutually prime.

Exercise 8.6. *Find* $lcm(5, 99)$, $lcm(25, 9)$.

Exercise 8.7. *Compute* $29 \times 31 \pmod 3$.

Solution. $29 = 3 \cdot 9 + 2$, $31 = 3 \cdot 10 + 1$, thus, $29 \times 31 \pmod 3 = 2 \cdot 1 = 2$.

Exercise 8.8. *Prove that* $m \times d \equiv 0 \pmod d$ *for every integer m.*

Exercise 8.9. *Check that* $49 \equiv 7 \pmod 7$, $349 \equiv 9 \pmod{17}$, *but* $249 \not\equiv 7 \pmod 7$. *Is* $5 \equiv -7 \pmod 3$?

Exercise 8.10 *Solve linear congruences.* $8x \equiv 3 \pmod{()13}$, $99x \equiv 1 \pmod{13}$; $99x \equiv 5 \pmod{13}$; $2x \equiv 4 \pmod{()13}$.

Exercise 8.11. *Check whether the integers* $12, 49, 73, 11$ *are multiplicative inverses of* $7 \pmod{61}$ *and of* $11 \pmod{()61}$.

Solution.   Compute $12 \times 7 \pmod{61} \equiv 84 \pmod{61} \equiv 23 \pmod{61}$,   but not 1. Hence, 12 and 7 are not inverses in this problem. However, $11 \times 11 \pmod{60} \equiv 1 \pmod{60}$. Thus, modulo 60, $11^2 \equiv 1$, or 11 is multiplicative inverse to itself, or $11 \equiv \sqrt{60} \pmod{60}$.

Exercise 8.12. *Compute the multiplicative inverses of the integers* $n = 1, 2, 3, 4, 5, 6, 7$ *in* $\mathcal{Z}_8$.

Solution. Now the even numbers do not have multiplicative inverses, while $1^{-1} = 1; 3^{-1} = 3$, since $3 \cdot 3 = 1 \cdot 8 + 1$; $5^{-1} = 5; 7^{-1} = 7$.

Exercise 8.13. *Calculate* $3^{50} \pmod 7$.

Exercise 8.14. *What sequence of pseudorandom numbers is generated by the congruence* $x_n = (9x_{n-1} + 15)(\text{mod } 9)$ *with the seed* $x_0 = 7$?

Exercise 8.15. *A farmer brought cucumbers to the market and was selling them by tens, but the last customer didn't have two cucumbers for the whole ten. Tomorrow he brought the same amount of cucumbers and was selling them by dozens, but now only 8 cucumbers were left for the last buyer. The farmer knew that each day he brought more than 300 but less than 400 items. How many vegetables did he brought every day?*

Exercise 8.16. *Prove that any load from 1 gram to 40 grams inclusive can be weighted by making use of the weights 1 g., 3 g., 9 g., and 27 g., if you can put the weights on both sides, and use any weight no more than once. In mathematical parlance, the problem says that the equations* $\pm 1 \cdot x \pm 3y \pm 9z \pm 27w = k$ *have a solution in natural* x, y, z, w *for every natural* $k, 1 \le k \le 40$. *What if* $k = 41$?

Exercise 8.17. *Find the smallest positive multiple of 7, such that after dividing over 2 the remainder is 1, after dividing over 3 the remainder is 2, after dividing over 4 the remainder is 3, after dividing over 5 the remainder is 4, after dividing over 6 the remainder is 5, and it is a multiple of 7. Find the smallest positive integer, if after dividing over 2 the remainder is 1, after dividing over 3 the remainder is 2, after dividing over 4 the remainder is 3, after dividing over 5 the remainder is 4, and after dividing over 6 the remainder is 5.*

Solution. Setting out the system of congruences and excluding step-by-step natural parameters, we arrive at the answer $x = 119$.

Exercise 8.18. *Find the smallest positive integer, which is a multiple of 2000, 2001, 2002, 2003, and 2004.*

Exercise 8.19. *(1) Seven friends bought together a lottery ticket and, all of a sudden, won a prize. Instead of dividing the prize into seven equal parts, they decided to play the following game, so that the winner would get the whole prize. They guess a natural number X, then stand in a circle, and the youngest among them counts them from 1 to X inclusive, starting from herself. The person X goes out and does not participate further in the game. Then the youngest starts from herself and again counts from 1 through X, and the new No. X goes out and also does not take part in the draw anymore. The last person, who obviously remains alone, wins the game and gets the* *prize. What is the smallest value of X, such that the youngest person gets the prize?*

(2) Generalize the problem for N, $2 \le N < \infty$, people.

Exercise 8.20. *Cathy goes out with her friends every 12 days, goes to her bank every 15 days, and cleans her apartment every 18 days. Today she does all the three things. When will be her next day like this?*

Exercise 8.21. *Compute, if they exist, the inverses $11^{-1} \pmod{37}$, $11^{-1} \pmod{33}$, $11^{-1} \pmod{371}$.*

Exercise 8.22. *Apply the encryption rule $m(n) = \chi n + \omega \pmod{26}$ with $\omega = 17$ and $\chi = 9$ to the sentence DDANKK!. Decrypt the cipher-word 2323060708.*

CHAPTER 9

BOOLEAN FUNCTIONS

9.1   BOOLEAN DEGREE: DNF AND CNF

We remind that the set $\mathcal{B}^n, n = 1, 2, 3, \ldots$, is a Cartesian power of n copies of the set $\mathcal{B} = \{0, 1\}$, and a map $f : \mathcal{B}^n \to \mathcal{B}$ is called a *Boolean function* (b.f.) of n variables. For example, $f(x_1, x_2, x_3) = x_1 \wedge (\neg x_2 \vee x_3)$ is a b.f. of three variables.

Definition 62. *An elementary conjunction (elementary disjunction) is a conjunction (disjunction) of the symbols of variables or the negations of variables, the negations can also be written as apperbars; for example, $p, \neg p, q, \neg s, \overline{y}$, etc.*

An elementary conjunction (elementary disjunction) is called complete, iff it contains all the n variables or, maybe, their negations, and each variable, without or with negation, occurs exactly once, A complete elementary conjunction is also called minterm.

A convenient way to write elementary conjunctions is by making use of the Boolean Powers.

Definition 63. *If x is a Boolean variable, then* $x^a = \begin{cases} x, & a = 1 \\ \neg x, & a = 0 \end{cases}$

If $x = (x_1, x_2, \ldots, x_n)$ is a variable vector and $a = (a_1, a_2, \ldots, a_n)$ is an exponent-vector of zeros and ones, then the Boolean power is

$$x^a = x_1^{a_1} \wedge x_2^{a_2} \wedge \cdots \wedge x_n^{a_n}.$$

Definition 64. *A disjunction of elementary conjunctions is called a Disjunctive Normal Form (DNF) of a Boolean function. A DNF is called Perfect (PDNF) if it contains only complete elementary conjunctions, that is, only minterms.*

Definition 65. *Give definitions of a Conjunctive Normal Form (CNF) and of the Perfect CNF (PCNF).*

For example, in the case of three variables x_1, x_2, x_3, the formula $x_1 \wedge x_2 \wedge x_3$ is a complete elementary conjunction (a minterm), and $\overline{x_1} \vee x_2 \vee \overline{x_3}$ is a complete elementary disjunction, while $x_1 \vee x_2$ is an elementary disjunction but not complete elementary disjunction. However, a simple transformation, based on an obvious formula $p \wedge \overline{p} \equiv 0$,

$$x_1 \vee x_2 \equiv x_1 \vee x_2 \vee (x_3 \wedge \neg x_3) \equiv (x_1 \vee x_2 \vee x_3) \wedge (x_1 \vee x_2 \vee \neg x_3)$$

results in an identical perfect conjunctive form; see the next Definition 63.

Problem 294. *(1) Prove the following identities for any Boolean variables and constants 0 and 1.*

$$p \vee p \equiv p; p \vee \neg p \equiv 1; p \wedge p \equiv p; p \wedge \neg p \equiv 0;$$

$$p \vee 1 \equiv 1; p \vee 0 \equiv p; p \wedge 0 \equiv 0; p \wedge 1 \equiv p$$

(2) Prove that any normal form can be transformed to an equivalent perfect normal form.

Lemma 10. *Since any Boolean function f is a map whose domain is the set $\mathcal{B}^n$ of n–vectors $(a_1, a_2, \ldots, a_n)$, where the arguments a_j are the n–vectors of zeros and ones, the Boolean function, that is, the map f can be represented as a truth table, containing 2^n rows of zeros and ones. The right-most vector of the table is the vector of the function values of f.*

Proof is obvious.

Example 38. *For example, consider the function $f(x_1, x_2) = x_2 \rightarrow (x_1 \vee \neg x_2)$. To simplify the computations, we replace the conditional as $a \rightarrow b \equiv \neg a \vee b$, and due to the idempotency of the disjunction, the function $f \equiv x_1 \vee \neg x_2$, whose truth table is given by Table 9.1.*

TABLE 9.1 The Truth Table of b.f. $f(x_1, x_2) \equiv x_2 \rightarrow (x_1 \vee \neg x_2) \equiv x_1 \vee \neg x_2$.

x_1	x_2	$\neg x_2$	$f(x_1, x_2) \equiv x_1 \vee \neg x_2$
0	0	1	1
0	1	0	0
1	0	1	1
1	1	0	1

Theorem 33. *Every b.f. f can be represented as a disjunctive normal form (DNF). Moreover, if f is not an identically zero, it can be represented by a perfect DNF. Similarly, every b.f. f can be represented as a conjunctive normal form (CNF) and moreover, if f is not identically one, as a perfect CNF.*

Proof. We prove only the first part of the theorem since the second part follows if we apply the first part to the b.f. $\neg f$. If $f \equiv 0$, it can be written as $f(x_1,...,x_n) \equiv x_1 \wedge \neg x_1 \vee \cdots \vee x_n \wedge \neg x_n$ or as any subset of this set of pairwise conjunctions. Thus, we suppose that f is not identically zero or, which is the same, not identically false.

Represent f by the truth table. Its right-most column is a vertical string of 2^n bits. In every row, the left n symbols, we denote them σ, represent the argument of the b.f. f in this row. Let us fix an n–tuple x and consider the disjunction $f_1(x) = \bigvee_{\sigma=0}^{2^n-1} f(\sigma) \wedge x^\sigma$. This disjunction contains 2^n addends, and each of them is a conjunction of a constant $f(\sigma)$ with the minterm x^σ. Since σ runs through all the 2^n elementary n-conjunctions, and fixed x is one of them, there is one term, where $x = \sigma$. In this term, $x^\sigma = x^x = x_1^{x_1} \wedge x_n^{x_n} = 1$, since each factor here, according to Def. 63, is either $1^1 = 1$, or $0^0 = \neg 0 = 1$. Thus, the whole of this term is $f(x) \wedge x^x = f(x) \wedge 1 = f(x)$.

If $x \neq \sigma$, then the elementary conjunction X^σ contains either term $1^0 = \neg 1 = 0$, or $0^1 = 0$. Therefore, now $f(x) \wedge x^\sigma = f(x) \wedge 0 = 0$. Therefore, always $f_1(x) = f(x)$. Finally, we can omit all but one zero terms $f(\sigma) \wedge x^\sigma = 0 \wedge x^\sigma = 0$. The remaining term, when $x = \sigma$ gives $f(x) \wedge x^x = f(x)$.

Example 39. *Find the normal forms for b.f.* $f(x_1,x_2,x_3) = x_1 \wedge (\neg x_2 \vee x_3)$.

Solution. We have the PDNF

$$f = x_1 \wedge x_2 \vee x_1 \vee x_3 = x_1 \overline{x_2} x_3 \vee x_1 \overline{x_2} \vee x_1 x_2 x_3.$$

At the same time, the right column of the truth table for f contains three ones, corresponding to $\sigma = (1,0,0)$, $\sigma = (1,0,1)$, and $\sigma = (1,1,1)$, giving the same PCNF of f.

TABLE 9.2 Truth Table for Example 39.

x_1	x_2	x_3	$f(\sigma)$
0	0	0	0
0	0	1	0
0	1	0	0

(Continued)

x_1	x_2	x_3	$f(\sigma)$
0	1	1	0
1	0	0	1
1	0	1	1
1	1	0	0
1	1	1	1

To calculate the DNF, we can either start with a DNF for the negation $\neg f$, or take the negation of the CNF we just computed. We proceed from the available truth table for f – see Table 9.2, but we build a DNF instead of CNF, that is, we follow the zeros in the right column of the table and get

$$f(x) = (0 \vee \neg x_1 \vee \neg x_2 \vee \neg x_3) \wedge (0 \vee \neg x_1 \vee \neg x_2 \vee x_3) \wedge (0 \vee \neg x_1 \vee x_2 \vee \neg x_3) \wedge$$
$$\wedge (0 \vee \neg x_1 \vee x_2 \vee x_3) \wedge (0 \vee x_1 \vee x_2 \vee \neg x_3) =$$
$$(\neg x_1 \vee \neg x_2 \vee \neg x_3) \wedge (\neg x_1 \vee \neg x_2 \vee x_3) \wedge (\neg x_1 \vee x_2 \vee \neg x_3) \wedge (\neg x_1 \vee x_2 \vee x_3) \wedge (x_1 \vee x_2 \vee \neg x_3)$$

which is a PCNF of f. The latter can be simplified to a (non perfect) CNF $\neg x_1 (x_2 \vee \neg x_3)$.

The next statements, which easily follow from the previous ones, are useful in general and also will be employed when we study complete systems of Boolean functions. The following claim is obvious.

Theorem 34. *A Boolean Function is identically true ($\equiv 1$) iff it has a CNF such that every elementary disjunction in it is identically true.*

In turn, an elementary disjunction is identically true iff it contains an indeterminate (an independent variable) together with its negation.

The next assertion directly follows from the previous one by duality.

Theorem 35. *A Boolean Function is identically false ($\equiv 0$) iff it has a DNF such that every elementary conjunction in it is identically false.*

In turn, an elementary disjunction is identically false iff it contains an indeterminate (an independent variable) together with its negation.

Proof of Theorem 33. Represent $f \equiv 1$ as a CNF. $f \equiv C_1 \wedge \cdots \wedge C_k$, where C_j are elementary disjunctions. If at least one elementary disjunction is not $\equiv 1$, then we can find the values of arguments, such that f is not identically true and get the contradiction. Going to the second statement, it is obvious that $z \vee \neg z \equiv 1$, where z is any variable.

Problem 295. *Find the identically true CNF for the syllogism*
$$[(a \to b) \wedge (b \to c)] \to (a \to c), \text{ and the identically false DNF for its}$$
negation.

Corollary 4. *If any CNF (or DNF) of any Boolean Function is identically true (is identically false), then any its CNF (DNF) has the same property.*

9.2 BOOLEAN POLYNOMIALS

Boolean functions can be written down not only as normal forms but also as special logical polynomials.

Definition 66. *Any conjunction of the variables is called a logical (Boolean) monomial. Therefore, a logical monomial is an elementary conjunction, which does not contain negations. The constant 1 is a logical monomial by definition. Any sum modulo 2 of logical monomials is called a logical polynomial (or Zhegalkin polynomial).*

Example 40. *For example, x_1 and $x_2 \oplus x_1 x_3 \oplus 1$ are logical polynomials but $x_2 \oplus x_1 \vee x_3 \oplus 1$ is not.*

Problem 296. *Prove that $x \oplus 1 \equiv \neg x$, $x \oplus x \equiv o$, $x_1(x_2 \oplus x_3) \equiv x_1 x_2 \oplus x_1 x_3$. Represent the function $x_2 \oplus x_1 \vee x_3 \oplus 1$ as a logical polynomial.*

Lemma 11. *Every Boolean function can be represented as a Boolean polynomial.*

Proof. We give a proof as an algorithm, where each step is either obvious or was justified earlier. Given a Boolean function f.

1. Represent f as either CNF or DNF, whichever is simpler.

2. De Morgan rules imply that $x \vee y \equiv \neg(\overline{x} \wedge \overline{y})$. Use this formula to exclude all the disjunctions from the formula in 1).

3. Use $\overline{x} \equiv x \oplus 1$ to exclude all the negations from the formula, derived in step 2).

4. Employ the distributive law to "open" all the parentheses.

5. Employ the rule $x \oplus x \equiv 0$ to combine like terms.

Example 41. *Compute the logical polynomial of the Boolean function*

$$f \equiv \overline{x_1 \wedge x_3} \vee (x_2 \oplus \overline{x_4}).$$

Solution. In examples, we can proceed in any logically correct order, which sometimes can be simpler than in the general proof. Thus, in the example, we, first of all, eliminate the disjunction, as

$$f \equiv \overline{x_1 \wedge x_3} \vee (x_2 \oplus \overline{x_4}) \equiv \neg(\neg x_1 \wedge x_3 \wedge \neg(x_2 \oplus \overline{x_4})).$$

We continue by eliminating the double negations and using $\neg x \equiv x \oplus 1$; also omit the sign of conjunction as $x \wedge y \equiv xy$. Hence,

$$f \equiv \neg(x_1 x_3 \overline{(x_2 \oplus x_4 \oplus 1)}) \equiv \neg x_1 x_3 (x_2 \oplus x_4 \oplus 1 \oplus 1)).$$

Since $1 \oplus 1 \equiv 0$, we end as

$$\equiv \neg x_1 x_3 (x_2 \oplus x_4) \equiv x_1 x_3 (x_2 \oplus x_4) \oplus 1 \equiv x_1 x_2 x_3 \oplus x_1 x_3 x_4 \oplus 1 -.$$

Now we can prove the important representation theorem.

Theorem 36. *Any Boolean function can be represented as a logical polynomial. Moreover, this polynomial can be unique, up to the order of its terms, written as the sum modulo 2 of the constant term a_0, of the sum of linear terms $a_i x_i$, $1 \leq i \leq n$, the sum modulo 2 of quadratic terms $a_{i,j} x_i x_j$, etc., up to the unique product (i.e., conjunction) $a_r x_1 x_2 \ldots x_n$. The coefficients a_I are certain Boolean constants from the algebra $\mathcal{B}$.*

Proof. Keeping in mind Lemma 15, we have to prove only the uniqueness. We compare the cardinalities of the set of Boolean functions and of the logical polynomials. Remind that there are 2^{2^n} Boolean functions of n variables, and compute the number of logical polynomials of n variables $x_1, x_2, \ldots, x_n$. To every monomial $x_{i_1} \cdots x_{i_k}$ one can put into a one-to-one correspondence a subset $x_{i_1}, \ldots, x_{i_k}$ of the set of variables $x_1, \ldots, x_n$; we mention that the constant 1 corresponds to the empty set $\varnothing$. Since the correspondence is one-to-one, there are 2^n Boolean monomials of n variables. But every polynomial can be considered as a subset of the set of monomials; hence there are 2^{2^n} polynomials, that is, exactly as many as there are Boolean functions. Moreover, an analog of Lemma 15 shows that different Boolean functions with different truth tables cannot have the same corresponding polynomial. This proves the uniqueness of the polynomials.

Problem 297. *Prove that the set of Boolean functions of n variables makes a Boolean algebra.*

Problem 298. *Find the Boolean polynomias for the Boolean function* $p_1 \oplus p_2$.

Problem 299. *Generalize Theorem 36 for a general Boolean algebra.*

9.3 BOOLEAN DERIVATIVES

In the classical definition of the derivative of a real-valued function $f(x)$ at a point x_0, f is defined over a dense neighborhood of the point x_0. However, the Boolean functions are defined over discrete sets. Therefore, *Boolean Derivatives*, or rather *Boolean Differences* are defined as follows.

Definition 67. *Let* $f = f(x)$, $x = (x_1, x_2, \ldots, x_n)$ *be an n–dimensional Boolean vector, and* $x^{[k]} = (x_{i_1} x_{i_2} \ldots x_{i_k})$ *be its k–dimensional sub-vector. The Boolean derivative (or Boolean difference) of* $f(x)$ *with respect to the component* x_i *is the Boolean function*

$$D_i f(x) = f(x_1, \ldots, x_i, \ldots, x_n) \oplus f(x_1, \ldots, x_i \oplus 1, \ldots, x_n).$$

It is useful to remind here that $x \oplus 1 = \overline{x}$.

Example 42. $D_1(x_1 \oplus x_2) = x_1 \oplus x_2 \oplus (x_1 \oplus 1) \oplus x_2 = 1.$

Problem 300. *Prove that* $D_1(x_2) = 0$ *and* $D_1(x_1 \wedge x_2) = x_2$.

Boolean derivatives have found applications in the theory of codes correcting errors, in cryptography, and in other areas of mathematics and engineering [14]. They preserve many properties of differential and integral operators in calculus. Proofs of the following properties of the Boolean derivatives are mostly straightforward and left to the Reader.

Problem 301. *1) Boolean derivatives with respect to different arguments commute, that is,* $\dfrac{d}{dx_j}\left(\dfrac{df(x)}{dx_i}\right) = \dfrac{d}{dx_i}\left(\dfrac{df(x)}{dx_j}\right).$

2) If a Boolean function f is a sum, then the derivative is also a sum, that is, $\partial(f \oplus g) = \partial f \oplus \partial g.$
3) The formulas for derivatives of conjunction or disjunction remind partially the classical ones, but may have an extra terms; for example,

$$\frac{d(f \vee g)}{dx_i} = f\frac{dg}{dx_i} \vee g\frac{df}{dx_i} \oplus \frac{df}{dx_i}\frac{dg}{dx_i}.$$

9.4   EXERCISES

Exercise 9.1. *Construct normal and algebraic forms of the propositions (Boolean functions)* $(1)\ a \rightarrow (b \rightarrow c)$ $(2)\ a \leftrightarrow (b \leftrightarrow c)$.

Exercise 9.2. *Find the Boolean polynomials for Boolean functions*

$$(1)\ (p_1 \rightarrow p_2) \rightarrow p_3 \quad (2)\ (p_1 \rightarrow p_2)(p_2 \leftrightarrow p_3).$$

10

HASHING FUNCTIONS AND CRYPTOGRAPHIC MAPS

10.1 ROSETTA STONE

If you have an encrypted message but do not have the key, you cannot read it. That was the case with ancient languages before 1799 when archeologists had many stone obelisks and clay tablets with texts on ancient languages but could not read them. However, at that year, a stone was found near a Rosetta village in Egypt, which was called the Rosetta stone. The stone contained parallel copies of the same text in three ancient languages. This was the key, which allowed historians and linguists to decipher one of the ancient languages. When we deal with any unknown language, ancient or alive or computer, we need a Rosetta Stone.

To use the databases efficiently, the latter must be properly stored in the computer memory, so that we can quickly extract the information, otherwise as we have mentioned before, we will drown in that information. There are potentially *infinitely many records*; therefore, we need a device, which would help us to *compress* certain information. One way to achieve that is to use the *hashing functions*. They accept *long records of variable length* and produce uniformly short images. This uniformly bounded image is called a *hash* or *digest*, or *signature* of the file.

10.2 HASHING FUNCTIONS

The hashing functions have numerous applications, for example, when we submit the electronic signature online, and at many other occasions. Their cryptographic applications will be considered in a later lecture.

Consider an example. Every person in this country is assigned a 9-digit Social Security Number, SSN. But that assignment cannot serve as a hashing function because there are as many digests as records. However, if we consider the remainders after dividing the SS numbers by, say, 1,000, the remainders can serve as digests. As always, this simple solution has drawbacks. Indeed, there are only 1,000 of the remainders, while the population counts many millions of people. That simplifies life since we do not need the large storage of the digests, but on the other hand, one constantly experiences the *collision* of digests, that is, one signature may belong to at least two records. In particular, this hashing function is *not injective*.

Of course, those collisions are unacceptable, must be *resolved*, and there are many recipes suggested for resolving the problem. The simplest approach is linear: If we come to place a digest into an already occupied address a, we look through the subsequent addresses $a + 1$, $a + 2$, etc., so long as we find an empty address. Surely, this solution might fail, and there are more sophisticated algorithms.

If we want to completely eliminate collisions, we must significantly increase the number of available locations. Consider, for example, the ISBNs, that is, the unique identificators, assigned to *every* printed book, and distinguishing the books, editions, etc. They are printed on the back covers and familiar to everyone but without details. The ISBN-10, consisting of 10 digits, were introduced in the 1970s. Here only the nine digits contained certain information about the book, and the tenth digit was the *check-digit* (CD). It was computed by making use of the initial nine digits and served, to some extent, to verify certain calculations. Since there are 10^9 nine-digit natural numbers, and the amount of printed books grows tremendously, the publishers in 2004-2006 introduced ISBN-13 with 12 significant digits and the 13th check-digit, CD.

We are discussing the example of the check-digits here because it directly relates to circular arithmetic. Indeed, for ISBN-13, if its 12 digits are $d_1 d_2 \ldots d_{12}$, then one must compute the quantity

$$D = 1 \times d_1 + 3 \times d_2 + 1 \times d_3 + \cdots + 1 \times d_{11} + 3 \times d_{12},$$

and the check-digit is $CD = 10 - D(\mathrm{mod}\ 10)$, where $D(\mathrm{mod}\ 10)$ is the remainder after dividing $D \div 10$.

Problem 302. *(1) For the ISBN-13, 9781548791094, check whether the CD is computed correctly.*
(2) Replace X with the correct CD in the ISBN-13, 978154810254X.

The hashing functions have important applications in cryptology. A detailed discussion of these issues would bring us way too far from the main topics of this book, and we just say a few words here addressing the reader, for instance, to [40, Chap. 10-13]. Hashing functions are used in almost every online procedure. For example, for the authentication of digital signatures, a special protocol was developed, and it is based on hashing functions. The hashing functions cannot be *injective*; for instance, in the case of databases, there are usually more records than keys. Thus, given two elements x_0 and y_0, satisfying $h(x_0) = y_0$, the equation $h(x_0) = y$ definitely has a solution $x_1 \neq x_0$. If a perpetrator would find that solution, the signature is compromised. That is why the hashing function must be a one-way (or *trap-door*) function so that it is *infeasible* to solve the equation above. Theoretically, a solution $x_1 \neq x_0$ does exist, but to compute it, one must spend, on average, so much time that the solution becomes useless.

One of the major occupations of many people these days is to exchange information, which requires coding. Since it is impossible to create a dedicated channel for every two people, we must use public, that is, *open* channels. Messages that go through open channels, are called *plaintext*. However, we often do not want other people to read our communication. Therefore, we want to make our texts unreadable to other people, that is, to *encrypt them*. Such encrypted message is called *ciphertext*. The discipline at the intersection of mathematics, computer science, and who knows what else, dealing with encrypted messages, is called *cryptology*.

10.3 SUBSTITUTION CIPHERS

The subject of cryptology exists for more than two millennia; still, Julius Caesar used the *substitution code*. Talking about encrypted messages, people often immediately imagine intelligence, spies, etc. It's not always the case. Even if we do not want to cover the information but just want to receive it in a correct, unperturbed way, we must use a kind of encryption. For example, the E. Berlekamp error-correcting codes have been systematically used in space communication because during the long travel through space, the information is almost inevitably corrupted, and at the receiving end, we want to be able to recover the unperturbed information. We start with a small example.

Example 43. *Every text, open or encrypted, must use an alphabet, for example, the Latin alphabet of 26 letters. We assume that in our small example, the alphabet consists of only five characters, $\{c, e, g, m, o\}$. The plain message is the word "come." However, to cover the invitation over, you substitute the characters in the plain text letter by letter by making use of Table 10.1: instead of "c," one must transmit "g," "o" stays the same, "e" must be replaced by "c," "g" by "m," and "m" by "e," thus, "come" becomes "goec," where only first two characters make a desired word. Table 10.1 is called the key to this system, since one needs it to encode the plain text, and after receiving the code-text, to decode it.*

TABLE 10.1 Substitution Table for the Baby Example.

Original Character	c	e	g	m	o
Substitution	g	c	m	e	o

Thus, to achieve a correct decoding, we need certain additional information. In this example, since the input and the corresponding output words may have different lengths, the output must sometimes be *padded*.

Example 44. *Find the ciphertext for the plain words "gem" and "cegmomeg."*

This example demonstrates characteristic features of any substitution cipher. One of the main problems is that because of padding we cannot easily return from the obtained ciphertext to the original plain text. Any solution, for instance, by including the dummy symbols into the basic alphabet has its difficulties.

In general, there are 26 characters to substitute instead of the first letter "A," but after that, the shift is always the same, the images for "B," "C," etc., are completely determined. Hence, after excluding a trivial shift for zero place, there are 25 keys in this shift cipher. We can say that the *key space* consists of 25 elements.

We will address these issues later on. Right now, we see that such a crypto-system can be described as a mapping of the input alphabet into itself.

To avoid some of these issues, we will require these maps to be bijections of the alphabet onto itself. If a map is not injective, one will meet the situation when a character in a codeword has at least two preimages; hence there is no unique decoding. If a map is not surjective, then decoding may not give a symbol at all. Hence, the coding is to be bijective, and then decryption is (theoretically!) achieved by applying the inverse map, which is also the bijection by Theorem 8. In this case, the crypto-system is called *symmetric* since both in the encryption and in the decryption, we apply the same key. Let us consider one more example, which combines the substitution and block methods.

Example 45. *Cipher the message* **HOW NOW, BROWN** *First, we introduce the substitution table, which we will use in the example. For this particular example, we use only two non-alphabetical symbols "," and "?". First, we code the letters with the integers, that is, we substitute the integers instead of the letters. We consider the correspondence*

TABLE 10.2 Substitution Table for the Example 45.

A	B	C	D	E	F	G	H	I	J	K	L	M	N
28	27	26	25	24	23	22	21	20	19	18	17	16	15
O	P	Q	R	S	T	U	V	W	X	Y	Z	,	?
14	13	12	11	10	9	8	7	6	5	4	3	2	1

Next, we split the message into three-letter blocks, thus we have **[HOW]**, **[NOW]**, **[BR]**, **[OWN]**. If the last block contains less than three symbols, we artificially pad it with extra symbols "Z" in the end. Denote these blocks as

$$B_1 = [21,14,6]^T, B_2 = [15,14,6]^T, B_3 = [2,27,11]^T, B_4 = [OWN]^T.$$

The superscript "T" indicates the transposition of the vectors. Furthermore, the block size is 3, so that we choose any non-singular 3×3 matrix; in this example, let us use

$$A = \begin{pmatrix} 1 & 2 & 3 \\ 0 & 1 & 2 \\ 1 & 0 & 4 \end{pmatrix};$$

its determinant is $\det(A) = 5$, and the inverse matrix is

$$A^{-1} = \frac{1}{125} \begin{pmatrix} 4 & -8 & 1 \\ 2 & 1 & -2 \\ -1 & 2 & 1 \end{pmatrix}.$$

To get the encoded message, we multiply the matrix A on the left by each three-symbol vertical block of the original message. Therefore, we obtain

$$A \cdot \boldsymbol{B}_1, A \cdot \boldsymbol{B}_2, A \cdot \boldsymbol{B}_3, A \cdot \boldsymbol{B}_4 =$$

$$= \begin{pmatrix} 67 \\ 26 \\ 45 \end{pmatrix}, \begin{pmatrix} 61 \\ 26 \\ 39 \end{pmatrix}, \begin{pmatrix} 89 \\ 49 \\ 46 \end{pmatrix}, \begin{pmatrix} 14 \\ 6 \\ 1 \end{pmatrix}, \tag{10.1}$$

which can be written as a string

$$67, 26, 45, 61, 26, 39, 89, 49, 46, 14, 6, 1. \tag{10.2}$$

To decode it, one must first rewrite string (10.1) into a string of 3×1 column-vectors (10.1), and then left-multiply the inverse matrix A^{-1} by (10.1).

Problem 303. *Decode string (10.2) and get the original message.*

The Substitution Table (10.2) is a bijective map between the two alphabets, in an upper row and a lower row of the table. Bijective maps are symmetric in the sense that both the function f and its inverse f^{-1} are bijections, but this is, so to say, theoretical equivalence. From the computational point of view, these functions can behave quite differently, one of them can work much slower than another. This distinction deserves a special definition.

Definition 68. *A bijective map f is called a one-way function, iff f computes in at the most polynomial time, while any computations of f^{-1} require more than polynomial time.*

This definition does not contain any numerical values, only the qualitative description. For example, a prime factorization of the integers is a one-way function. Indeed, a multiplication of several prime numbers, even very big, is straightforward and can be done very quickly. However, if an integer contains several hundred digits, then the search for its prime factors with the current technology is almost always extremely time-consuming. The RSA crypto algorithm is based on this observation.

Problem 304. *Is solving quadratic equations a one-way algorithm?*

We finish this section by describing the brute-force attacks on a crypto-message. Suppose that by eavesdropping or by any other way, we have a crypto-message, and we want to decrypt it. "Brute-force" means that we go through all the possible substitutions until we get a plausible decoded text. Since we choose a substitution and proceed at random, the answer can, by chance, occur at the very first attempt but can also occur at the very last one. Hence we assume the *worst-case scenario* and suppose that the proper result will appear at the very last trial. Therefore, we must calculate how many different substitutions there are at all. The calculation is easy, though.

The English alphabet consists of 26 letters. For the first symbol, we can use all 26 characters. But since we want to use the one-to-one substitutions, at the second step, we have only 25 symbols since we cannot use the initially chosen symbol again. In the same way, at the third step, we have only 24 symbols, then 23, then 22, 21, ... , 2, 1. Next, every way to select a letter at step one can be combined with every possibility to choose a letter at the second step, and with every possibility to choose a letter at the third step, etc. The reader will find more on this in a chapter about combinatorics, and we immediately recognize here a permutation of 26 different symbols without repetition.

Hence, there are $26 \times 25 \times 24 \times \cdots \times 2 \times 1 = 26!$ permutations, or 26! ways to make substitutions, and the problem immediately leads us to the *factorials* of the integers. In general, there are 26! ways to perform substitutions in this problem. The factorials were defined in Definition 2, where we evaluated that $26! \approx 4.02 \times 10^{26}$. Let us compute how long will it take to try all the substitutions. Suppose your computer can try a trillion, that is, 10^{12} substitutions a second, which sounds huge, but is not too big for modern technology, see, for example, [13, p. 33]. So that, it will take almost 13,000 years to verify all the possible substitutions.

That leads us to another important issue of modern cryptology. Theoretically, we solved the problem: indeed, after *finitely many* trials, we will get the answer. However, in 13,000 years, our solution will definitely be useless to anyone then existing. So that, our solution shows that the substitution code should be considered very safe – as long as there is no other decoding algorithm. But this algorithm does exist, and the substitution codes that were in use for many centuries are not in use anymore. To finish this section, we consider the so-called Caesar code, which has just a historical value, but will lead us to important mathematical concepts.

Caesar's cipher is a simple substitution code. First of all, one enumerates all the characters in the used alphabet by the consecutive natural numbers, starting with 0. Thus, "a" becomes Nos. 0, "b" gets Nos. 1, etc., "z" is Nos. 25. A substitution is determined by an explicit rule. For instance, Caesar himself liked the rule $n \mapsto n + 3$, that is, the symbol, having Nos. k, must be replaced by the character Nos. $k + 3$. Therefore, "m," which is Nos. 11, must be replaced with the character with Nos. $11 + 3 = 14$, that is, with "p." It is also important, that to decipher a ciphertext, one must apply the same algorithm with the opposite shift. Thus, in Caesar cipher, for deciphering one must use the rule $n \mapsto n - 3$. This rule, though, does not tell us how to proceed if, for example, $n = 25$, since $25 + 3 = 28$, and no letter Nos. 28 exists in English or Latin alphabet. That difficulty, however, has been overcome centuries back, when people realized that we must just wrap up the alphabet after "z." Hence, the letter Nos. 26 becomes again Nos. 0, then Nos. 27 does, that is, "a" becomes "b," etc. So that, we are led directly to the circular or *modular arithmetic*, which was discussed in Chapter 8.

This cipher is called a group cipher, because the original word was broken into *groups*, and after that, every group was dealt with separately. More power ciphers occur if we generalize the encryption function, since we can vary group sizes and use different matrices for each group.

The *affine ciphers*, which we considered in Section 8.1, are based on a different approach.

10.4 EXERCISES

Exercise 10.1. *Use the information in Example 45 to encode and then decode the message WE MEET AS USUAL.*

Exercise 10.2. *Compose Caesar ciphertext for the plain text "Meet tomorrow at our place" and find the plain text for the ciphertext "Ablablah."*

CHAPTER

11

GENERATING POLYNOMIALS AND INVERSION FORMULAS

While deciphering a coded message, we actually try to invert (decipher) some mappings. In this chapter, we consider the inversion problem in general since it occurs in discrete mathematics, in calculus, and everywhere else. Many quantities, occurring in mathematics and elsewhere, depend upon one or more numerical parameters. For instance, factorials n! depend upon a natural parameter n = 0, 1, 2, 3, ..., binomial coefficients C(n, k) depend on two parameters n and k, etc. These parameters take values in the *discrete* sets of the integers; therefore, we cannot immediately apply the well-developed powerful techniques of classical analysis. To employ these tools, various methods were developed. In this chapter, we consider the method of *generating functions*.

For a sequence $\{a_n\}_{n=0}^{\infty}$, its generating function (GF) is a series $\sum_{n=0}^{\infty} a_n t^n$. However, to consider infinite series, we must make sure it is *convergent*, thus assuming certain requirements regarding the analytical background of the readers. Of course, we can forget about the convergence and consider the *formal* power series, which in turn, requires a certain algebraic background of the readers. To avoid these additional prerequisites for the readers, instead of infinite sequences, we often consider finite sequences $\{a_n\}_{n=0}^{Q}$ and the corresponding *generating polynomials* $\sum_{n=0}^{Q} a_n t^n$, thus circumventing any convergence issues.

Definition 69. *Given a sequence $\{a_0, a_1, a_2, \ldots, a_n, \ldots\}$, its generating polynomial of degree k is*

$$p_{a,k}(z) = p_k(z) = a_0 + a_1 \cdot z + a_2 \cdot z^2 + \cdots + a_k z^k.$$

Problem 305. *Find generating polynomials (GP) for the number sequence* $u_0 = \{1, 1, 1, \ldots\}$.

Solution. The generating polynomial of degree k for the sequence u_0 is

$$p_k(t) = 1 + 1 \cdot t + 1 \cdot t^2 + \cdots + 1 \cdot t^k.$$

This is a finite geometric progression, thus, $p_k(t) = \dfrac{1 - t^{k+1}}{1 - t}$ for $t \neq 1$ and $p_1(1) = 1 + k$. The polynomial is called GP, because it *generates* the ith term of the sequence as the coefficient of power t^j, $j = 0, 1, 2, \ldots, k$. In particular, the polynomials are $p_0(t) = 1$, $p_1(t) = 1 + t$, $p_2(t) = 1 + t + t^2$, $p_3(t) = 1 + t + t^2 + t^3$, etc.

These are the simplest examples of the GPs. Now let us study their simple applications.

11.1   PARTITIONS OF THE INTEGERS

Problem 306. *The postage for a letter is a dollar and 10 cents. In how many ways can you buy the stamps to send out the letter?*

Solution. Assume that the Post Office has an unlimited supply of regular 55-cent stamps and of 1-cent stamps. Since we can buy either (1) 2 stamps of 55 cents, or (2) one big stamp and 55 penny stamps, or (3) 110 penny stamps, there are three ways to buy the postage.

It is worth repeating that the integer 110 in the problem was represented as a linear combination of two other integers, 55 and 1, in three different ways, namely,

$$110 = 2 \times 55 = 1 \times 55 + 55 \times 1 = 110 \times 1.$$

If we want the linear combinations to contain both 55 and 1, then the first and the last sums can be written as

$$110 = 2 \times 55 + 0 \times 1 = 1 \times 55 + 55 \times 1 = 0 \times 55 + 110 \times 1,$$

These sums are called *partitions* of the integer 110. The first factor in every product indicates the *multiplicity* (the number of repetitions) of the second factor of the product, which is a building block of the sum.

Now we give another solution of Problem 306, which is more formal and not so obvious but has many applications. We start with a definition.

Definition 70. *Given two natural numbers n and k, a set of ordered pairs of natural numbers*

$$\coprod(n, k) = \{(n_1, k_1), (n_2, k_2), \ldots, (n_l, k_l)\}$$

is called a partition of the integer n into k terms $n_1, n_2, \ldots, n_l$ with multiplicities $k = k_1 + k_2 + \cdots + k_l$ and $n = n_1 k_1 + n_2 k_2 + \cdots + n_l k_l$, that is, the term n_j in the sum repeats k_j times. Since the addition is commutative, we always assume that $1 \leq n_1 < n_2 < \cdots < n_l$. We emphasize that the partition contains l different terms $n_1, n_2, \ldots, n_l$, which cannot repeat, and the partition contains k, maybe repeating terms.

For example, in Problem 306 we found three partitions of the integer $n = 110$, with $k = 1$ (twice) or $k = 2$, hence either $110 = 2 \times 55$, or $110 = 1 \times 55 + 55 \times 1$, or $110 = 110 \times 1$. These partitions can be written as $\coprod(110, 2) = \{(55, 1), (1, 55)\}$, or $\coprod(110, 1) = \{(55, 2)\}$, or $\coprod(110, 1) = \{(1, 110)\}$.

Problem 307. *A girl bought a small toy for 75 cents and paid for it with a $1 bill. In how many ways can she get the change?*

Even for small n, it may be cumbersome to verify all the possible partitions. The GFs and GPs can be of use in these problems.

Theorem 37. *The GF for the number of partitions $\coprod(n, k)$ is*

$$(1 + t + t^2 + t^3 + \cdots)(1 + t^2 + t^4 + \cdots) \cdots (1 + t^k + t^{2k} + t^{3k} + \cdots) \cdots \quad (11.1)$$

$$= \{(1 - t)(1 - t^2) \cdots (1 - t^k) \cdots\}^{-1}.$$

Proof. All series and the infinite products here are absolutely convergent if $|t| < 1$, whether t is real or complex. If we multiply out the parentheses in (11.1), we see that for every term t^n, the exponent represents exactly a partition of n into k addends.

If in a particular problem, a certain degree of t cannot appear in the final answer, we can safely cross out these degrees from any power series occurring hereafter in the solution.

The second line in (11.1) follows immediately if we apply the formula for the sum of infinite geometric progression to every series in the first line.

For example, a payment in Problem 306 can include at most two 55-cent stamps and at most 110 cents; therefore, the GP for the partitions in this problem is

$$\coprod(110, 2) = (1 + t + t^2 + \cdots + t^{110})(1 + t^{55} + t^{110}).$$

When one multiplies out these polynomials term-wise, the monomial $t^j t^{i \cdot 55}$ indicates that the payment consists of j "singles" and i 55-cent stamps, where i can be only $i = 0$, $i = 1$ or $i = 2$. If $i = 2$, that is, $2i = 110$, then it must be $j = 0$; if $i = 1$, then $j = 55$, and if $i = 2$, then $j = 110$. Therefore, again only three exponents are possible, and only three monomials satisfy the problem. Thus, the GF, which is GP as well, is $f(t) = 1 \cdot t^{110 \times 1} + 1 \cdot t^{55 \cdot 1 + 1 \times 55} + 1 \cdot t^{2 \cdot 55}$.

Problem 308. *Prove that the GF for partitions with different addends, that is, with $k_1 = k_2 = \ldots = k_1 = 1$ is*

$$(1 + t)(1 + t^2)(1 + t^3) \cdots.$$

11.1.1 Compositions

All the partitions, we considered so far, were non-ordered. Next, we study *ordered* partitions, called compositions. The following notation of a *natural segment* $\mathbf{N}_m = \{1, 2, \ldots, m\}$ is convenient hereafter. Repeat that $\mathbf{N} = \{0, 1, 2, \ldots\}$ while $0 \notin \mathbf{N}_m$.

Definition 71. *A composition of a natural number n, containing m parts, is a map $f : \mathbf{N}_m \to \mathbf{N}$, such that $f(1) + f(2) + \ldots + f(m) = n$.*

Example 46. *Consider a partition $\coprod(13, 6)$, which has at least two representations, $\{(1, 3), (3, 2), (4, 1)\}$ or also $\coprod(13, 6) = \{(1, 3), (2, 2), (6, 1)\}$, that can be conveniently displayed as Ferrers diagrams or Young tableaus*

FIGURE 11.1 The Ferrers diagrams of partitions $\coprod(13,6)$.

These two diagrams of the partition $\coprod(13, 6)$ generate two compositions of six with six parts, namely $13 = \{1, 1, 1, 3, 3, 3, 4\}$ and $13 = \{1, 1, 1, 2, 2, 6\}$. The first composition can be realized as a mapping $f_1 : \{1, 2, 3, 4, 5, 6\} \to \mathbf{N}$, $f_1(1) = f_1(2) = f_1(3) = 1, f_1(4) = f_1(5) = 3$ and $f_1(6) = 4$.

The second composition can be given as a mapping f_2 with the same domain and codomain, such that $f_2(1) = f_2(2) = f_2(3) = 1, f_2(4) = f_2(5) = 2$ and $f_2(6) = 6$.

Exactly the same way we convince ourselves that the GF for the compositions of a number n, consisting of m parts is

$$(t + t^2 + \cdots)^m = t^m (1 - t)^{-m}.$$

This equation immediately proves the next theorem.

Theorem 38. *The amount of compositions of number n with m parts is* $C(n-1, m-1)$. *The same number is the number of combinations with unlimited repetitions from the elements of m types, containing at least one element of each type.*

Corollary 5. *If we allow a composition to contain zero terms. than there are* $C(n+m-1, m-1)$ *such compositions.*

11.1.2 Inversion Formulas

The generating functions, GP in particular, may be useful in problems about the *Inversion Formulas*. For instance, let us solve the following problem.

Problem 309. *Let two infinite sequences* $\{a_n\}_{n=0}^{\infty}$ *and* $\{b_n\}_{n=0}^{\infty}$ *be connected by the system of equations* $a_n = \sum_{k=0}^{n} (-1)^k C(n,k) b_{n-k}$, $\forall n = 0,1,2,...$, *where* $C(n, k)$ *are, of course, the binomial coefficients. Then these sequences are also connected by the inverted formulas* $b_n = \sum_{k=0}^{n} C(n,k) a_{n-k}$, $\forall n = 0,1,2,....$ *Hence, the equations for* a_n *can be solved (inverted), giving us equations for* b_n.

Solution. Introduce the generating polynomials $A_n(t) = \sum_{j=0}^{n} a_j t^j$ and $B_n(t) = \sum_{j=0}^{n} b_j t^j$ for the sequences $\{a_n\}$ and $\{b_n\}$ above. Inserting the expression for a_n into $A_n(t)$, after simple identical transformations and using the Problem 114, 4), we derive an expression for b_n. The opposite inversion formula can be proved in exactly the same way.

For the method of GF (and GP) to be really useful, there must be a correspondence between linear operations over sequences and over GF. The correspondence between the addition/subtraction and the multiplication by a scalar is obvious. However, how to multiply sequences? It turns out that an appropriate linear operation is the *convolution* of sequences.

Definition 72. *The convolution of the sequences* $a = \{a_0, a_1, a_2, ...\}$ *and* $b = \{b_0, b_1, b_2, ...\}$ *is the sequence* $c = a * b = \{c_0, c_1, c_2, ...\}$, *such that* $c_n = a_0 \cdot b_n + a_1 \cdot b_{n-1} + a_2 \cdot b_{n-2} + \cdots + a_{n-1} \cdot b_1 + a_n \cdot b_0$, $n = 0,1,2,....$

For example, let $a = \{a_0, a_1, ...\}$ and $\mathbf{1} = \{1, 0, 0, ...\}$, then one can easily verify that $a * \mathbf{1} = \mathbf{1} * a = a$, hence the sequence $\mathbf{1}$ is the unit element for the convolution.

Problem 310. *Prove that the set of sequences equipped with the common termwise addition/subtraction and multiplication by constants, the* *convolution for multiplication, and the sequence* **1** *as the unity element is the unital commutative ring.*

Consider again the sequence $S = \{1, 1, 1, \ldots\}$, and its GF $f_S(t) = \sum_{k=0}^{\infty} t^k$. It is well known that this series converges in the open unit disc to $1 / (1 - t)$,

$$\frac{1}{1-t} = 1 + t + t^2 + t^3 + \cdots + t^n + \cdots.$$

Squaring both sides of this equation and applying the Cauchy rule to the square of the infinite series on the right, we deduce the following formula,

$$\frac{1}{(1-t)^2} = 1 + 2t + 3t^2 + 4t^3 + \cdots + (n+1)t^n + \cdots.$$

Let us notice that $\sum_{k=0}^{n} 1 = n + 1$, that is, the coefficient of t^n in the last series is the *sum* of the first $n + 1$ terms of the sequence being squared.

Problem 311. *As another example, consider*

$$(1 + 2t - 3t^4 + 9t^{18} + 6t^{40})(1 - t)^{-1} =$$

$$= 1 + 3t + 3t^2 + 3t^3 + 0 \cdot t^4 + \cdots + 0 \cdot t^{17} + 9t^{18} + 9t^{19} + \cdots + 9t^{39} + 15t^{40} + \cdots.$$

Here $1 + 2 + 3$, $1 + 2 - 3 = 0$, $1 + 2 - 3 + 9 = 9$, etc. For $n \geq 40$, all the coefficients are equal to the sum $1 + 2 - 3 + 9 + 6 = 15$.

Because of this feature, the sequence S is called the *summator*.

Problem 312. *Use summator's properties to prove, for every natural n, the equations* $\sum_{k=0}^{n} k^2 = (1/6)n(n+1)(2n+1)$ *and* $\sum_{k=0}^{n} k^3 = (1/4)n^2(n+1)^2$.

11.2 EXERCISES

Exercise 11.1. *Find GPs for the number sequences* $u_1 = \{1, 0, 1, 0, 1, \ldots\}$ *and* $u_2 = \{0, 1, 0, 1, \ldots\}$.

Exercise 11.2. *(1) Prove that the GF for partitions with all terms not exceeding a fixed natural number q, is*

$$(1 + t + t^2 + \cdots) \cdot (1 + t^2 + t^4 + \cdots) \cdots (1 + t^q + t^{2q} + \cdots) =$$

$$= \left((1-t)(1-t^2) \cdots (1-t^q)\right)^{-1}.$$

(2) Prove that the GF for partitions containing exactly k terms is

$$t^k \left((1-t)(1-t^2) \cdots (1-t^k)\right)^{-1}.$$

For example, If $k = 2$, the last GF is $\dfrac{t^2}{(1-t)(1-t^2)} = t^2 + t^3 + O(t^4)$. Thus, the numbers 0 and 1 have, obviously, no two-part partitions, 2 and 3 have only one such partition, $2 = 1 + 1$ and $3 = 1 + 2$.

Exercise 11.3. *(1) Prove the identity, for $|t| < 1$,*

$$\frac{1}{1-t} = (1+t+t^2+\cdots+t^9) \times (1+t^{10}+t^{20}+\cdots+t^{90}) \times$$

$$\times (1+t^{100}+t^{200}+\cdots+t^{900}) \times \cdots$$

(2) Use this identity and the summator sequence to prove that every integer can be written, and uniquely, in the decimal numerical system.
(3) Prove that every integer can be, and uniquely, written in the binary, ternary, etc., number systems.

Exercise 11.4. *Prove the pair of the Inversion Formulas*

$$a_n = \sum_{k=0}^{n} C(n+p, k+p)b_k, \; n = 0,1,2,\ldots, \; and$$

$$b_n = \sum_{k=0}^{n} (-1)^{n-k} C(n+p, k+p)a_k, \; n = 0,1,2,\ldots.$$

12

SYSTEMS OF REPRESENTATIVES

Example 47. *Preparing to the new school year, 16 first-graders created five committees, A, B, C, D, and E, such that every student participated in at least one committee, and the memberships at some committees can be identical. Before the year starts, the Principal called a meeting, where every committee had to be represented by one member. Is it possible that all the committees will be represented by different students?*

12.1   SDRs

Such problems often appear not only in mathematics, hence it is important to learn how to determine whether the answer is positive or negative. The students' body in the problem is a universal set; the membership of any committee is a subset. It might happen that two or more committees consist of the same students, that is, these committees make the *same set*. In this case, we attach subscripts (indices), thus making these subsets *distinct*. Even if the subsets are element-wise identical but are treated as different, we speak about the *Systems of Distinct Representatives (SDR)*.

Example 48. *Consider the set $U = \{1, 2, 3\}$ and its subsets $X_1 = \{1, 2\}$, $X_2 = \{1, 3\}$ and $X_3 = \{1, 2, 3\}$. This system has more than one SDR, for example, $\{2, 3, 1\}$. However, the system $X_1 = \{1, 2\}$, $X_2 = \{1, 3\}$, $X_3 = \{1, 2, 3\}$ and $X_4 = \{2, 3\}$ with the same universal set U does not have a SDR, since just three elements cannot represent four sets one-by-one.*

As P. Hall showed almost a century ago, this simple necessary condition is at the same time sufficient. It should be mentioned that this theorem about the SDRs is equivalent to many other statements, for example, to the maximal flow theorem, or to the theorem about the disjoint chains in graphs, or to the theorem about matching in bipartite graphs, etc. Proofs of the equivalence of these and other results can be found elsewhere, for example, in [28].

Theorem 39. *(Hall) The set U and its subsets X_1, ..., X_k have at least one SDR iff the union of any family of the subsets contains at least as many elements as the number of subsets in this family, that is, iff for every natural k and every family of k subscripts $\{i_1, i_2, ..., i_k\}$ the inequality holds,*

$$|X_{i_1} \cup X_{i_2} \cup \cdots \cup X_{i_k}| \geq k.$$

Problem 313. *Follow Example 48 above almost word-by-word to prove the necessary part of this theorem.*

A proof of sufficiency is more elaborate and can be found in many sources; we skip it here. In particular, an inductive proof with an estimate of the number of SDRs that follows an original proof, can be found in [28]; a constructive proof is given in Halmos [24], who coined the term "A theorem on the village marriages."

Next, we show a few applications of the SDRs.

Problem 314. *Every girl, attending a dance party, is familiar with at least m boys there, while every boy at the party knows at most m girls who are present. Prove that every girl can call a familiar boy for a dance.*

Partially ordered sets (posets) were introduced in Definition 40. Consider, for instance, a 5-element poset $X = \{a, b, c, d, f\}$, where a binary relation of partial order ρ is assigned so as this set of ordered pairs is $\rho = \{(a,c),(a,d),(b,d),(d,f)\}$, that is, $a \prec c, a \prec d, b \prec d, d \prec f$. Then X can be written as the union of chains in several ways, for example, as

$$X = \{a,c\} \cup \{b,d\} \cup \{f\}$$

or as

$$X = \{a,d\} \cup \{b,f\} \cup \{c\}$$

or as

$$X = \{a,c\} \cup \{b\} \cup \{d\} \cup \{f\}.$$

The last union contains four chains; however, there are unions with three chains, and one can easily check that it is impossible to reduce this number of chains to 2; thus, the minimum number of the disjoint chains, containing all the elements of X is 3. On the other hand, there also are three pairwise, noncomparable elements of X, and this number cannot be decreased. That is the statement of the Dilworth Theorem (Theorem 40), whose proof is also left to the reader (or to any appropriate reference, e.g., [28]).

Theorem 40. *Let X be a finite poset. The minimum number of disjoint chains containing all the elements of X, is equal to the maximum number of pairwise noncomparable elements of X.*

Problem 315. *Prove that the theorems of Hall and Dilworth are equivalent.*

Problem 316. *Does the poset X above, in addition to the three given ones, have other chain decompositions?*

Problem 317. *Prove that if a finite set and a family of its subsets satisfy the Hall condition of the existence of SDRs, then the family has the unique SDR iff the cardinality of this union of the sets in the family is equal to the cardinality of the family.*

The topic of SDRs is connected with the next one, the Block Schemes, and in particular, the Triple Systems. We again start with a typical example. Suppose we want to study the effect of three different fertilizers on two various kinds of corn. We can use for the experiments the three available fields; thus, we have $3 \times 2 \times 3 = 18$ different combinations (fertilizer, field), and if we study them one pair at a year, it would be way too long to get any conclusion. However, researchers developed the *scheduling theory* and the *experiment planning*, and using these theories, one can proceed much faster and more efficiently. We consider the following model.

Example 49. *Organizers of a soccer tournament invited nine teams $T_1, T_2, \ldots, T_9$ and rented three fields for the games. Each team must play every other team once. What should be a schedule for the shortest possible tournament?*

Solution. In total, $C(9, 2) = 36$ games are necessary; thus, $36 \div 3 = 12$ games are to be scheduled at every field. If the games are scheduled in two shifts, morning and evening, the tournament will last 6 days. We try to split the teams into three groups, by three teams in each group. These groups are called *blocks*. Then we arrange mini-tournaments of the three teams in every block, and finally, shuffle the blocks. There should be $C(3, 2) = 3$ games within every block. If this scheme works, there should be $36 \div 3 = 12$ blocks $B_1, B_2, \ldots, B_{12}$.

It is not clear at all that this method works; hence we must construct the actual blocks. Since initially there is no restriction on the teams or blocks, we suppose that the first shuffle consists of the blocks

$$B_1 = (T_1, T_2, T_3), \;\; B_2 = (T_4, T_5, T_6), \;\; B_3 = (T_7, T_8, T_9).$$

Now we have to shuffle teams, that is, blocks. To introduce the certain systematic method, for the next block, we pick the initial elements of blocks $B_1 - B_3$, for the next block, choose the second elements, and finally, for the following block, select the third elements; thus, we get the second shuffle

$$B_4 = (T_1, T_4, T_7), \;\; B_5 = (T_2, T_5, T_8), \;\; B_6 = (T_3, T_6, T_9).$$

It is clear now how to choose the third shuffle,

$$B_7 = (T_1, T_5, T_9), \;\; B_8 = (T_3, T_4, T_8), \;\; B_9 = (T_2, T_6, T_7),$$

and the last one,

$$B_{10} = (T_3, T_5, T_7), \;\; B_{11} = (T_1, T_6, T_8), \;\; B_{12} = (T_2, T_4, T_9).$$

One can verify immediately that this arrangement of 9 elements T_1, ..., T_9 into 12 blocks B_1, ..., B_{12} satisfies all the stipulations of the *Balanced Incomplete Block Design* (*BIBD*) with the parameters $(9, 12, 3, 4, 1)$, that is, there are nine elements (teams) arranged in 12 blocks of three elements each, such that every team belongs to 4 shuffles and every pair of elements (of teams) appears together exactly in one block.

This can be generalized as the following.

Definition 73. *Let X be any v–element set, whose subsets are called blocks. The same block can appear (repeat) more than once, but with different subscripts. Let this family of blocks contain b blocks. The family of blocks is called BIBD(v,b,k,r,λ) iff every block contains k elements, every element belongs to r blocks, and every pair of elements meets in λ blocks.*

Problem 318. *Given certain entities, say the natural numbers $\{1, 2, 3, 4, 5, 6, 7\}$ prove that the family of blocks*

$$\{B_1, B_2, B_3, B_4, B_5, B_6, B_7\}$$

where the blocks are $B_1(1, 3, 7)$, $B_2(1, 2, 4)$, $B_3(2, 3, 5)$, $B_4(3, 4, 6)$, $B_5(4, 5, 7)$, $B_6(1, 5, 6)$, and $B_7(2, 6, 7)$, is a block-scheme BIBD($7, 7, 3, 3, 1$).

12.1.1 Systems of Triples

The simplest and the most common block designs are the systems of triples, that is, the $BIBD(v, b, 3, 3, 1)$, called Steiner triple systems. See Problem 324 and Example 49. Unlike the systems with $k \geq 5$, for the systems with $k = 3$

and $k = 4$, there are known the necessary and sufficient conditions for the parameters to satisfy.

Theorem 41. *The system of triples BIBD(v, b, 3, 3, 1) exists iff $v = 6t + 1$ or $v = 6t + 3$ with any natural t. Proofs can be found in more extensive treatments of combinatorics.*

Problem 319. *Let $S(w)$ be a Steiner subsystem of the Steiner system of triples $S(v)$. Prove that $2w \leq v - 1$.*

Problem 320. *Nine professors must proctor 12 exams in 4 days, such that each test must be observed by a committee of 3 professors. Compose a schedule of the tests so that every pair and every triple of the professors do not meet more than once during the tests.*

12.2 EXERCISES

Exercise 12.1. *A bridal shop has dresses of three designs and of three colors. Can the owner choose for the display exactly three dresses, which demonstrate all the three fashions and all the three colors? Is it possible to choose two dresses if there are two designs and two colors?*

Exercise 12.2. *Four kindergarten students go out in a 2×2 formation. For how many days can they go this way, so that every day each student has a new neighbor in her row? The same question for 9 students and 3×3 square? For 25 students and 5×5 square?*

Exercise 12.3. *An ice hockey team has nine forwards. The team plays four games a week. Prove that the coach can set up the triples of field players so that no two forwards play twice in the same triple.*

Exercise 12.4. *Construct BIBD(13, 26, 3, 6, 1).*

13

BOOLEAN ALGEBRAS

In Chapter 2, we studied the propositions and operations with them; in Chapter 3, we considered quite different objects, sets and corresponding operations. However, if we compare the formulas in Chapters 2 and 3, we observe a lot of similarities. Such coincidence cannot occur by chance, and there should be a certain algebraic structure underlying both theories. Such structure indeed exists and is called *Boolean algebra*.

13.1 OPERATIONS AND AXIOMS

Definition 74. *A set B with a negation or complementation unary operation ⁻, with two binary operations +, called the sum or disjunction or binary addition, and ·, called conjunction or multiplication, and with two different elements **0** and **1**, called zero and unity, respectively, is called a Boolean algebra, iff the following properties are valid for all the elements $x \in B$.*

0 *is the neutral element for the addition, that is, for all $x \in B$, $x + \mathbf{0} = x$, and **1** is the unit element for multiplication, that is, $x \cdot \mathbf{1} = \mathbf{x}$, respectively. If possible, the symbol · will be omitted. Moreover, $x + \overline{x} = \mathbf{1}$ and $x \cdot \overline{x} = \mathbf{0}$.*

Both operations, the addition "+" and the multiplication "·" are commutative, associative, and are connected with two symmetric[1] distributive laws, that is,

$$x + y = y + x, \, x \cdot y = y \cdot x$$
$$(x + y) + z = x + (y + z), \, (x \cdot y) \cdot z = x \cdot (y \cdot z)$$
$$x \cdot (y + z) = x \cdot y + x \cdot z, \, x + (y \cdot z) = (x + y) \cdot (x + z).$$

---

[1] From this remark we observe an essential distinction between the algebra that we studied at high school and the Boolean algebra.

The simplest example of a Boolean algebra[2] is the two-element set $\mathcal{B}_2 = \{\mathbf{0},\mathbf{1}\}$ with the standard logical operations $\neg, \vee, \wedge$.

Problem 321. *Prove that one can define a division in this Boolean algebra. Thus, this is an example of the two-element Boolean Field. Describe the neutral and opposite elements in this field. Can one compute radicals (square roots) in this ring? Solve linear and quadratic equations?*

Problem 322. *Prove that the Boolean 2^X of any finite set X and the set of n–tuples $(x_1, x_2., ..., x_n)$, where $n \geq 2$ is any natural number, are the Boolean Algebras; if X is an infinite set, the conclusion is the same, but a proof requires more mathematical technique. In the case of the Boolean, the operations are the operations with the subsets, the null element is the empty set $\varnothing$, and the unity is the Boolean 2^X itself. In the case of n–tuples, the operations are coordinate-wise Boolean operations, and the neutral elements are $(0, ..., 0)$ and $(1, ..., 1)$, respectively. The set of n–tuples for $B = \{0, 1\}$ is denoted as B^n.*

The value of these abstract definitions is that, in particular, we do not have to prove the same feature again and again in every particular instance – if we prove, say, the duality law in the general Boolean algebra, we can apply that rule to every instance of this structure; this is, by the way, an example of the deduction scheme of proof. As an example, we consider *the Duality Principle* in a Boolean algebra.

Let $\mathcal{B}$ be a Boolean algebra with finitely many variables (*indeterminates*) $x_1, x_2, ..., x_n$, standard operations $\neg, \vee, \wedge$, and standard separators. It is, of course, possible to introduce the conditional and biconditional operations as *abbreviations*, that is, $p \rightarrow q \equiv \neg p \vee q$ and $p \leftrightarrow q \equiv (p \rightarrow q) \wedge (q \rightarrow p)$, but we do not need that now. Using these symbols, we develop proper formulas as described above in Lecture 2, and call them *Boolean expressions*, for example, $A(x, y, z) = \neg x \vee z \wedge y$.

For any properly built Boolean expression A, we can construct its *dual* A^* by replacing every sum $a + b$ with the product $a \cdot b$ and vice versa, every product $a \cdot b$ with the sum $a + b$, and similarly, interchanging zeros $\mathbf{0}$ with ones $\mathbf{1}$. It is important *to preserve* the same order of operations.

Example 50. $A \equiv \neg x \cdot 1 + y + \ \neg z; A^* \equiv \neg x + 0 \cdot y \cdot \ \neg z.$

Theorem 42. *Given any Boolean algebra, the dual of every properly built Boolean equality is also Boolean equality.*

---

[2] Hence, the term "Boolean algebra" has two meanings: this is a technical term in the sense of Definition 74, and a generic name of a set with Boolean operations. This duality does not lead to any misunderstanding.

Proof. The proof is obvious, since all the defining equations in any Boolean algebra appear in *dual* pairs.

Problem 323. *Consider a three-element set* $\left\{0, \dfrac{1}{2}, 1\right\}$. *Define proper Boolean operations, making it a Boolean algebra, in which, however, the equivalences* $x \vee \overline{x} \equiv 1$ *and* $x \wedge \overline{x} \equiv 0$ *are invalid.*

13.2 BOOLEAN RINGS

The *Boolean rings*, such as B^n, are important in mathematical logic and elsewhere.

Definition 75. *An unital ring K is called a Boolean ring, iff in addition to the six ring axioms $K_1 - K_6$ (Definition 15) and to the axiom K_7 of the existence of the unity in a ring, there holds one more axiom; the operations with this property are called idempotent operations,*

$$K_8 : \forall a \in K(a \cdot a = a).$$

For example, the conjunction and disjunction are idempotent operations, since in B, $a \vee a \equiv a$ and $a \wedge a \equiv a$, while the negation is not idempotent, because $\neg(\neg p) \not\equiv \neg p$.

Definition 76. *Two rings R_1 and R_2 are called isomorphic, iff there exists a one-to-one correspondence between their elements, which preserves the results of the ring operations, that is, the sum corresponds to the sum and the product to the product.*

Let $\mathcal{A}$ be a set of cardinality $n, |\mathcal{A}| = n \geq 1$. One can straightforwardly check that the Boolean $P(\mathcal{A})$ is isomorphic to both the Cartesian degree B^n and to the power-set $B^{\mathcal{A}}$. To this end, it is convenient to employ again the characteristic function χ_A of a subset $A \subset \mathcal{A}$. It is clear that $\chi_A \in B^{\mathcal{A}}$, and any function $f \in B^{\mathcal{A}}$ is a characteristic function for a certain subset $A' \subset \mathcal{A}$, namely, for $A' = f^{-1}(\{1\})$. Therefore, the isomorphism we sought for, is given by the correspondence

$$P(A) \ni A \sim \chi_A \in B^{\mathcal{A}}.$$

If $\mathcal{A} = \{a_1, \ldots, a_n\}$, then the correspondence between $B^{\mathcal{A}}$ and B^n is established by

$$B^{\mathcal{A}} \ni f \sim (f(\alpha_1), f(\alpha_2), \ldots, f(\alpha_n)) \in B^n.$$

Problem 324. *Prove that these correspondences indeed establish an isomorphism, that is, they are bijective and preserve the ring operations.*

Let $\mathcal{A}$ be an arbitrary set and $P(\mathcal{A})$ its Boolean. The operations of addition and multiplication on this Boolean are defined as follows.

$$A + B = (A \cup B) \setminus (A \cap B)$$

and

$$A \cdot B = A \cap B.$$

Problem 325. *Check that the Boolean with these operations is a ring.*

We prove below that any *finite* Boolean ring is isomorphic to the ring of subsets of a certain finite set; thus, all finite Boolean rings have a similar structure. First of all, we study in more details the ring of subsets $P(\mathcal{A})$.

Definition 77. *Let $\mathcal{A}$ be a Boolean ring and element $A \in \mathcal{A}$. Its complement is called the element $\bar{A} = \mathcal{A} \setminus A$, or which is the same due to the axiom K_r, $\bar{\bar{A}} = A$, $\bar{A} = A + \mathcal{A}$.*

The complement, clearly, is the complement to the universal set, is idempotent, and satisfies $A \cdot \bar{A} = \varnothing$. Moreover,

Problem 326. $A \cup B = A + B + A \cdot B$

Problem 327. *Prove also the following transitivity property. If $A \subset C$ and $B \subset C$, then $A \cup B \subset C$.*

Next, we remark that the equation

$$A \cup B = A + B + A \cdot B = A + B + A \cap B$$

implies that the union of disjoint sets $A \cap B = \varnothing$ coincides with their ring sum $A + B$. However, every nonempty set is a union of its one-element subsets. Therefore, if $A = \{a_1, \ldots, a_l\}$, this set can be represented as

$$A = \cup_{j=1}^{l} \{a_j\} = \sum_{j=1}^{l} \{a_j\} = \sum_{a \in A} \{a\}.$$

Remark 11. *The one-element subsets $\{a\}$ are minimal nonzero elements for the binary relation $\subset$, see Definition 42.*

We have shown above that the ring operations on a Boolean can be expressed through the set-theory operations $\vee$, $\wedge$, and $\setminus$. On the other hand, we have shown just now that the inverse is also true: set-theory operations can be expressed through the ring ones. It turns out that this correspondence is valid for any Boolean ring – given any Boolean ring, one can put into a one-to-one correspondence another structure on the same set, called the Boolean lattice. Essentially, this is the same Boolean algebra. We continue the study of rings, and first of all, give a definition of a complement of an element, but only for unital rings.

Definition 78. *Let K be a ring with a unity e. A complement of an element $a \in K$ is an element $a + e$; it is denoted as $\bar{a}$.*

Definition 79. *A binary relation in a ring K is called a partial order, iff for any elements a, b in K, the inequality $a \leq b$ is equivalent to the equation $a \cdot b = a$. The partial order is called a strong order $a < b$, iff $(a \leq b)$ and $a \neq b$.*

Definition 80. *Let K be a ring with a partial order $\leq$. A minimal nonzero element a is called an atom, iff for every $b \in K$, an inequality $b \leq a$ implies either $b = 0$ or $b = a$.*

Lemma 12. *The following equations are valid in any ring K for every elements $a, b, c \in K$:*

1. $a + a = 0$

2. $a + b = 0 \Rightarrow a = b$

3. $ab = ba$

4. $\bar{\bar{a}} = a$

5. $a + \bar{a} = e$

6. $a \cdot \bar{a} = 0$

7. $a \leq a$

8. $(a \leq b) \wedge (b \leq a) \Rightarrow a = b$

9. $(a \leq b) \wedge (b \leq c) \Rightarrow a \leq c$

10. $(a \leq c) \wedge (b \leq c) \Rightarrow a + b \leq c$

11. $ab \leq a$ and $ab \leq b$

12. $a \geq 0$.

Proof. Many of these properties are now well familiar. For example, property (3) means that a Boolean ring is commutative, (4) states that the complement in a Boolean ring is idempotent, etc. Properties (7)–(9) mean that the binary relation ≤ is a partial order, that is, any Boolean algebra is a Poset. We just collected all these features together for convenience.

To prove Property (1), we set

$a: = a + e$ in the Axiom 8 to get $(a + e) \cdot (a + e) = a + e$. Therefore, $a \cdot a + e \cdot a + a \cdot e + e \cdot e = a + e$,

or

$a + (a + a) + e = a + e$. But $a + e + 0 = a + e$, and since the difference is unique, $a + a = 0$.

(2) $a + b = 0$ by the condition, and we have just proved that $a + a = 0$. The uniqueness of the difference again implies $a = b$.

(3) By axiom K_8, $(a + b) \cdot (a + b) = a + b$. Distributing and keeping in mind that the Boolean ring multiplication is idempotent, and the equation $ab + b \cdot a = 0$ due to the uniqueness of the difference, we can apply just proved property (2).

(4) $\overline{\overline{a}} = (a + e) + e = a + (e + e) = a + 0 = a$.

(5) $a + \overline{a} = a + (a + e) = (a + a) + e = 0 + e = e$.

(6) $a \cdot \overline{a} = a \cdot (a + e) = a \cdot a + a \cdot e = a + a = 0$.

(7) The inequality $a \le a$ is equivalent to the equation $a \cdot a = a$, which is the axiom K_7.

(8) Similarly, $a \le b \Leftrightarrow a \cdot b = a$ and $b \le a \Leftrightarrow b \cdot a = b$. Due to the commutativity of multiplication, $a = b$.

(9) As before, $a \le b \Rightarrow a \cdot b = a$ and $b \le c \Rightarrow b \cdot c = b$. Now,

$$a \cdot c = (a \cdot b) \cdot c = a \cdot (b \cdot c) = a \cdot b = a.$$

But the equality $a \cdot c = a$ means what we want, $a \le c$.

(10) By condition, $a \le c \Leftrightarrow a \cdot c = a$ and $b \le c \Leftrightarrow b \cdot c = b$. By adding these inequalities, we get $(a + b) \cdot c = a + b$, that is, $a + b \le c$.

(11) Since $(a \cdot bb) \cdot a = a(b \cdot a) = a \cdot (a \cdot b) = (a \cdot a) \cdot b = a \cdot b$, we get $a \cdot b \le a$, and the inequality $a \cdot b \le b$ is now obvious.

(12) By definition, the inequality $0 \le a$, that is, equivalent to the evident equation $a \cdot a = 0$.

Up to now, we have studied any Boolean rings. Hereafter, we consider only *finite* Boolean rings.

Lemma 13. *If K is a finite Boolean ring, then for every $0 \ne b \in K$, there exists an atom $a \in K$, such that $a \le b$.*

Proof. If the element b itself is an atom, it is enough to set $a = b$. Otherwise, there is an element $b_1 < b$ and $b_1 \ne 0$, that is, $0 < b_1 < b$. If b_1 is an atom, we are done; otherwise, there exists an element b_2 such that $0 < b_2 < b_1 < b$, and so on. Since the ring is finite, after finitely many steps, this chain ends, and an element b_l, where $0 < b_l < b_{l-1} < \ldots < b_1 < b$, must be an atom.

Lemma 14. *Different atoms are orthogonal, that is, if $a \neq b$, then $a \cdot b = 0$.*

Proof. By Property (11), $a \cdot b \leq a$, and $a \cdot b \leq b$. Since $a \cdot b \neq 0$, we conclude that $a \cdot b = a$ and $a \cdot b = b$, since an atom can be preceded by zero only. Therefore, $a = b$, and we get the contradiction.

We are ready to prove the major statement of this chapter.

Theorem 43. *Every finite Boolean ring K is isomorphic to the ring B_n, where the index n depends upon the ring.*

Proof. Let us arbitrarily number the atoms in K, $a_1, a_2, \ldots, a_n$, there could be only finitely many of them, and let $b \neq 0$ be any non-zero element of K. Consider the element $b' = \sum_{a_i' \leq b} a_i$, where the sum is taken over all the atoms a_i, preceding b. Moreover, $b' \leq b$. We show now that as a matter of fact, $b' = b$. Otherwise, $b' < b$, hence, there exists an element $d \neq 0$ such that $b' + d = b$. We multiply the latter by b', and since the inequality $b' \leq b$ is equivalent to equation $b' \cdot b = b'$, we get $b' + d \cdot b' = b'$, that is, $d \cdot b' = 0$. By the same property, there exists an atom $a \leq d$. We prove now that this element $a \leq b$. It is sufficient for that to check inequality $a \leq b$, that is, $d \cdot b = d$. However, $d \cdot b = d \cdot (b + d) = d \cdot b + d \cdot d = d$, therefore, $a \leq b$.

We have now $a \cdot b' = a \cdot \sum_{a_i \leq b} a_i = \sum_{a_i \leq b} (a \cdot a_i)$. Since $a \leq b$, the atom a occurs in the latter sum, say, $a = a_{i_0}$, hence $a \cdot b' = \sum_{a_i \leq b, i \neq i_0} a \cdot a_i + a \cdot a_{i_0}$. Here, all $a_i \neq a$, and due to the orthogonality of atoms, in this sum, $a \cdot a_i = 0$. At the same time, $a \cdot a_{i_0} = a \cdot a = a$. Hence, $a \cdot b' = a \neq 0$.

On the other hand, since $a = a \cdot d$, which is equivalent to the inequality $a \leq d$, we have $a \cdot b' = (a \cdot d) \cdot b' = a \cdot (d \cdot b') = a \cdot 0 = 0$. The contradiction shows that $d = 0$, therefore, $b' = b$.

For any element $b \in K$, we got a representation $b = \sum_{a_i \leq b} a_i$. If we add to that sum all the other atoms with zero coefficients, we get the sum, which contains all the atoms in the ring K,

$$b = \sum_{i=1}^{n} \mu_i \cdot a_i. \tag{13.1}$$

Thus, n in this theorem is the amount of the atoms in the ring, and is the same for every b. The coefficients μ_i here are either 0 or e. If an element b has another representation, then subtracting one from another, we will find $0 = \sum_{i=1}^{n} (\mu_i + v_i) a_i$. Multiplying this sum by $a_1, \ldots, a_n$ and using the orthogonality of the atoms, we get that all the coefficients of this sum are zeros, that is, $\mu_i = v_i$, and the presentation is unique.

To conclude the proof, we take two elements, $b = \mu_1 a_1 + \ldots + \mu_n a_n$ and $c = v_1 a_1 + \ldots + v_n a_n$. It is clear that in ring B_n the operations are done coordinate-wise, namely, $b + c = (\mu_1 + v_1) \cdot a_1 + \cdots + (\mu_n + v_n) \cdot a_n$ and $b \cdot c = (\mu_1 \cdot v_1) \cdot a_1 + \cdots + (\mu_n \cdot v_n) \cdot a_n$. Finally, we define a function

$$f(\mu) = \begin{cases} 1 & \text{if } \mu = e \\ 0 & \text{if } \mu = 0, \end{cases} \quad \text{and put into a correspondence to an element } b \in K$$

a string $(f(\mu_1), \ldots, f(\mu_n)) \in B^n$. It is clear that we now establish the isomorphism $K \sim B$, we sought for.

Problem 328. *Let X and Y be nonempty sets. Consider the set of binary relations on the Cartesian product X × Y, with the operations complementation, intersection and union. Prove that this is a Boolean algebra with respect to these operations. What are its neutral elements?*

Remark 12. *Pay attention to a similarity between the structure of the finite Boolean rings and vector spaces.*

Problem 329. *Let P_n be the family of algebraic predicates (Chapter 5) with the fixed domain $M = M_1 \times M_2 \times \ldots \times M_n$. Prove that P_n is a Boolean algebra with respect to operations $p + q = (p \wedge q) \vee (q \wedge p)$ and $p \cdot q = p \wedge q$. What are the unity, the zero, the atoms of this algebra? When is it finite?*

Problem 330. *State and prove the duality principle in Boolean algebra.*

13.3 EXERCISES

Exercise 13.1. *For any element x in a Boolean algebra, prove that $x + x = x$. Hence, in any Boolean algebra, all the coefficients in equations are equal to 1.*

Solution. It directly follows from the axioms.

Exercise 13.2. *What other Boolean operations are idempotent? Are there idempotent operations in elementary algebra?*

Exercise 13.3. *Prove that if K is a Boolean ring, then K^n is also a Boolean ring. In particular, B and B^n are Boolean rings. What are the zero and the unity in B^n?*

Exercise 13.4. *Prove that $A \subset B$ iff $A \cdot B = A$, or iff $A \cup B = B$.*

CHAPTER 14

COMBINATORIAL CIRCUITS

14.1   GRAPHS AND SCHEMES

In this chapter, we describe a relatively old, while still important and widely used application of the theory of Boolean functions. This application is based on a simple observation that many technical devices, in particular, those used in computers, can be in two states only; therefore, it may be useful to describe their mathematical models in terms of the two-valued Boolean logic. The simplest among these models are contact schemes, which are used in technical devices. We do not state the features of the schemes in detail, referring the reader, for example, to [20]. The schemes, for us, are *weighted graphs*, considered in more detail below, in Chapter 16 and the followings.

A graph is a collection of points in a plane, called the vertices of a graph or the poles of the scheme, connected to one another with *straight segments*, called the edges of the graph (scheme). The edges are marked with symbols from a certain alphabet, $x_1, \overline{x}_1, \ldots, x_n, \overline{x}_n$, not all of them should appear on the scheme. The symbols are called the contacts of the scheme; a contact x_i is called *connecting* and a contact $\overline{x}_i$ *disconnecting*. If contacts x_i are interpreted as Boolean (two-valued) variables and edges as wires connecting the contact x_i or $\overline{x}_i$, then the piece of a circuit containing x_i is connected iff $x_i = 1$, and it is disconnected iff $\overline{x}_i = 0$.

On the other hand, the part of a circuit containing $\overline{x}_i$ is connected iff $x_i = 0$, and it is disconnected iff $\overline{x}_i = 1$. Furthermore, it is obvious that the conjunction of the corresponding variables corresponds to the consecutive connection of the contacts, and the disjunction of the corresponding variables corresponds to the parallel connection of the contacts, see Figure 14.1.

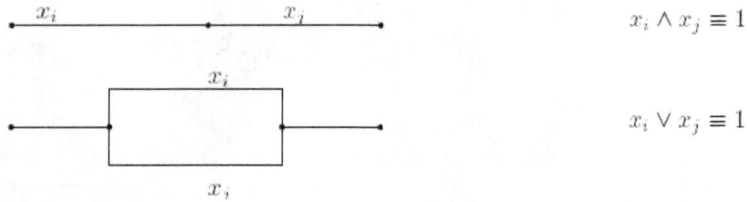

FIGURE 14.1 The upper chain is closed iff $x_i \wedge x_j \equiv 1$. The lower chain is closed iff $x_i \vee x_j \equiv 1$.

Therefore, we put a Boolean function into correspondence with every pair "in–out" of the poles of a network (or "entrance–exit"); this function is called the conductivity function of a scheme. It should be mentioned that the conductivity function is derived in its DNF. It is obvious that the converse problem, that is, given a conductivity Boolean function, to construct the corresponding scheme, is solved exactly as easy as the previous problem. If a conductivity function is constructed, we can use the well-developed theory of Boolean functions, for example, to simplify the scheme.

Problem 331. *Consider the scheme in Figure 14.2:*

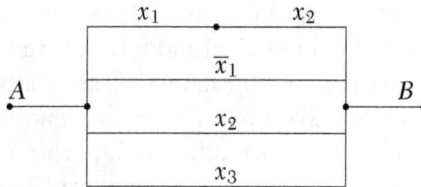

FIGURE 14.2 The original scheme for Problem 331.

The conductivity function between the points $A - B$ can be immediately written down as

$$f_{A-B} = x_1 \wedge x_2 \vee \overline{x}_1 \vee x_2 \vee x_3.$$

Since $ab \vee b = b$ and $ab \vee \overline{a} = b \vee \overline{a}$, we can simplify it as $f_{A-B} = \overline{x}_1 \vee x_2 \vee x_3$, and the scheme for this function (Figure 14.3) is equivalent to, but noticeably simpler than the original scheme in Figure 14.2.

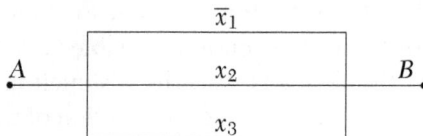

FIGURE 14.3 The reduced scheme for Problem 331.

It is equivalent to the original scheme in Figure 14.2 in the sense that their parts between the contacts $A - B$ in both circuits are always either closed or open simultaneously. Of course, we assume that the contacts, denoted with the same characters, work synchronously; they are either both open or both closed.

14.2 LOGICAL PROBLEMS

In the next well-known example, we show how Boolean functions can be used to solve logical problems. We follow the exposition of S. Epp [15, p. 56].

Problem 332. *There are two doors at the opposite ends of a long hallway, and only one electrical bulb in the middle of it. If nobody is in the hallway, the light is to be down. There is a switch by every door, that turns the bulb on or off. When nobody is in the hallway, the bulb is off, but anyone entering the hallway through either door, can turn the light on and then, while exiting through the opposite door, turn it off. Design a circuit controlling the light in the hallway.*

Solution. Introduce three Boolean variables, B for the bulb, s_1 and s_2 for the switches, and construct the truth table for the variable B as a function of s_1 and s_2. It is natural to assume that the circuit was installed so that initially there is no light and both switches are *off*, that is, initially, $B \equiv 0$ and $s_1 \equiv s_2 \equiv 0$.

If one enters the hallway through any door, say, the first one, and turns the corresponding switch s_1 *on*, the light must turn *on*, hence, $B = 1$, $s_1 = 1$, while still $s_2 = 0$. After that, exiting through another door, one has to turn the light *off* by the switch s_2, while the switch s_1 was not changed. Thus, it is now $B = 0$, $s_2 = 1$ and yet $s_1 = 1$. After that, if anyone enters through another door, she generates the string $B = 1$, $s_2 = 1$, $s_1 = 0$. Therefore, we designed the following truth table, Table 14.1, for the variable $B = B(s_1, s_2)$.

TABLE 14.1 The Truth Table for Problem 332.

s_1	s_2	B
0	0	0
0	1	1
1	0	1
1	1	0

We immediately recognize here the truth table for the binary addition, thus, $B \equiv s_1 \oplus s_2$.

14.2.1 Binary Adders

Now we design a logical circuit for adding two n-digit binary integers. Independently upon a specific technical realization, the binary adder is one of the major parts of any computer or any electrical scheme. There is no theoretical difficulty here, we just want to demonstrate again a simple approach to such a problem. Thus, we have to remind how to add binary numbers. While adding in binary system, we follow the similar procedure as with decimals, and write sometimes the subscript 2 or 10 to specify the base of the number system used.

Example 51. *Compute* $1101_2 + 111_2$.

Solution. We start by adding the two right-most digits. In the binary system, $1 + 1 = 10_2$. Therefore, we write $1 \oplus 1 = 0$, which is the digit of units in the binary sum, and *carry* the 1 one step to the left, to the binary "tens." Now, in the binary "tens", that is, in the column of 2s, one has $0 + 1 + 1$, where the second 1 is *the carry*. Since $(0 + 1 + 1)_{10} = 10_2$, the digit of the 2s is 0, and the next carry is again 1. One more step to the left, in the column of $2^2 = 4s$ in the binary system; in the decimal system, it would be the column of hundreds. Hence, this binary digit is 1 and the carry is also 1. Only the first addend contributes 1 in the next column to the left (the second addend actually gives 0, and also, we have the carry 1). The sum is 10100_2. Just for curiosity, computations can be verified in the decimal system, indeed, $1101_2 = 13_{10}$, $111_2 = 7_{10}$, $13_{10} + 7_{10} = 20_{10} = 10100_2$.

Remark 13. *The sum operation of binary numbers above is not the addition modulo 2, denoted as $a \oplus b$. In particular, if $a = (a_1, a_2, ..., a_n)$ and $b = (b_1, b_2, ..., b_n)$ are binary vectors, then the addition modulo 2 is being done termwise, that is, $a \oplus b \equiv (a_1 \oplus b_1, a_2 \oplus b_2, ..., a_n \oplus b_n)$.*

Problem 333. *Compute $a = (1,0,1,1,0) \oplus (0,0,1,1,1)$.*

Solution. $a = (1, 0, 0, 0, 1)$

The binary numbers are added exactly as the decimal integers. In a decimal place value system, the integers are represented as sums of the powers of 10, which is the base of the number system; therefore, we need 10 digits. In the binary system, any integer is written as the sum of powers of 2, and we need only two digits, 0 and 1. We use the powers of 2 written as decimal numbers, that is,

$$2^0 = 1, 2^1 = 2, 2^2 = 4, 2^3 = 8, 2^4 = 16, 2^5 = 32, 2^6 = 64, 2^7 = 128,$$
$$2^8 = 256, 2^9 = 512, 2^{10} = 1024, 2^{11} = 2048, ...$$

What is more, since the difference $1 - 1 = 0$ and the sum $1 + 1 = 10_2$, the binary difference is, in the same column, the same as binary addition, and binary multiplication is just the shift to the left. So that, to construct a circuit performing binary arithmetic, it is enough to materialize the binary adder $s_1 \oplus s_2$ given by Table 14.1. By making use of the table, we immediately construct the DNF

$$s_1 \oplus s_2 \equiv \neg s_1 \wedge s_2 \vee s_1 \wedge \neg s_2.$$

Hence, we must have circuits for the standard Boolean operations: disjunction, conjunction, and negation. Hereafter, we assume that these three elementary Boolean functions are available "in store," and only use them to design larger circuitry. They are often called logical *gates* and may have different notations.

Let us denote the addends as $x_n x_{n-1} \ldots x_2 x_1$ and $y_n y_{n-1} \ldots y_2 y_1$, their sum $z_{n+1} z_n z_{n-1} \ldots z_2 z_1$, and the *carry-ons* as $p_n, p_{n-1}, \ldots, p_2, p_1$. Hence, the binary addition of two n-digit integers can be represented by Tables 14.2–14.4.

TABLE 14.2 The Binary Adder.

	p_n	p_{n-1}	$\cdots$	p_2	p_1
$-$	$\leftarrow$	$\leftarrow$	$\cdots$	$\leftarrow 0$	$\leftarrow$
$-$	x_n	x_{n-1}	$\cdots$	x_2	x_1
$-$	y_n	y_{n-1}	$\cdots$	y_2	y_1
$-$	$-$	$-$	$\cdots$	$-$	$-$
z_{n+1}	z_n	z_{n-1}	$\cdots$	z_2	$-$

Since there are only two digits in the binary system, it is convenient to treat x_i and y_i as Boolean variables, where $z_{n+1} = p_n$, z_1 and p_1 are Boolean functions of two arguments x_1 and y_1, while $p_2, \ldots, p_n, z_2, \ldots, z_n$ are Boolean functions of three arguments, given by the truth tables 14.2–14.3.

TABLE 14.3 The Truth Table for z_1 and p_1.

x_1	y_2	z_1	p_1
0	0	0	0
0	1	1	0
1	0	1	0
1	1	0	1

We see immediately from these tables that $z_1 = x_1 \oplus y_1$, $p_1 = x_1 \wedge y_1$, $z_2 = x_2 \oplus y_2 \oplus p_1$, $p_2 = x_2 \wedge y_2 \vee x_2 \wedge p_1 \vee y_2 \wedge p_1$, $z_3 = x_3 \oplus y_3 \oplus p_2, ..., z_{n+1} = p_n$.

Problem 334. *Construct the contact schemas for $z_2, ..., z_n$ and $p_2, ..., p_n$.*

We remark that $z_2, z_3, ..., z_n$ are computed with the same Boolean function; the same is valid for $p_2, p_3, ..., p_n$. Therefore, we can define a functional element with three entrances and two exits, shown in Figure 14.4 as triangles, and use it to build an n-digit binary adder.

To end this section, we return to logical problems and puzzles.

TABLE 14.4 The Truth Table for z_i and p_i, $i \geq 2$.

x_i	y_i	p_{i-1}	z_i	p_i
0	0	0	0	0
0	0	1	1	0
0	1	0	1	0
0	1	1	0	1
1	0	0	1	0
1	0	1	0	1
1	1	0	0	1
1	1	1	1	1

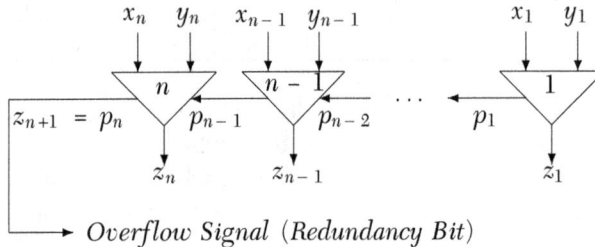

FIGURE 14.4 *n*-Digit adder.

Problem 335. *Mother, father, their son and two daughters are resting at the beach. They follow the next schedule. (1) If a father is swimming, the mother and the son follow him. (2) If the son is swimming, then the smaller sister also goes with him. (3) The older daughter and mother are in water only together.*

(4) Every morning, at least one of the parents is swimming. On Sunday, only one of the daughters went to the beach. Who else was there that Sunday?

Solution. First of all, we need good notation. Let the following symbols M, F, S, D_o, D_y stand for the propositions "The mother is on the beach," "The father is on the beach," "The son is on the beach," "The older daughter is on the beach," "The younger daughter is on the beach," respectively. In these notations, the problem can be concisely written as the following propositions, which by the problem must be all true:

1. $F \to M \wedge S$

2. $S \to D_y$

3. $M \equiv D_o$

4. $F \vee M$.

Having good notations, the solution is almost "mechanical." We transform every proposition to the CNF and make their conjunction, which must be *true*, thus

$$(\overline{F} \vee M) \wedge (\overline{F} \vee S) \wedge (\overline{S} \vee D_y) \wedge (\overline{M} \vee D_o) \wedge (M \vee D_o) \wedge (M \vee F) \equiv 1.$$

Next we transform the left-hand side to DNF. Since $D_y \wedge D_o \equiv 0$, we conclude that $M \wedge D_o \wedge \overline{F} \wedge \overline{S} \equiv 1$, so that only the mother and the older daughter were on the beach that Sunday.

Out of the endless collection of logical puzzles, we show only the following one, belonging to R. Smullyan [43, p. 19]. This is a good example to demonstrate a purely formal solution without resorting to the human language.

Problem 336. *The population of an island consists of two kinds of people, knights who always tell the truth and only the truth, and knaves, who always lie. You encounter two people, A and B. As a greeting, A says, "B is a knight," and B says, "The two of us are opposite types." Can you determine who is who?*

Solution. Let a and b be the propositions "A is a knight" and "B is a knight," respectively; each of them can be either true or false, independently on one another. Then, if A is a knight, his claim can be stated as $a \to b$, which must be true. If A is not a knight, the conditional $a \to b$ has a false premise; thus, it is true in this case as well. By the same token, the conditional $(\neg a) \to (\neg b)$ also is a tautology. Therefore, the conjunction $(a \to b) \wedge ((\neg a) \to (\neg b))$ is identically true.

Similar reasoning applies to B. The statement that A and B are of opposite types can be written as $((\neg a) \wedge b) \vee (a \wedge (\neg b))$, hence the conditional

$$b \to (((\neg a) \wedge b) \vee (a \wedge (\neg b)))$$

is a tautology. Quite similarly,

$$(\neg b) \to ((a \wedge b) \vee ((\neg a) \wedge (\neg b))) \equiv 1.$$

Whence, this puzzle claims that the following conjunction of the four conditionals is a tautology,

$$(a \to b) \wedge ((\neg a) \to (\neg b)) \wedge$$

$$\wedge(b \to (((\neg a) \wedge b) \vee (a \wedge (\neg b)))) \wedge ((\neg b) \to ((a \wedge b) \vee ((\neg a) \wedge (\neg b)))).$$

Replacing here conditionals as $a \to b \equiv \neg a \vee b$ and simplifying the conjunctions by making use of the distributive, absorption, and deletion laws, we derive the proposition

$$(\neg a) \wedge (\neg b) \equiv 1,$$

which is possible iff $\neg a \equiv \neg b \equiv 1$, that is, iff $a \equiv b \equiv 0$. Therefore, both A and B are knaves.

Problem 337. *There are five people on the committee, four regular members, m_1 - m_4 and the chairman C. The decision is taken by the majority, but the chairman can veto a positive decision. To vote, every member presses a button. Design a scheme, such that if the decision is positive, the lamp is "on." Solve the same problem for the committee of three members without the chairman.*

Solution. For the first question, an obvious solution is given by the following formula

$$(m_1 m_2 \vee m_1 m_3 \vee m_1 m_4 \vee m_2 m_3 \vee m_2 m_4 \vee m_3 m_4) \wedge C$$

where the symbols of conjunction in $m_i \wedge m_j \equiv m_i m_j$ are omitted. This Boolean function can be realized by the scheme in Figure 14.5 and can be shortened to

$$(m_1(m_2 \vee m_3 \vee m_4) \vee m_2(m_3 \vee m_4) \vee m_3 m_4)C.$$

Problem 338. *Construct a scheme for the shortened Boolean function for Problem 337.*

Problem 339. *At the exam, a student gets three "true-false" questions. Design a scheme such that a student presses a button if her answer is "true," and then the scheme shows the number of correct answers.*

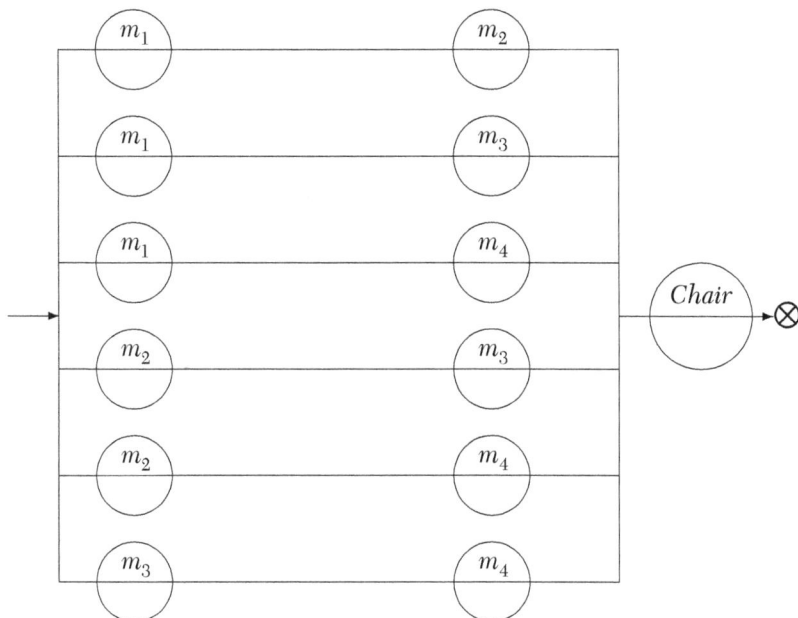

FIGURE 14.5 A scheme to Problem 337. Here ⊗ stands for the lamp.

14.3 EXERCISES

Exercise 14.1. *Perform the arithmetic operations with binary numbers:*
$(1, 0, 1, 1, 0)_2 + (0, 0, 1, 1, 1)_2; (1, 0, 1, 1, 0)_2 - (0, 0, 1, 1, 1)_2; (1, 0, 1, 1, 0)_2 \times (0, 0, 1, 1, 1)_2.$

Exercise 14.2. *Find the binary representations of the (decimal) numbers, and verify your computations by returning back from the binary numbers to the decimal system.* $0_{10} = 0_2; 1_{10} = 1_2; 2_{10} = 10_2; 3_{10} = 11_2; 4_{10} = 100_2; 5_{10} = 101_2; 6_{10} = 110_2; 7_{10} = 111_2; 8_{10} = 1000_2; 9_{10} = 1001_2; 10_{10} = 1010_2; 11_{10} = 1011_2; 78_{10} = 1001110_2.$

Exercise 14.3. *Change* $(6304)_7$ *to the decimal system.*

Exercise 14.4. *Find the base 4 expansion of the decimal number* 817_{10}.

Exercise 14.5. *Draw the schemes for the Boolean functions* $x\bar{z} \vee y\bar{t}$ *and* $x \to (y \to z)$.

Exercise 14.6. *You meet someone but do not know whether this person is a knight or a knave. How to determine this by asking only one question?*

Exercise 14.7. *In a big hall, there are four light switches, one at each of the four walls. Each switch must turn the light off if it was on, and vice versa. Construct the scheme.*

CHAPTER 15

COMPLETE SYSTEMS OF BOOLEAN FUNCTIONS AND BASES

In Chapter 9, we proved that Boolean functions could be written as a normal form. In other words, any Boolean function can be represented as a superposition of some simple functions, namely, conjunction, disjunction, and negation. A natural question is whether one can find a simpler superposition, and how to determine what system of Boolean functions can be used for that? We study these questions in this chapter.

Given two functions $f(x, y, \ldots)$ and $g(\ldots)$, their superposition, or composition, or composite function $f(g, \ldots)$ is a function derived by substitution of g instead of some of the indeterminates into f. We will soon see many examples of the superpositions.

15.1 COMPLETE SYSTEMS AND BASES

Definition 81. *A system of Boolean functions $\mathcal{M}$ is called functionally complete iff every Boolean function can be written as a superposition of finitely many functions from this system. Functions in that representation can repeat.*

Definition 82. *A (functionally) complete system is called a basis iff, after removing any function from the system, it is not complete anymore.*

Example 52. *Prove that the systems $\{\wedge, \vee, \neg\}, \{\rightarrow, \wedge, \vee, \neg\}, \{\wedge, \neg\}, \{\vee, \neg\},$ $\{\wedge, \oplus, 0, 1\}$ are complete. What about the system $\{\wedge, \vee\}$? About the system $\{\wedge\}$? Which of these systems make a basis? Make no basis?*

Definition 83. *Given a system of Boolean functions* $\mathcal{M}$ *the family of all finite superpositions of functions of the system* $\mathcal{M}$ *is called the closure of this system and is denoted as* $[M]$. *The system is complete iff its closure is the set of all the Boolean functions with any finitely many arguments.*

Thus, the system of Boolean functions $\mathcal{M}$ is complete (with respect to the superposition) if it coincides with its own closure, that is, $\mathcal{M} = [\mathcal{M}]$.

Problem 340. *Prove that* (1) $[[\mathcal{M}]] = [\mathcal{M}]$
(2) *If* $\mathcal{M} \subset \mathcal{N}$, *then* $[\mathcal{M}] \subset [\mathcal{N}]$
(3) $[\mathcal{M}] \cup [\mathcal{N}] \subset [\mathcal{M} \cup \mathcal{N}]$.

Definition 84. *A Boolean function* f *is said to preserve* 0 *iff* $f(0, \ldots, 0) \equiv 0$, *and to preserve* 1 *iff* $f(1, \ldots, 1) \equiv 1$. *We denote these two classes as* $\mathcal{M}_0$ *and* $\mathcal{M}_1$, *respectively.*

Problem 341. *How to determine whether a Boolean function* f *belongs to* $\mathcal{M}_0$ *or* $\mathcal{M}_1$ *by the truth table of* f?

Problem 342. *Prove that the cardinality* $|\mathcal{M}_0| = 2^{2^{n-1}}$. *What is the cardinality of* $\mathcal{M}_1$?

Definition 85. *Given a Boolean function* f *of* n *variables, the function*

$$f^*(x_1, \ldots, x_n) = \overline{f(\overline{x_1}, \ldots, \overline{x_n})}$$

is called the dual of f. *A Boolean function is called self-dual, iff* $f \equiv f^*$. *The class of self-dual Boolean functions is denoted as* S; *if we want to emphasize the number of variables, we will write* $S(n)$.

Problem 343. *For every Boolean function in Chapters 9 and 10, find the dual function. Which of these functions are self-dual?*

Problem 344. *Prove that if a Boolean function* f *can be written as a formula, containing the symbols of variables, two constants* 0 *and* 1, *and the symbols of operations* $\neg, \vee, \wedge$, *then to derive a formula for the dual function* f^* *it is enough to substitute* 0 *in this formula instead of* 1, 1 *instead of* 0, *the symbol* $\vee$ *instead of* $\wedge$ *and* $\vee$ *instead of* $\vee$.

Problem 345. *Prove that the Boolean function is self-dual iff for all the values of the arguments, the following equation holds good,*

$$\overline{f(x_1, \ldots, x_n)} \equiv f(\overline{x_1}, \ldots, \overline{x_n}).$$

The set of Boolean vectors B^n can be made a poset as well.

Definition 86. *A Boolean vector* $(b_1, ..., b_n)$ *precedes a vector* $(c_1, ..., c_n)$ *of the same length, iff* $b_i \le c_i$ *for all* $1 \le i \le n$.

We will denote this relation by $\prec$, thus, $(0,1) \prec (1,1)$ but $(1, 0)$ is not comparable with $(0, 1)$. This example shows that $\prec$ is a binary relation of partial but not linear order.

Problem 346. *Compare, pairwise, all the vectors from* B^3 *and* B^4. *Determine the largest, the smallest, maximal, and minimal elements, if they exist, of* B^n *with respect to the relation* $\prec$.

Definition 87. *A Boolean function* f *is called monotone iff* $x \prec y$ *implies* $f(x) \le f(y)$ *for all the Boolean vectors* x, y *in the domain of* f. *The class of monotone Boolean functions is denoted as* $M = M(n)$.

Problem 347. *Characterize the monotone functions through their truth tables.*

Problem 348. *Find the upper and lower bounds for the cardinality of* $M = M(n)$.

Problem 349. *Prove that the majority function (or quorum function)*

$$MAJ_n(x_1, x_2, ..., x_n) = \begin{cases} 1, & \sum_{i=1}^{n} x_i \ge n/2, \\ 0, & otherwise \end{cases}, n \ge 3,$$

is a monotone Boolean function. What if $n = 1$ *or* $n = 2$?

In cryptography, the following *XOR* function is often used, which for three variables is given as $f_2(x,y,z) \equiv x \oplus y \vee x \oplus z \vee y \oplus z$.

Problem 350. *Prove that the function* f_2 *is a monotone function. Find its dual function* f_2^*.

A study of other cryptographically important properties of the monotone Boolean functions would move us too far from our major topic; for that issue, we address the reader to [11].

Finally, we define linear Boolean functions.

Definition 88. *A Boolean function is called affine iff its Zhegalkin polynomial does not contain any conjunction of the variables (contains no minterm). Another way, a Boolean function is called affine iff it can be written as*

$$f(x_1, ..., x_n) \equiv a_0 \oplus a_1 x_1 \oplus a_2 x_2 \oplus \cdots \oplus a_n x_n,$$

where multiplication means the conjunction, and the coefficients a_j *are Boolean constants 0 or 1. An affine Boolean function is called linear iff in* *its representation above $a_0 \equiv 0$. The class of linear functions is denoted as $L = L(n)$.*

Problem 351. *Characterize the affine and linear Boolean functions through their truth tables.*

Problem 352. *Find the cardinality of the class of linear functions $L(n)$ and of the class of affine functions.*

Another way to determine whether a Boolean function is linear is to apply the method of undetermined coefficients, demonstrated in the following problem.

TABLE 15.1 Five Precomplete Systems of Boolean Functions.

f	T_0	T_1	S	M	L
0	$+$	$-$	$-$	$+$	$+$
1	$-$	$+$	$-$	$+$	$+$
$f(x) \equiv x$	$+$	$+$	$+$	$+$	$+$
$f(x) \equiv \neg x$	$-$	$-$	$+$	$-$	$+$
$x \wedge y$	$+$	$+$	$-$	$+$	$-$

Problem 353. *Is the disjunction $x \vee y$ an affine function? Is it a linear?*

Solution. If this function is affine, it can be written as $x \vee y \equiv a_0 \oplus a_1 x \oplus a_2 y$, where a_0, a_1, a_2 are undetermined coefficients yet. Since the latter equation must fulfill identically in B^2, it leads to four equations in three unknowns, namely,

$$\begin{cases} 0 \equiv a_0 \\ 1 \equiv a_0 \oplus a_2 \\ 1 \equiv a_0 \oplus a_1 \\ 1 \equiv a_0 \oplus a_1 \oplus a_2 \end{cases}$$

From the first three equations (equivalences) we consecutively find $a_0 \equiv 0$, $a_2 \equiv a_1 \equiv 1$; however, these values do not satisfy the last equivalence. This contradiction proves that one cannot find the coefficients a_0, a_1, a_2, a_3; hence the disjunction is not an affine function, and moreover, it is not a linear Boolean function.

Let us consider Table 15.1, where the signs + or – show that the Boolean function belongs or does not belong to the corresponding class. Thus, the negation is self-dual, since $\neg\neg\neg x \equiv \neg x$, and linear, since $\neg x \equiv x \oplus 1$.

Problem 354. *Prove that the negation function* $\neg$ *does not belong to the other precomplete classes.*

Problem 355. *What Boolean function is represented by the scheme in Figure 15.1?*

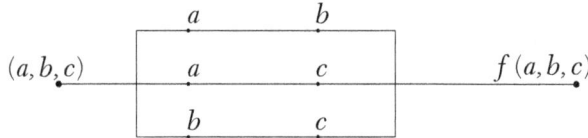

FIGURE 15.1 The scheme for Problem 355.

The five classes of Boolean functions, namely, those preserving zeros, preserving ones, self-dual, monotone, and linear functions, called *precomplete classes*, are important for the study of the complete systems of Boolean functions; this study was undertaken by Emil Leon Post (1897-1954).

Since every column of Table 15.1 contains a "-" sign, we see that none of these five classes exhausts the set of all the Boolean functions. On the other hand, there are Boolean functions, which do not belong to any of these classes. The two well-known examples of these functions are the Sheffer Stroke, denoted as $x \mid y$, and the Peirce Arrow, also called the Lukasiewicz symbol, denoted as $x \downarrow y$. These two connectives are defined in Table 15.2.

TABLE 15.2 Definitions of the Sheffer Stroke $x \mid y$ and the Peirce Arrow $x \downarrow y$.

$f(x, y)$	$(0, 0)$	$(0, 1)$	$(1, 0)$	$(1, 1)$
$x \mid y$	1	1	1	0
$x \downarrow y$	1	0	0	0

We see that the Sheffer Stroke is the negation of the conjunction $x \mid y \equiv \neg(x \wedge y)$, that is why it is also called the NAND gate, while the Peirce Arrow $x \downarrow y \equiv \neg(x \vee y)$, and that is why it is also called the NOR connective, or the NOR gate.

Problem 356. *Prove that the Sheffer Stroke, as well as the Peirce Arrow do not belong to any of the five classes T_0, T_1, S, M, L.*

Problem 357. *Prove the following equivalences:*

$$(x \mid x) \equiv (\neg x),$$
$$(x \mid x) \mid (y \mid y) \equiv (x \vee y),$$

so that any Boolean formula can be expressed through the Sheffer Stroke. Moreover, express the conditional $x \rightarrow y$ and the conjunction $x \wedge y$ through the Sheffer Stroke.

Problem 358. *Prove that any Boolean formula can be expressed through the Peirce Arrow.*

Problem 359. *Prove that a non-constant Boolean function is monotone iff it can be written as a superposition of only two functions - conjunction and disjunction.*

In Problem 352, we had to find that there are 2^{n+1} linear functions with n arguments; hence the class of linear Boolean functions is relatively small and is not of much interest for cryptography. More useful from this point of view are *partially linear Boolean functions*, which systematically appear in cryptography; they were considered, in particular, by Beale and Monaghan [3] and O'Connor [39].

An *ad hoc* definition can be done as follows. Let $f(x_1, \ldots, x_n)$ be a Boolean function, whose n indeterminates can be split into two disjoint groups $X = \{x_{i_1}, \ldots, x_{i_k}\}, |X| = k$, and the rest arguments $Y, |Y| = n - k, X \cap Y = \varnothing$. Suppose that with respect to every argument in X, f is (independently) linear, that is, there is a Boolean function g with the set of indeterminates Y and k binary constants $b_1, b_2, \ldots, b_k$, such that

$$f(x_1, \ldots, x_n) \equiv b_1 x_{i_1} \oplus b_2 x_{i_2} \oplus \cdots \oplus b_k x_{i_k} \oplus g(Y)$$

identically over all the variables $(x_1, \ldots, x_n)$. The following lemma is obvious due to our previous results and the Inclusion-Exclusion Principle.

Lemma 15. *The number of partially linear Boolean functions with n arguments, such that $k, 1 \leq i \leq n$, among them are independently linear, is*

$$\sum_{k=1}^{n} (-1)^{k-1} \binom{n}{k} 2^{k+1} \times 2^{2^{n-k}},$$

where $\binom{n}{k}$ are the binomial coefficients - see Chapter 7.

Definition 89. *A system or class of Boolean functions is closed with respect to substitution (or the superposition, which is the same), iff all possible finite superpositions of functions of the class into themselves belong to the system. A system of Boolean functions is complete if every Boolean function can be constructed as a superposition of the functions in the system.*

For instance, the system of all Boolean functions is complete, as well as the two-function systems $\{\neg, \vee\}$ and $\{\neg, \wedge\}$, while the two-function system $\{\vee, \wedge\}$ is not complete. Below we prove the criterion for a system to be complete, which was given almost a century ago by E. L. Post.

15.2 POST THEOREM

We have already proved that any Boolean function can be represented as a superposition of the three functions, $\neg, \vee, \wedge$, and we even can reduce this system to two functions, either $\neg, \vee$, or $\neg, \wedge$. However, if we try to express a Boolean function, for example, the negation as a superposition of only the conjunction and the disjunction, we fail, and the question is, why is it wrong? Is it impossible at all, that is, the system $\{\wedge, \vee\}$ is not complete, or can we just not solve this problem? This question is crucial in the production of digital and electronic schemata; indeed, what elements must be "physically" made to include in the system? The answer is given by the following statement.

Theorem 44 *(Post). A system of Boolean functions* **X** *is complete with respect to the superposition, iff the system contains a function not belonging to the class* $\mathbf{T}_0$*, a function not belonging to the class* $\mathbf{T}_1$*, a function not belonging to the class* **M**, *a function not belonging to the class* **S**, *and a function not belonging to the class of linear Boolean functions* **L**.

These five classes are called precomplete, because, as we will prove soon, they are not complete, but if one adds to either of these classes a Boolean function external to the class, the expanded class becomes complete.

For example, both the disjunction and conjunction preserve 0 and 1, therefore, the system consisting of just two functions, $\{\vee, \wedge\}$, does not satisfy the theorem. We will prove this theorem later on in this chapter after some auxiliary results.

Since we deal with functions, in this section we systematically write the equality sign = together with the equivalence ≡. To prove Post Theorem (Theorem 44), we are to study certain features of the precomplete classes.

Lemma 16. *Each of the five precomplete classes is closed with respect to the superposition.*

Proof. We must prove that the closure of each of these classes coincides with the same class. We start with T_0. Let the functions $f, f_1, \ldots, f_k \in T_0$, that is, $f(0, \ldots, 0) \equiv 0$, and the same is valid for every f_j, $1 \leq j \leq k$.

Set $F \equiv f(f_1, \ldots, f_k)$, hence,
$F(0, \ldots, 0) \equiv F(f_1(0, \ldots, 0), f_2(0, \ldots, 0), \ldots, f_k(0, \ldots, 0)) \equiv f(0, \ldots, 0) \equiv 0$.

Therefore, class T_0 is closed with respect to the superposition; computations for class T_1 are similar; hence, class T_1 is also closed.

Lemma 17. $|T_0| = |T_1| = 2^{2^n-1}$.

Proof. It is enough to remark that the truth table of any Boolean function of n arguments has 2^n rows, and in the case of these two classes, one of these rows is fixed.

Lemma 18. $|S| = 2^{2^{n-1}}$.

Proof. It is enough to note that the column of the values of a self-dual function is symmetric with respect to its middle.

Lemma 19. $|L| = 2^{n+1}$.

Proof. It is enough to note that a linear function is determined by its $n + 1$ Boolean coefficients.

To prove the Post theorem, we have to study more properties of these functions. Next, we prove that the class of self-dual Boolean functions S is closed with respect to the superposition; thus its closure is the same class S. It is enough to employ the same notation $F(f_1, \ldots, f_k)$ as above, and straightforwardly calculate the dual F^* of this Boolean function.

To prove that the class M of monotone Boolean functions is closed, we consider Boolean vectors α and β, such that $\alpha \prec \beta$, and the sets $\xi = (f_1(\alpha), \ldots, f_k(\alpha))$ and $\eta = (f_1(\beta), \ldots, f_k(\beta))$. Since $\alpha \prec \beta$ and $f_i \in M$, $1 \leq i \leq k$, then $f_i(\alpha) \leq f_i(\beta)$, that is, every component of vector ξ does not exceed the corresponding component of vector η. Therefore, $\xi \prec \eta$. Now, since $f \in M$, this implies that $F(\alpha) = f(\xi) \prec f(\eta) = F(\beta)$.

Problem 360. *Prove the duality principle for Boolean functions, namely, prove that if*

$$F(x_1, \ldots, x_n) = f(f_1(x_1, \ldots, x_n), \ldots, f_n(x_1, \ldots, x_n)),$$

then

$$F^*(x_1, \ldots, x_n) = f^*(f_1^*(x_1, \ldots, x_n), \ldots, f_n^*(x_1, \ldots, x_n)).$$

Finally, to prove that the class of linear Boolean functions is closed, we straightforwardly compute a superposition of linear functions, considering that the conjunction is distributive with regard to the binary addition, and conclude that a superposition of linear functions is a linear function, that is, the class L of linear functions is closed.

Lemma 20. *(On a non-self-dual function). A constant can be deduced as a superposition of an arbitrary non-self-dual function, any argument, and its negation.*

Proof. Let $f \notin S$. Then there exists a vector $\alpha = (\alpha_1, \ldots, \alpha_n)$ such that $f(\alpha) = f(\bar{\alpha})$, where $\bar{\alpha} = \bar{\alpha}_1, \ldots, \bar{\alpha}_n$. Define a Boolean function $\phi(x) = f(x^\alpha)$, where x^α is a Boolean power defined earlier. By the definition, $f(\alpha) = f(\bar{\alpha})$, or $\phi(0) = \phi(1)$, that is, ϕ is a constant Boolean function.

Problem 361. *What constant, 0 or 1, will be deducted from a given Boolean function $f \notin S$?*

Problem 362. *Deduce both constants 0 and 1, if this is possible, from the conjunction and from the disjunction. Can we determine in advance which constant will be generated by the algorithm in the lemma?*

Lemma 21. *(On a nonmonotone function). The negation can be represented as a superposition of an arbitrary nonmonotone function and constants. More specifically, given any nonmonotone function f, the constants, and an argument x, one can get its negation $\neg x$, that is, the simplest and the only one nonmonotone function of one variable.*

Proof. Let $f \notin M$. That means that there are two Boolean tuples $\alpha \prec \beta$, such that $f(\alpha) > f(\beta)$. Hence, $\alpha \neq \beta$, that is, the set of subscripts where $\alpha_{i_j} \neq \beta_{i_j}$ is nonempty. In turn, that implies the equations $\alpha_{i_1} = \cdots = \alpha_{i_k} = 0$ while $\beta_{i_1} = \cdots = \beta_{i_k} = 1$.

Now we design the Boolean vectors $\tilde{\alpha} \neq \tilde{\beta}$, and $\tilde{\alpha} \prec \tilde{\beta}$, which are neighbors just in one coordinate, and such that $f(\tilde{\alpha}) > f(\tilde{\beta})$. To this end, we compare the vectors $\alpha = (\alpha_1, \ldots, \alpha_n)$ and $\alpha^1 = (\alpha_1, \ldots, \alpha_{i_1-1}, 1, \alpha_{i_1+1}, \ldots, \alpha_n)$, which are the neighbors at the coordinate i_1. We remind that $\alpha_{i_1} = 0$. If $f(\alpha^{(1)}) = 0$, we choose $\tilde{\alpha} = \alpha$, $\tilde{\beta} = \alpha^1$, and the claim is proven, since $f(\alpha) = 1$.

On the other hand, if $f(\alpha^{(1)}) = 1$, we let

$$\alpha^{(2)} = (\alpha_1, \ldots, \alpha_{i_1-1}, 1, \alpha_{i_1+1}, \ldots, \alpha_{i_2-1}, 1, \alpha_{i_2+1}, \ldots, \alpha_n,$$

which was derived from $\alpha^{1)}$ by changing i_2 – component from 0 to 1, exactly as $\alpha^{1)}$ was derived from α by changing the component α_{i_1} from 0 to 1. The construction implies that $\alpha^1 \prec \alpha^2, \alpha^{(1)} \neq \alpha^{(2)}$, and these two vectors are neighbors at the coordinate i_2. If $f(\alpha^{(2)}) = 0$, then the proof is again done, since we can set $\alpha = \alpha^1$ and $\beta = \alpha^{(2)}$, and $f(\alpha^1) = 1$. If, however, $f(\alpha^2) = 1$, we proceed as before, that is, we construct the vector $\alpha^{(3)}$, and so on. Since $\alpha^{(k)} = \beta$, while $f(\beta) = 0$, in no more than $k-1$ steps, we arrive at the vectors $\tilde{\alpha}$ and $\tilde{\beta}$, we sought for; they are different while neighboring in i – coordinate, $\tilde{\alpha} \prec \tilde{\beta}$, $f(\tilde{\alpha}) = 1$ and $f(\tilde{\beta}) = 0$.

To complete the proof of the lemma, we set $\tilde{\alpha} = (\alpha_1, ..., \alpha_{i-1}, 0, \alpha_{i+1}, ..., \alpha_n)$, then $\tilde{\beta} = (\alpha_1, ..., \alpha_{i-1}, 1, \alpha_{i+1}, ..., \alpha_n)$, and consider a Boolean function of one variable x, $\phi(x) = (\alpha_1, ..., \alpha_{i-1}, x, \alpha_{i+1}, ..., \alpha_n)$. We have $\phi(0) = f(\tilde{\alpha}) = 1$, while $\phi(1) = 0$, meaning $\phi(0) > \phi(1)$, therefore, $\phi(x) = \neg x$.

So far, we considered 4 out of 5 precomplete classes, and we have to study the linear/nonlinear functions. It should be noted that all the four Boolean functions of one variable are *linear* functions, since in particular, $\neg x \equiv x \oplus 1$. Thus, we have to consider a nonlinear function with at least two variables; the conjunction $\wedge$ will do; we also fix, without any loss of generality, the variables x and y.

Lemma 22. *(On a nonlinear function) The conjunction $x \wedge y$ can be deduced as a superposition of an arbitrary nonlinear Boolean function with the constants 0, 1, and negations of these arguments. In other words, from any nonlinear function with at least two arguments, substituting instead of its arguments their negations, the two constants, and negating, it this is necessary, the function itself, one can get the conjunction of the two selected variables.*

Proof. Let f be written as a logical polynomial. Since $f \notin L$, it must contain at least one conjunction of the arguments; renumbering the variables, if necessary, we can assume that f contains $x \wedge y \equiv xy$. The polynomial f can be split into four groups of monomials:

1. containing $x \wedge y$

2. containing x but not containing y

3. containing y but not containing x

4. containing neither x nor y.

The last three groups can be empty, but the first one contains at least one monomial. Factoring x, y, and xy, we get the polynomial $f(x_1, x_2, ..., x_n) = xy f_{1,2}(x_3, ..., x_n) \oplus x f_1 \oplus y f_2 \oplus f_0$, where f_0, f_1, f_2 and $f_{1,2}$ are

Boolean polynomials, which do not depend upon x and y; moreover, $f_{1,2} \neq 0$, thus, there is a tuple $\omega = (x_3^0, \ldots, x_n^0)$ such that $f_{1,2}(\omega) = 1$.

Denote $f_1(\omega) = \alpha$, $f_2(\omega) = \beta$, $f_0(\omega) = \gamma$, and consider a Boolean function ψ of two variables $\psi(x,y) = f(x,y,\omega) = xy \oplus \alpha x \oplus \beta y \oplus \gamma$, which is the conjunction, we sought for, and maybe, a linear term. To eliminate the latter, we introduce one more Boolean function $\phi(x,y) = \psi(x \oplus \beta, y \oplus \alpha) \oplus \alpha \wedge \beta \oplus \gamma$. After simple algebra, we get $\phi(x,y) = x \wedge y$. Let us note also that $x \oplus \beta = x$ if $\beta = 0$, and $x \oplus \beta = \bar{x}$ if $\beta = 1$, Finally, if $\alpha\beta \oplus \gamma = 1$, then $\phi \oplus \alpha\beta \oplus \gamma = \neg\phi = \neg(x \wedge y)$.

We finish this section by proving the criterion of the completeness of a system of Boolean functions.

Theorem 45. *In order for a system of Boolean functions $\mathcal{M}$ to be complete, it is necessary and sufficient that the system does not belong entirely to neither of the five precomplete classes T_0, T_1, S, M, L. In other words, the system must contain Boolean functions $f_0 \notin T_0, f_1 \notin T_1, f_m \notin M, f_s \notin S, f_l \notin L$.*

These do not have to be five different Boolean functions, some of them or even all of them can coincide. For instance, in the system $\mathcal{M} = \{\neg, \wedge\}$ we have $f_0 = f_1 = f_m = \neg$ and $f_s = f_l = \wedge$.

Proof of Necessity. We have to prove that if the system $\mathcal{M}$ is complete, it cannot be a subset of any of the five precomplete classes. On the contrary, let us assume that $\mathcal{M} \subset T_0$. Then, as we showed, its closure $[\mathcal{M}] \subset [T_0]$. But $[T_0] = T_0$ due to the closeness of any precomplete class. Since we assume that the system $\mathcal{M}$ is complete, its closure is the set of all the Boolean functions; thus, the inclusion $\mathcal{M} \subset T_0$ means that any Boolean function preserves zero, which is a contradiction. The reasonings in the case of the other four classes are exactly the same.

Proof of Sufficiency. We have to prove that the given system $\mathcal{M}$ is complete. To this end, it is enough to show that just two Boolean functions, the negation, and the conjunction, can be represented as superpositions of the functions of the system Hence, one has to find only two formulas,

$$\neg x = f_{neg}(\ldots, x, \ldots) \tag{15.1}$$

$$x \wedge y = f_{con\,j}(\ldots, x\ldots, y, \ldots), \tag{15.2}$$

where all the functions that appeared in these formulas, are taken from the system $\mathcal{M}$. In the right part of (15.1), there is only one argument x, and in the right part of (15.2), there are only two arguments x and y. Indeed, if we have (15.1) and (15.2), we can prove the completeness of system $\mathcal{M}$ as follows. Let f be an arbitrary Boolean function. We represent it as a DNF and exclude from the latter all disjunctions by making use of the property $p \lor q \equiv \neg(\overline{p} \land \overline{q})$. Then we replace every negation in the derived formula by formula (15.1) and every conjunction by formula (15.2), since these formulas contain only the functions from the system $\mathcal{M}$; thus, we arrive at the representation of an arbitrary Boolean function f as a superposition of the functions from the system $\mathcal{M}$.

Finally, we take up formulas (15.1)-(15.2), and first, consider the negation. By the statement, there exists a Boolean function $f_0 \in \mathcal{M} \setminus T_0$. We identify all the arguments of f_0, that is, consider a function of only one argument $\phi(x) = f_0(x, ..., x)$. By condition, $\phi(0) = 1$, but $\phi(1)$ can be either 0 or 1, and we consider these two cases separately. Let $\phi(1) = 1$, then the function ϕ is a constant, $\phi(0) = \phi(1) = 1$. Next, we use $f_1 \in \mathcal{M} \setminus T_1$ and make substitutions $\psi(x) = f_1(\phi(x), ..., \phi(x)) = f_1(f_0(x, ..., x), ..., f_0(x, ..., x))$. Since $\phi(x) = 1$, we have $\psi(x) = f_1(1, ..., 1) = 0$. Thus, we get both constants as superpositions of the functions of our system. Now considering Lemma 21 on nonmonotone function, by a superposition of this nonmonotone function f_m and the constants, that is, functions f_0 and f_1, we can derive the negation, that is, we construct formula (15.1).

Consider another case, $\phi(1) = 0$. Since $\phi(0) = 1$, the function $\phi(x) = \neg x$. Now by Lemma 21 we can construct from ϕ (actually, from f_0 and $f_s \in \mathcal{M} \setminus S$) a constant function. The other constant is given by a superposition $\phi(\psi) = \neg \psi$.

In both cases, we derived the two constants and the negation in our system. To deduce formula (15.2), that is, to get the conjunction, it is enough to employ function $f_l \in \mathcal{M} \setminus L$ and use Lemma 22 on nonlinear function.

Remark 14. *It is worth emphasizing, that all the proofs in this chapter are constructive, that is, every proof contains an algorithm, explaining what substitutions must be done to prove the claim.*

15.2.1 Bases

If we analyze the proof of the sufficiency of Theorem 45, we observe that in the first case, when $\phi(1) = 1$, we have used four functions of the system (f_0, f_1, f_m, f_l) out of five, and in the second case, when $\phi(1) = 0$, we have used only three functions out of five. Therefore, a complete system of more than four functions cannot be a basis. Leaving out all the functions of the system, not used in the proof, we still have a complete system. This proves the first part of the concluding theorem of this chapter.

Theorem 46. *A basis cannot contain more than four Boolean functions. This number cannot be reduced, since there are bases of one, of two, of three, and of four Boolean functions.*

Proof. To complete the proof, we have to show the appropriate examples of bases. We have checked already (see Table 15.2) that any of the Sheffer Stroke and of the Peirce Arrow does not belong to either of the five precomplete classes; therefore, every of these two functions makes a basis. We have also proved that the system $\{\neg, \wedge\}$ is a basis. The system $\{1, \oplus, \wedge\}$ is an example of a basis of three functions, and a system $\{0, 1, \vee, x \oplus y \oplus z\}$ is an example of a basis of four functions.

15.3 EXERCISES

Exercise 15.1. *Consider 16 propositions in Table 2.3. For each of the Boolean functions equivalent to these propositions determine to what Post classes these functions belong or do not belong. Apply the Lemmas 20-22 on nonmonotone function, or non-self-dual function, or nonlinear function to any of these 16 functions, if possible, and determine what Boolean functions are produced.*

Exercise 15.2. *Prove that $\{\neg, \vee\}, \{1, \oplus, \wedge\}, \{0, 1, \vee, x \oplus y \oplus z\}$ are bases in the set of all Boolean functions. Can we exclude any function from these systems?*

Exercise 15.3. *Describe the truth tables of the self-dual functions.*

Exercise 15.4. *Prove that another majority function $f_1(x, y, z) \equiv x \wedge y \vee x \wedge z \vee y \wedge z$ is a monotone function. Find its dual function f_1^*.*

Exercise 15.5. *Give an example of a linear and non-self-dual Boolean function. What linear Boolean functions are self-dual?*

16

INTRODUCTORY GRAPH THEORY, EULER'S FORMULA, AND UNBREAKABLE CIPHERS

16.1 GRAPHS AND DIAGRAMS

Both the kids and adults like drawing doodles, and those *dessin d'infant* give sometimes a valuable information, which can be translated into more rigorous terms. For instance, an observation of the Koenigsberg bridges led Leonhard Euler in 1736 to the notion and theory of *Eulerian graphs*, which are useful both in pure mathematics and in applications, and will be discussed later in this chapter. Below we study initial notions of the *Graph Theory*. Graphs can be defined in two different, but to some extent, equivalent ways, similarly to the profound concept of the Riemann surface.

Definition 90. *A geometric graph or a diagram $G(V, E)$ is a pair of two sets satisfying an incidence relation: a non-empty set $V \neq \varnothing$ of points (on paper or on the blackboard) called vertices of the graph, and a set of smooth arcs E connecting some pairs of vertices; these arcs are called edges of the graph. Each edge has two end-points, which belong to V; these end-points can merge into one, in which case the edge is called a loop. An actual shape of an edge is immaterial. The set of edges can be empty, in which case every vertex is isolated. If a vertex v is an endpoint of an edge e, these vertex and edge are called incident to each other.*

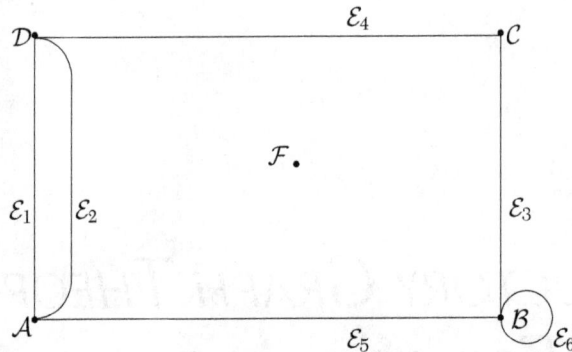

FIGURE 16.1 Graph $G(V, E)$ has the set of vertices $V = \{A,B,C,D,F\}$ and the set of edges $E = \{\mathcal{E}_1,\mathcal{E}_2,\mathcal{E}_3,\mathcal{E}_4,\mathcal{E}_5,\mathcal{E}_6\}$, among them two parallel edges $\mathcal{E}_1$ and $\mathcal{E}_2$ connecting the same pair of vertices A and D, and a loop $\mathcal{E}_6$ at vertex B. Vertex F is called isolated, since it is incident to no edge.

The number of vertices, $p = |V|$, is called the order of the graph, and the number of edges $q = |E|$ is the size of the graph. For instance, the graph $G(V, E)$ (Figure 16.1) has five vertices and six edges, thus, it is of order 5 and of size 6; vertex A is incident to three edges $\mathcal{E}_1$, $\mathcal{E}_2$, and $\mathcal{E}_5$. In turn, these three edges are incident to the vertex A. Two vertices are called *adjacent* if they are the endpoints of the same edge.

Thus, in Figure 16.1 the vertices A and B are adjacent vertices but the vertices A and C are not. For a vertex v, the total number of its incident edges is called the degree $deg(v)$ of this vertex: in Figure 16.1, $deg(A) = 3$, $deg(B) = 4$, $deg(C) = 2$, $deg(D) = 3$, $deg(F) = 0$; keep in mind that a loop is counted twice, it has *two* endpoints. The incidence relation i can be formalized as the following map: $i(\mathcal{E}_1) = i(\mathcal{E}_2) = \{A,D\}$, $i(\mathcal{E}_3) = \{B,C\}$, $i(\mathcal{E}_4) = \{C,D\}$, $i(\mathcal{E}_5) = \{A,B\}$, $i(\mathcal{E}_6) = \{B\}$.

The two endpoints of any edge are connected by that edge. Two vertices $v_1, v_2 \in V$ can also be connected with a chain of intermittent edges and vertices, such that v_1 and v_2 are the endpoints of the chain. Depending upon whether the chain is closed or open, whether it contains repeating edges or vertices, there are many specialized terms: ways, trails, circles, routes, paths, loops, etc., see, for example, [28, Chap. 2]. In this short introduction, we always write *path* or *loop*, adding if necessary, more clarification.

Two endpoints of any edge are symmetric, this is the same edge since the points coincide, $e_1 = e_2$. When we have to distinguish these two edges, as having *different directions*, we must consider *directed* graphs or digraphs; we will do that in Chapter 17. The premise that a loop has two end-points, allows us to state and prove the following claim.

Lemma 23. *In any graph, the sum of the degrees of all the vertices is even. If a graph has the size of q, then*

$$\sum_{v \in V} deg(v) = 2q. \tag{16.1}$$

Proof. We use in this proof a *double counting*, which means that we count the same quantity twice, in two different ways, and compare the results. This is a common device in combinatorics. In this case, we count twice the total number of the endpoints in a graph. On the one hand, we just sum up all the endpoints in the graph. On the other hand, we observe for the same quantity, that every edge, including loops, has two ends.

Remark 15. *This statement is called* The Handshaking Lemma. *Indeed, if we imagine participants of a party, draw each of them as a point, that is, a vertex of a graph, and connect the two vertices iff the corresponding partici-pants exchange a handshake, we will have exactly the corresponding graph where every hand appears twice. For example, the graph in Figure 16.1 has q = 6 vertices and the sequence of degrees is 3 + 3 + 2 + 4 + 0 = 12.*

Corollary 6. *In any graph, the number of odd vertices is even.*
    The same graph can be drawn in many ways, which may look differently, but exhibit the same incident relationship among the elements of a set V. We reiterate, that not the shape of the edges, but the connectivity of the vertices is important. We associate a few matrices with every graph, which can reveal its certain important features.

Definition 91. *Let G(V, E) be a graph of order $p = |V|$ and of size $q = |E|$. The $p \times p$ square matrix A(G) is called the adjacency matrix of G iff every its element p_{ij} is equal to the number of edges connecting the vertices v_i and v_j.*

    A $p \times q$ matrix B(G) is called the incidence matrix of G iff the element b_{ij} is 2 iff e_{ij} is a loop at the vertex v_i, $b_{ij} = 1$ iff e_{ij} is an edge with an endpoint at v_i, and $b_{ij} = 0$ iff the vertex v_i is not an endpoint at the edge e_j.

Example 53. *Thus, the elements of these matrices are natural numbers. For the graph G in Figure 16.1, the adjacency matrix is the square matrix of order 5,*

$$A(G) = \begin{pmatrix} 0 & 1 & 0 & 2 & 0 \\ 1 & 2 & 1 & 0 & 0 \\ 0 & 1 & 0 & 1 & 0 \\ 2 & 0 & 1 & 0 & 0 \\ 0 & 0 & 0 & 0 & 0 \end{pmatrix}$$

and the incidence matrix is 5 × 6 matrix

$$B(G) = \begin{pmatrix} 1 & 1 & 0 & 0 & 1 & 0 \\ 0 & 0 & 1 & 0 & 1 & 2 \\ 0 & 0 & 1 & 1 & 0 & 0 \\ 1 & 1 & 0 & 1 & 0 & 0 \\ 0 & 0 & 0 & 0 & 0 & 0 \end{pmatrix}.$$

The adjacency matrix is symmetric, because the graph is undirected, the 2s in the first row and in the first column mean that the vertices v_1 and v_4 are connected with two parallel edges, and the 2 in the main diagonal indicates the loop at the vertex v_2; we count a loop as having two ends, the agreement that is convenient in many problems.

The incidence matrix $B(G)$ shows what edges are incident with what vertex. The graph matrices give a suitable device for storing graphs in computer memory. We consider more applications of these matrices in Chapter 17.

Problem 363. *Does there exist a simple graph with 13 vertices of degree 3 each?*

Problem 364. *Find the numbers of vertices, edges, and the degree of every vertex in the graph G in Figure 14.2. Identify, whether it is a simple graph, a multigraph, or a pseudo-graph. Find the sub-graphs, its isolated and pendant vertices. Find the adjacency and the incidence matrices of G. Is G a bipartite graph?*

Problem 365. *Given the adjacency matrices*

$$A = \begin{pmatrix} 0 & 1 & 0 & 1 \\ 1 & 0 & 0 & 0 \\ 0 & 0 & 0 & 1 \\ 1 & 0 & 1 & 0 \end{pmatrix}$$

and

$$B = \begin{pmatrix} 0 & 1 & 1 & 0 \\ 1 & 0 & 0 & 1 \\ 1 & 0 & 0 & 1 \\ 0 & 1 & 1 & 0 \end{pmatrix}$$

of two simple graphs; are these graphs isomorphic? Is either of them bipartite?

16.2 CONNECTIVITY IN GRAPHS

Graph theory provides an appropriate language for formalizing the idea of connectivity of different objects.

Definition 92. *Two vertices v_1 and v_2 of a graph are connected iff there is an intermittent sequence of edges, separated by vertices, which starts with v_1 and ends with v_2, whose ordering is irrelevant. If any two vertices of a graph are connected, the graph is called connected. Otherwise, the graph is called disconnected. If every two vertices of a graph are the end-points of an edge in this graph, and there are no parallel edges or loops (the condition we usually assume) the graph is called complete and is denoted as K_n, where n is the order of the graph, that is, the number of its vertices. Thus, we can imagine K_n as an n–gon such that every two vertices are connected with one and only one segment, either a side or a diagonal. A connected subgraph of a disconnected graph is called its connected component. An edge of a graph is called the bridge, iff the removal of that edge increases the number of the connected components.*

Problem 366. *If two vertices of a graph are connected, do they have to be adjacent? What about the converse implication?*

Problem 367. *Sketch complete graphs K_n for n = 1, 2, 3, 4, 5, 6.*

Definition 93. *An abstract graph is a triple G = (V, E, f), where $V \neq \varnothing$ is a non-empty set, whose elements are called vertices of the graph G, E is a set (maybe empty), whose elements are called edges, and $f : E \to V \cup C(V, 2)$ is an incidence function; hence the image f(e) of every edge contains either one or two vertices. If $f(e) \in C(V, 2), e \in E$, the vertices $v_1, v_2, v_1 \neq v_2$, are called the end-points of the edge e. If $f(e) \in V, e \in E$, the edge e is called a loop. Similarly to geometric graphs, we often denote an abstract graph by G = (V, E) without mentioning explicitly its incidence function. A graph is called finite, iff both sets V and E are finite. Two different edges $e_1 \neq e_2 \in E$ are called parallel, if they have the same pair of end-points.*

Remark 16. *Any geometric graph can be thought of as an abstract graph. In this case, the sets V and E consist of usual geometric objects - points and smooth arcs.*

Vice versa, any abstract graph can be visualized in different ways, but some of them cannot be regularly embedded in the plane. We study this issue later on, while considering planar graphs. Therefore, from now on we mostly say "graph" and specify whether it is a diagram or an abstract graph only if we must distinguish them.

Example 54. *The graphs in Figures 16.1–16.4 have no bridges, while every edge in the directed graph in Figure 16.5 is a bridge.*

Remark 17. *In many problems, we must consider directed graphs – see, for example, the next chapter, but for now all the graphs considered are non-directed.*

Problem 368. *Prove that if the degree of every vertex of a finite graph is at least two, then the graph contains a cycle, and moreover, a simple cycle.*

Problem 369. *A county has several townships. Any two of them are connected either by a road or by a trail, but not both. From every township there originate two roads and three trails. How many are there townships, roads, and trails in the county?*

16.2.1 Unbreakable Ciphers

In Chapter 8, we discussed certain basic notions of ciphering. All of those simple ciphers could be more or less easily deciphered if one has enough ciphered text for analysis. The natural question has arisen whether does exist an "unbreakable cipher," which could not be deciphered, whatever time and material resources are available? The answer is positive, indeed, such a cipher does exist, and we design it now.

As the alphabet, we use 26 Latin characters, 10 decimal digits, and a few symbols for the punctuation marks. Thus, we assume that the working alphabet has $64 = 2^6$ characters, that is, we have to cipher texts consisting of 6–digit binary words. To cipher such a text T, we specify an auxiliary text M, called a *mask*, also consisting of the same 6–digit binary characters. This new text does not have to be meaningful, its only goal is to mask the text T, which we want to transmit. Here, we apply the standard binary addition $\oplus$ without the carries, that is, $0 \oplus 0 = 1 \oplus 1 = 0$ and $0 \oplus 1 = 1 \oplus 0 = 1$. Several words are added by inserting these characters position-wise without carries.

Problem 370. *For any words x and y of 6–digit characters, prove the following properties of the binary addition:*

$$x \oplus y = y \oplus x \text{ and } x \oplus x = 0, \text{ so that } x \oplus y \oplus y = x.$$

Now let us use the binary addition $\oplus$ to add, character-wise, two texts T and M, both consisting of 6–digit binary characters. Denote the sum $T \oplus M = S$. Here T is the transmitted text, and the text M is used to cover, to mask the text T. The crucial observation is that if we know only the sum (the received text) S, but do not know the mask M and the sent text T, we cannot recover the latter. However, if we know the received text S and the mask M, we can straightforwardly recover the sent text T just by adding the mask again to the result, since by Problem 370,

$$S \oplus M = T \oplus M \oplus M = T.$$

However, if the mask were used repeatedly, the problem of deciphering becomes much simpler. Indeed, if we apply the same mask twice and both results would be eavesdropped, then we can write the result as a system of two equations, $T_1 \oplus M = A$ and $T_2 \oplus M = B$, where both A and B are known binary numbers. This leaves very little space for guesses, and even small computer experiments can recover the mask and all the other unknowns.

In this section, we study graph-traversing problems. The next question should remind an old puzzle to the reader.

Problem 371. *Can you draw either of the two graphs in Figure 16.2 without traversing any edge twice and without interrupting the drawing (that is, a pencil must not leave the paper)?*

Definition 94. *A closed path (circuit) in a graph is called Eulerian iff it contains every edge of the graph exactly once. A graph is called Eulerian if it contains an Eulerian circuit. A graph is called semi-Eulerian iff it contains an Eulerian trail.*

It is obvious that to be Eulerian or semi-Eulerian, the graph must be connected. Therefore, this necessary condition is included in all the statements in this section by default and we do not necessarily repeat it.

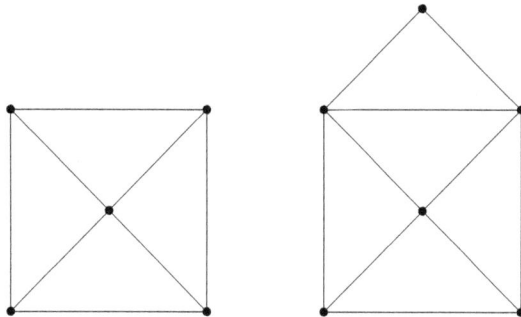

FIGURE 16.2 Is any of these "envelopes" (closed and open) Eulerian? semi-Eulerian?

The results of this section essentially depend on the *parity* of the vertex degrees of a graph, that is, whether the degree of a vertex is even or odd. We call a vertex even (odd) if its degree is even (odd).

Problem 372. *Is there a graph with just one odd vertex? One even vertex?*

Problem 373. *Is any of the graphs in Figure 16.3 – 16.6 Eulerian? Semi-Eulerian?*

Theorem 47. *A connected graph is Eulerian iff it has only even vertices. A connected graph is semi-Eulerian iff it contains exactly two odd vertices.*

Proof. The necessity of these conditions, including the connectedness, is obvious. Indeed, if we begin to traverse an Eulerian circuit and remove behind ourselves every edge traversed, after passing through any vertex its degree decreases by 2, so that the parity of any vertex's degree does not change. After completing the route, we must arrive at the initial vertex after traversing and removing behind ourselves every edge. Thus, the degree of each vertex gradually reduces to zero in steps of two, so that initially, the degree was even. In the case of semi-Eulerian graphs, the same argument works if we begin at either one of the two odd vertices; we must finally arrive at another odd vertex.

The sufficiency will be now proved by induction on the size q of the graph. Begin again with the Eulerian case. If $q = 1$, the graph consists of one vertex with an attached loop, so the statement is obvious. Suppose now that for all connected graphs of the size $|E| = q, q \geq 2$, with all the even vertices, the statement is correct, and consider a connected graph $G = (V, E)$ with $|E| = q + 1$. By Problem 231, G contains a cycle C. If this cycle includes all the edges of G, then there is nothing more to prove. Otherwise, we remove all edges of the cycle C from G, which can result in decomposing G into several connected components $G_1, \dots, G_l$, of smaller sizes.

Since G is connected, every its component G_i contains some vertex $v_i \in C$, whose degree in G must be at least 4; indeed, the cycle C gives two sub-components, and another two subcomponents, which cannot belong to the cycle C, are given by the connected component, attached to this vertex v_i. Therefore, the degree of v_i in G_i is at least 2. By the inductive assumption, each of G_i, $1 \leq i \leq l$, has an Eulerian circuit C_i, and we conclude that $v_i \in C_i$ as well. It is now obvious how to assemble all cycles $C, C_1, \dots, C_l$ back in an Eulerian cycle in graph G and finish the proof in the case of Eulerian graphs.

To consider the case of semi-Eulerian graphs, we connect the two existing odd vertices by an additional edge, thus making the graph Eulerian, and apply the statement we have just proved.

Problem 374. *(Veblen) Prove a generalization of Theorem 47 onto not necessarily connected graphs: A set of edges of a graph can be partitioned into cycles iff every vertex of the graph has an even degree. [6, p. 5]*

Problem 375. *Draw the floor plans of the buildings on your campus, where you are or were taking classes. Draw a graph representing each room with a vertex, such that two vertices are connected by an edge if the corresponding rooms have a common wall. Are these graphs Eulerian? Semi-Eulerian? Find, if any, semi-Eulerian trails or Eulerian cycles in these graphs.*

Problem 376. *Prove that the following procedure, called Fleury's algorithm, returns an Eulerian circuit in any Eulerian graph.*

Fleury's algorithm. *Start at any vertex and pass any edge incident to this vertex. Remove the edge passed and go through any other edge incident to the vertex reached, subject to the only restriction: a bridge can be used only if there is no other edge available.*

Problem 377. *Apply Fleury's algorithm to those graphs in the figures above, which are Eulerian or semi-Eulerian, and find semi-Eulerian trails or Eulerian circuits in those graphs.*

We see that the edge transversal problem allows for an effective and relatively simple solution – Fleury's algorithm. To consider vertex traversal problems, that in general, are more complicated, we introduce Hamiltonian graphs. This definition and the two following exercises contain some properties of the Hamiltonian graphs. More details about the Hamiltonian graphs can be found, for example, in [10, p. 167].

Definition 95. *A path (circuit) without repeating vertices in a graph G is called Hamiltonian iff it contains every vertex $v \in V$ of G. A graph is called Hamiltonian if it has a Hamiltonian path.*

We mention, without a precise statement, that the problem of finding a Hamiltonian cycle of the smallest possible weight is equivalent to another classical problem called the *Traveling Salesperson Problem*.

Problem 378. *Prove the following* necessary *condition of the existence of a Hamiltonian circuit: If a graph G contains a Hamiltonian circuit, then it contains a connected spanning subgraph H, which has the equal order and size,* *such that the degree of every vertex of H is 2. Here, "spanning" means that H contains all the vertices of G. Spanning trees are studied in more detail in Chapter 17.*

In the opposite direction, prove the following sufficient condition: If in a simple graph G = (V, E) of order p ≥ 3,

$$deg(v) \geq p/2, \forall v \in V,$$

then G has a Hamiltonian circuit.

Problem 379. *Find a Hamiltonian circuit in the complete graph K_5.*

Problem 380. *Is any of the graphs in Figure 16.1-16.4 Hamiltonian?*

The existence of Hamiltonian paths is related to many other problems, in particular, to the standard Gray code and the Wallet problem. More specifically, the coding is any change of the representation of information and will be studied in more detail in the following chapters. For instance, the Morse code is the traditional way to transmit alpha-numerical information. Here the secrecy is not necessarily an issue; we may do that just for convenience, for example, for reliability, that is, for the ability to resist disruptions. We say now a few words about the Gray codes, relevant to the Graph Theory. For an excellent detailed survey of that topic, see, for example, C. Savage survey [44]. As always, we begin with a motivating problem.

Problem 381. *You have bills of $1, $2, $5, $10, and $20 in your wallet, one bill of each denomination, and you want to pay your expenditures as much as possible with these bills, but you can spend no more than M = 29 dollars. What bills should you use?*

Solution. Since there are bills of $k = 5$ denominations, we can describe the problem by making use of *Boolean* vectors $b_1 b_2 \ldots b_5$ of the length 5, where $b_i = 1$ if and only if the ith bill was used in the payment; thus, if the payment includes the $5 bill, which is the third in the list, it must be $b_3 = 1$, otherwise, $b_3 = 0$. For example, vector (00101) shows that the payment was made by making use of $5 and $20 bills; the total payment was $5 + $20 = $25; this amount is called the *weight* of the payment.

Now we list the binary 5–vectors and the corresponding weights; say, the third row $(02) \sim 3 : (0,0,0,1,1) = 10 + 20 = 30$ of the array below is marked by (02), since the numbering started with zero, but if we start counting with a natural 1, it is the third row. After that, the row means that this payment includes the 4th and 5th bills, that is, $10, and $20, and the sum is

10 + 20 = \$30. However, we list them not in the standard increasing order of the natural numbers, but in another order, called the standard Gray code.

It must be said here that a Gray code is not a computer programing code, nor is it a coding of certain words by making use of a different alphabet. Gray code is an *arrangement* of several Boolean words in a special order, such that any word can be derived from its neighbor by changing exactly one symbol, that is, either 0 to 1, or vice versa, 1 to 0 (i.e., the Hamming distance, see Definition 118, between any two neighboring Boolean words, including the very first and the very last row, is exactly one).

First we show the code, that is, the arrangement (the ordering) of Boolean 5–vectors in the case when $r = 5$, of $2^5 = 32$ Boolean 5–vectors or code words. After that the algorithm will be described. All $2^5 = 32$ code words are listed below.

$(00) \sim 0 : (0,0,0,0,0) = \$0$

$(01) \sim 1 : (0,0,0,0,1) = \$20$

$(02) \sim 3 : (0,0,0,1,1) = 10 + 20 = \$30$

$(03) \sim 2 : (0,0,0,1,0) = \$10$

$(04) \sim 6 : (0,0,1,1,0) = 5 + 10 = \$15$

$(05) \sim 7 : (0,0,1,1,1) = 5 + 10 + 20 = \$35$

$(06) \sim 5 : (0,0,1,0,1) = 5 + 20 = \$25$

$(07) \sim 4 : (0,0,1,0,0) = \$5$

$(08) \sim 12 : (0,1,1,0,0) = 2 + 5 = \$7$

$(09) \sim 13 : (0,1,1,0,1) = 2 + 5 + 20 = \$27$

$(10) \sim 15 : (0,1,1,1,1) = 2 + 5 + 10 + 20 = \$37$

$(11) \sim 14 : (0,1,1,1,0) = 2 + 5 + 10 = \$17$

$(12) \sim 10 : (0,1,0,1,0) = 2 + 10 = \$12$

$(13) \sim 11 : (0,1,0,1,1) = 2 + 10 + 20 = \$32$

$(14) \sim 9 : (0,1,0,0,1) = 2 + 20 = \$22$

$(15) \sim 8 : (0,1,0,0,0) = \$2$

$(16) \sim 24 : (1,1,0,0,0) = 1 + 2 = \$3$

$(17) \sim 25 : (1,1,0,0,1) = 1 + 2 + 20 = \$23$

$(18) \sim 27 : (1,1,0,1,1) = 1+2+10+20 = \$33$

$(19) \sim 26 : (1,1,0,1,0) = 1+2+10 = \$13$

$(20) \sim 30 : (1,1,1,1,0) = 1+2+5+10 = \$18$

$(21) \sim 31 : (1,1,1,1,1) = 1+2+5+10+20 = \$38$

$(22) \sim 29 : (1,1,1,0,1) = 1+2+5+20 = \$28$

$(23) \sim 28 : (1,1,1,0,0) = 1+2+5 = \$8$

$(24) \sim 20 : (1,0,1,0,0) = 1+5 = \$6$

$(25) \sim 21 : (1,0,1,0,1) = 1+5+20 = \$26$

$(26) \sim 23 : (1,0,1,1,1) = 1+5+10+20 = \$36$

$(27) \sim 22 : (1,0,1,1,0) = 1+5+10 = \$16$

$(28) \sim 18 : (1,0,0,1,0) = 1+10 = \$11$

$(29) \sim 19 : (1,0,0,1,1) = 1+10+20 = \$31$

$(30) \sim 17 : (1,0,0,0,1) = 1+20 = \$21$

$(31) \sim 16 : (1,0,0,0,0) = \$1$

Slightly repeating ourselves, we explain what is the standard Gray code and the algorithm behind the chart. The standard Gray code works with words in the alphabet $B = \{0, 1\}$. It starts with the word B, but writes it as a column $\begin{matrix} 0 \\ 1 \end{matrix}$. Next let us imagine a horizontal mirror line below the bottom symbol in this column, and reflect the column in this mirror, to get a twice bigger column $\begin{matrix} 0 \\ 1 \\ 1 \\ 0 \end{matrix}$. Moreover, we attach a column of the same size on the left, but in this new column the upper half is filled with zeros, and the bottom half with ones. Thus, we get the chart $\begin{matrix} 0 & 0 \\ 0 & 1 \\ 1 & 1 \\ 1 & 0 \end{matrix}$. At the next step, we do in the same fashion and derive the chart of the Boolean words of the length 3,

$$\begin{matrix} 0 & 0 & 0 \\ 0 & 0 & 1 \\ 0 & 1 & 1 \\ 0 & 1 & 0 \\ 1 & 1 & 0 \\ 1 & 1 & 1 \\ 1 & 0 & 1 \\ 1 & 0 & 0 \end{matrix}$$

$\begin{matrix} 0 & 1 & 0 \\ 1 & 1 & 0 \end{matrix}$, etc. This procedure explains why the algorithm is also called *the*

Reflected Binary Code. After two more reflexions, we derive a 32×5 table of the Boolean 5–vectors, shown above. This is the standard Gray code representing all the Boolean vectors of length 5. The right-most column in the chart represents the weight (payment) of that row. Since the sum cannot exceed \$29, the row $(22) \sim 29 : (1,1,1,0,1) = 1 + 2 + 5 + 20 = \$28$ contains the answer. Namely, the payment must consist of the bills of \$20, \$5, \$2, and \$1, totaling \$28; in the problem no payment gives exactly \$29. We also remark that every two Boolean words in the code, including the first one and the last line, differ at exactly one place, which is an essential feature of any Gray code.

Problem 382. *List the binary vectors of length 4 and 6, by making use of the Standard Gray Code.*

Problem 383. *Find all Gray codes for binary vectors of length three starting with the vector* (0, 0, 0).

Problem 384. *Which truth tables in Lecture 2 are developed by using the Gray codes?*

The Wallet problem, which is similar to the Knapsack problem[1][49], shown above, can be stated as follows: Given the denominations and the quantity of bills in your wallet, find the largest amount you can spend, subject to specified restrictions.

It is convenient to represent the information we have as binary vectors. The binary vectors are listed in certain order, called the standard Gray code. This code represents information with Boolean words of length k. These issues are demonstrated by the problem below, where $k = 5$.

Problem 385. *Solve the Wallet problem, using the multiple checks of \$25, \$50, \$100, and \$200, and the threshold of M = \$500.*

---

[1] For a cryptographic application of the Knapsack problem see, for example, [45], Chapter 15.

238 • DISCRETE MATHEMATICS WITH CRYPTOGRAPHIC APPLICATIONS

Problem 386. *The Gray codes are also useful for constructing the truth tables of Boolean Functions. Of course, the order of vectors in the table differs from the first algorithm. For example, let us represent a Boolean function* $f(x,y,z) = x + \overline{y} + \overline{z}$ *by the table Gray code.*

Solution. The regular truth table for a function or a proposition with three arguments has $2^3 = 8$ rows, see Table 16.1,

TABLE 16.1 Truth Table for a Boolean Function of Three Variables $f(x_1, x_2, x_3)$.

x_1	x_2	x_3	$f(x_1, x_2, x_3)$
0	0	0	0
0	0	1	1
0	1	0	1
0	1	1	0
1	0	0	1
1	0	1	0
1	1	0	0
1	1	1	1

However, this table is not generated by a Gray code, since to get from the second to the third row, one must change two symbols. The table generated by a Gray code, is Table 16.2. The reader can directly verify that to move from any row to the next one, including from the last to the firs one, one must change exactly one symbol.

TABLE 16.2 Truth Table for a Boolean Function of Three Variables $f(x_1, x_2, x_3) = x + \overline{y} + \overline{z}$.

x_1	x_2	x_3	$x_1 + \neg x_2 + \neg x_3$
0	0	0	0
0	0	1	1
0	1	1	0
0	1	0	1
1	1	0	0
1	1	1	1
1	0	1	0
1	0	0	1

We saw in the preceding chapter, that it is necessary in many problems to distinguish a graph, as an abstract object, and a drawing, that is, its geometric realization. These drawings are called embeddings of a graph and consist of points drawn in a plane, or more generally, points in $\mathbb{R}^n, n \geq 2$. Remind that $C(V, 2)$ denotes the set of all 2-element subsets of a set V. For instance, if V is a three-element set $V = \{a, b, c\}$, then $C(V, 2) = \{\{a, b\}, \{a, c\}, \{b, c\}\}$.

Definition 96. *A (geometric) graph $G(V, E)$ is said to be regularly embedded in $\mathbb{R}^n$, iff its edges have no common points except, maybe, at the vertices, that is, at the end-points of its edges.*

Definition 97. *A graph is called planar, iff it can be regularly embedded in $\mathbb{R}^2$, that is, its edges can have common points only at the vertices, or to put that another way, iff two edges cannot intersect at a point, which is not a vertex of the graph.*

Thus, the graph $K_{3,3}$, Figure 16.4, p. 242, is not planar, see Corollary 9 on p. 241. On the other hand, consider the complete graph K_4, Figure 16.3. The left diagram is not regularly embedded in $\mathbb{R}^2$, while the right diagram embedded in $\mathbb{R}^2$ regularly; hence, the graph is clearly planar. Here, the pairs of the diagonals $v_1 - v_4$ and $v_2 - v_3$ intersect at an interior point, but this point is not a vertex of the graph. The latter can be redrawn as that on the right, where the right-most point is not a vertex, which is clearly planar.

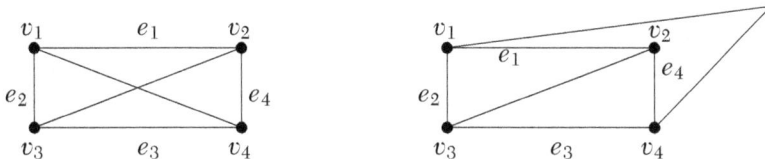

FIGURE 16.3 Graph K_4.

It is worth repeating that if a graph is regularly embedded in $\mathbb{R}^2$, its edges cannot have a common point except for, maybe, a vertex.

Planar graphs are important not only in entertainment. When engineers invented printed design, that is, schemes printed on insulated plates with colored curves, which could conduct electrical current, they immediately arrived at a mathematical problem: namely, how to draw a diagram without any intersection of edges, that is, the colored curves? Whence, to avoid short circuits, they need planar graphs. And as it often happens in history, the planar graphs in mathematics were considered before any industrial applications.

Let G be a graph with v vertices and d edges. Of course, we suppose that G was drawn on a black board or on a white paper. Its edges split the plane into several *bounded* domains $D_1, D_2, \ldots, D_{f-1}$ and an *external* domain D_f, which we consider as a vicinity of the point at infinity. These domains are called the *faces* of G, so that G has v vertices, d edges, and f faces. We repeat that the boundary of every face consists of several whole edges separated by the vertices of the graph. The following is called Euler's formula for polyhedrons.

Theorem 48. *If G is connected planar graph, then* $v + f - d = 2$.

Proof. Before doing a formal proof, we discuss some heuristics. Suppose we add an edge to the graph. An edge has two ends-vertices. If we attach this edge to an existing vertex, another end of the edge gives an extra vertex for the graph; thus, the difference $v - d$ does not change, since both v and d are increased by 1. If we attach this edge by making use of both its ends, we create a new face, and again the difference $f - d$ remains the same. We conclude that the expression $\kappa(G) = v + f - d$ should be constant. The expression $\kappa(G)$ is called the *Euler* (1707-1783) or Euler-Poincare (1854-1912) characteristic of graph G. Now we prove by induction over g that under the conditions of the theorem, $\kappa(G) = 2$.

Let the graph contain one edge only, $d = 1$. Hence, it has only two vertices - its ends, it cannot have isolated vertices due to the connectedness, thus, $v = 2$. What is more, there is only one face – the entire plane cut along the edge, hence $f = 1$, and we directly verify $v + f - d = 2$, that is, $2 + 1 - 1 = 2$.

Now we can repeat our heuristic reasoning at the beginning of the proof. Given a $G(v, f, d)$ graph, let us add to it another edge d'; thus, the enlarged graph has $d + 1$ edges. If we attach the new edge to G with only one end, we do not change the number of faces and increase by one both v and d; thus, the linear combination $v + f - d$ does not change and remains equal to the same 2. If we attach both ends d' to the existing vertices, then the new edge must split an existing face into two; hence, both f and d increase by one, but their difference does not change. In any case, the characteristic remains $v + f - d = 2$, as before.

For example, if G is a tree with v vertices, then $d = v - 1$ and $f = 1$, therefore, $v - (v - 1) + 1 = 2$ in agreement with Theorem 48.

Corollary 7. (**Euler formula for polyhedrons**). *For every convex polyhedron, $V + F - E = 2$, where E, V, F are the numbers of edges, vertices, and faces of the polyhedron, respectively.*

Sketch of the Proof. Fix any interior point of any face F, and a point, which is close enough to the face but is outside of the polyhedron. Then project the polyhedron onto the plane of the face F. The projection is connected planar graph, and we can apply Theorem 48.

Corollary 8. *If the initial convex polyhedron is regular, that is, all its faces are regular polygons, and all dihedral angles between faces are equal, then we conclude that there are only five regular polyhedra, called the Platonic solids, known from the ancient times: tetrahedron ($E = 6$, $F = 4$, $V = 4$), cube ($E = 12$, $F = 6$, $V = 8$), octahedron ($E = 12$, $F = 8$, $V = 6$), dodecahedron ($E = 30$, $F = 12$, $V = 20$), and icosahedron.($E = 30$, $F = 20$, $V = 12$).*

Corollary 9. *Graph $K_{3,3}$ is not planar.*

Problem 387. *Is there a planar graph with 6 vertices, each of them of degree 3? Of degree 4? Of degree 5?*

Problem 388. *Give an example of a graph, whose Euler characteristic is not 2. On what surface is it located?*

Corollary 10. *If G is a simple connected planar graph with $v \geq 3$ vertices and d edges, then $d \leq 3v - 6$.*

Proof. A simple graph cannot have parallel edges; hence every face of this graph is bounded with at least three edges. Moreover, every edge connects two faces, so that, $3f \leq 2v$. Combining this inequality with Theorem 48, we derive the corollary.

Problem 389. *Give another proof of Theorem 48, now by making use of the induction with respect to the number of faces f.*

Lemma 24. *Prove that the complete graph K_5 is not planar. Complete graphs K_n were defined at the previous chapter.*

Proof. Corollary 10 immediately leads to contradiction, since for K_5, $v = 5$ and $d = 10$.

16.2.2 Bipartite Graphs

A graph $G(V, D)$ is called *bipartite* if the set of its vertices is the union $V = V_1 \cup V_2$, where $V_1 \cap V_2 = \varnothing$, and no two vertices within the same component V_i. $i = 1, 2$, are connected with an edge. A bipartite graph is called complete and denoted $K_{m,n}$, where $m = |V_1|$ and $n = |V_2|$ $p1$, iff each vertex in V_1 is connected with all of the vertices in V_2, and vice versa, each vertex in V_2 is connected with all of the vertices in V_1.

Problem 390. *The complete bipartite graph* $K_{3,3}$ *gives rise to a known puzzle, included by L. Carroll in his book about Alice's adventures. The statement of Corollary 9 or Lemma 25 means that one cannot draw Figure 16.4 in plane without having at least one intersection of the edges.*

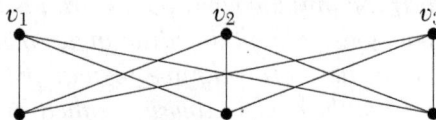

FIGURE 16.4 Graph $K_{3,3}$.

Problem 391. *Prove that any cycle in a bipartite graph contains an even number of edges.*

Lemma 25. *The graph* $K_{3,3}$ *is not planar.*

Proof. In Corollary 10, we proved the inequality $3f \le 2d$. Now, since every cycle contains at least 4 edges by Problem 391, this inequality can be strengthened to $4f \le 2d$. This inequality, together with the major inequality for planar graphs and Corollary 9, gives $d \le 2v - 4$ for every bipartite graph. In $K_{3,3}$, $v = 6$ and $d = 9$, which is contradiction if $K_{3,3}$ were planar.

We considered the graphs K_5 and $K_{3,3}$ in some details because all the non-planarity in our world boils down, in a sense, to these two graphs.

Theorem 49. *A graph is planar iff it does not contain a subgraph, which can be constructed from either* K_5 *or* $K_{3,3}$ *by subdividing some of its edges by inserting additional vertices[2].*

A proof of this theorem can be found in many places, for example, in a beautiful book [6].

Problem 392. *Give an example of a bipartite graph, which shows that the converse of the statement of previous problem is false. For what m and n, the complete bipartite graphs* $K_{m,n}$ *are trees?*

The following is a variant of the ancient puzzle about Wolf, Goat, and Cabbage.

Problem 393. *Three students, C, G, and W, came to the Instructor office to have make-up tests. The Instructor knows that the students C and G, as well*

---

[2] This theorem was proved by K. Kuratowski, and independently, by L. Pontryagin, O. Frink, K. Menger, and P. A. Smith.

*as G and W, if left alone, cannot be trusted because of cheating, while C and W
can. So that, if the Instructor has to leave the office, she will take a student(s)
with her. Describe the logistics of the uncompromised make-up test.*

Solution. We depict the initial state of affairs as $(C,G,W,I-\varnothing)$, meaning
that all four people are outside the office, which is empty. Since C and W
can be trusted, at the first step, the Instructor takes the student G into the
office, then gives the test to the student and goes out, leaving the student in
there. Thus, we get the first two edges (Figure 16.5) of the future graph of
the CGW – puzzle.

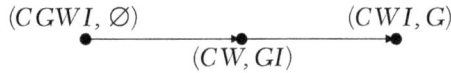

$$(CGWI, \varnothing) \xrightarrow{\quad (CW, GI) \quad} (CWI, G)$$

FIGURE 16.5 Graph of the CGW – puzzle. First two steps.

The next steps are clear from Figure 16.6, which shows that the puzzle
has two solutions.

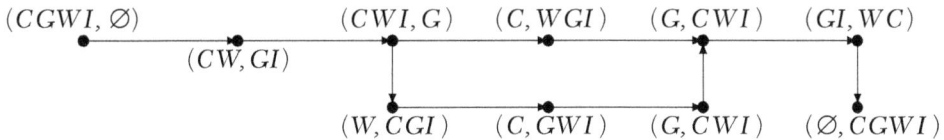

FIGURE 16.6 Two solutions of the CGW – puzzle.

We see that one solution goes straight from vertex (CWI,G) to (C, WGI),
while another makes detour from (CWI,G) to (W,CGI). In general, these
two paths have different weights.

16.3 EXERCISES

Exercise 16.1. *Draw two more diagrams representing the same incidence
relation as that represented in Figure 16.1.*

Exercise 16.2. *Does there exist a graph with three vertices of degree 5, two
vertices of degree 4, one vertex of degree 1, and a total of 11 edges?*

Exercise 16.3. *Prove that a complete graph K_n has $n\,(n-1)/2$ edges. What
is the degree of any vertex of K_n?*

Exercise 16.4. *Draw a planar graph, which is (connectivity) equivalent to the graph in Figure 16.3.*

Exercise 16.5. *Prove that any diagram can be regularly embedded in* $\mathbb{R}^3$.

Exercise 16.6. *A connected planar graph has 6 vertices; among them, there are 5 vertices of degree 3 and a vertex of degree 1. In how many regions does the graph divide the plane? Draw the diagram.*

Exercise 16.7. *Sketch a graph* $K_{2,3}$ *and prove that it is a planar graph.*

Exercise 16.8. *Prove that the complete bipartite graph* $K_{m,n}$ *has* $m \cdot n$ *edges.*

17

TREES AND DIGRAPHS

Trees are special kinds of graphs.

Definition 98. *An acyclic graph, i.e., a graph without cycles, is called a* forest. *A connected forest is called a* tree. *A rooted* tree *is a tree, which has a singled out vertex, called the root of the tree.*

Thus, a forest is a family of trees, and a tree is a connected graph without cycles.

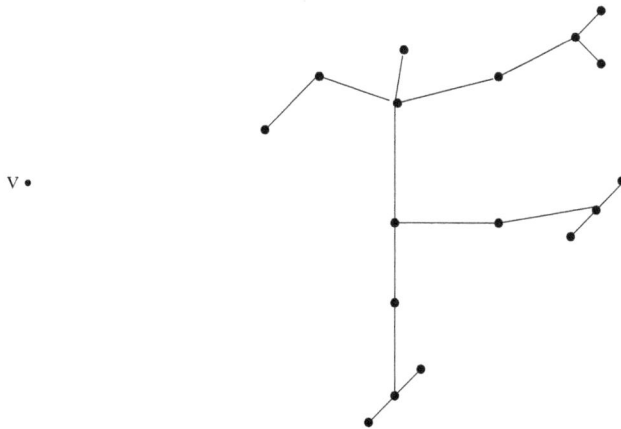

FIGURE 17.1 An example of a forest consisting of two trees. The isolated vertex *v* is a tree with no edge.

The following theorem gives several equivalent properties of the trees.

Theorem 50. *Let $G = (V, E)$ be a finite graph of order $|V| = p$. Then the following statements are equivalent.*

1. *G is a tree.*

2. *G is a connected graph, and each of its edges is a bridge.*

3. *G is an acyclic graph, and its size is* $|E| = p - 1$.

4. *G is a connected graph and* $|E| = p - 1$.

5. *For any pair of vertices of G there is a unique path connecting them.*

6. *G is acyclic but any new edge added to G generates precisely one cycle.*

Proof. We follow R. Wilson [53] and establish the following chain of implications,

$$1) \Rightarrow 2); 1) \& 2) \Rightarrow 3) \Rightarrow 4) \Rightarrow 5) \Rightarrow 6) \Rightarrow 1).$$

First, we prove that condition 1) implies 2). Since a tree is connected by definition, it suffices to prove that every edge is a bridge. On the contrary, if we assume that some edge $e \in E$ is not a bridge, then we can remove it and still get a connected graph G'. Hence the end-vertices of e are connected in G' by a walk, which cannot contain e. Now, if we return the edge e into the latter walk, we would generate a cycle in G, which is impossible since G is a tree and cannot contain cycles.

Next, we prove that 1) and 2) together imply 3). Indeed, if we remove any edge from a cycle, the remaining graph is still connected, implying that G must be acyclic. To prove that G has $p - 1$ edges, we apply mathematical induction on $p = |V|$. If $p = 1$, this single vertex must be isolated and the assertion is trivial. Suppose the assertion is valid for all trees of the order less than some $p > 1$ and consider a graph $G = (V, E)$ of order p. Let $e \in E$, then e is a bridge by the assumption. Thus, if we remove e and denote the remaining graph by G' and the number of connecting components of G by $cc(G)$, then $cc(G') = cc(G) + 1 = 2$. Let $G_1 = (V_1, E_1)$ and $G_2 = (V_2, E_2)$ be two connected components of G'. They cannot be empty, cannot have cycles, and by the inductive assumption, $|E_1| = |V_1| - 1$, $|E_2| = |V_2| - 1$. Adding up these equations and keeping in mind that $E = E_1 \cup E_2 \cup \{e'\}$, $V = V_1 \cup V_2$, where both unions are disjoint, we arrive at a conclusion.

To prove that statement 3) implies 4), we again assume the contrary, i.e., that G is not connected. Thus, it consists of $k \geq 2$ connected components. Each of these components is a tree, and by the assumption, for each of them $|E'| = |V'| - 1$. Adding up $k \geq 2$ such equalities leads to $|E| = |V| - k = p - k < p - 1$, which contradicts the premise.

The implication 4) $\Rightarrow$ 5) follows readily if we notice that if there are two vertices connected by two different paths, then these paths together make up a cycle. Removing any edge e off this cycle, we get a *connected* graph $G'(V, E')$, where $E' = E\backslash\{e\}$, such that $|V| = p$ and $|E'| = p - 2$; contradiction.

Next, we prove the implication 5) ⇒ 6). If we can add an edge and generate two cycles, this would mean that the end-vertices of the new edge were connected by two paths in the original graph, which is impossible.

Finally, to prove that 6) ⇒ 1), we have to prove that the graph is connected, which is obvious; indeed, a new cycle connects any two of its vertices twice; thus, one connection must have existed before we added the edge.

To study other properties of the trees, we compare the graph in Figure 17.1 and the tree $T = (\{v_1, v_2, v_3, v_4\}, \{e_1, e_2, e_5\}, f_T)$, see Figure 17.2.

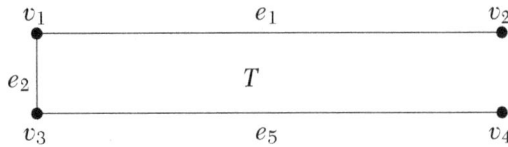

FIGURE 17.2 Tree T has four vertices and three edges.

Problem 394. *Write down explicitly the incidence function* f_T *of graph* T *in Figure 17.2.*

Solution. From Figure 17.2 we observe that $f_T(e_1) = \{v_1, v_2\}, f_T(e_2) = \{v_1, v_3\}$, and $f_T(e_5) = \{v_3, v_4\}$.

17.1 SPANNING TREES

The tree T in Figure 17.2 is a connected spanning subgraph of the complete graph K_4 in Figure 16.3 – it contains all its vertices and some of its edges.

Definition 99. *If a spanning graph of a graph G is a tree, this tree is called a* spanning tree *of G.*

Thus, the tree T (Figure 17.2) is a spanning tree of graph K_4. A graph may have several spanning trees. The next claim follows easily from Theorem 50.

Corollary 11. *Every graph has a spanning forest. Every connected graph has a spanning tree.*

It is often useful to supply edges of a graph with an additional piece of information, often called the *weight* of this edge. The weight can be a number like the length of an edge or a symbol like a traffic sign indicating the height of the overpass. If every edge of a graph carries weight, the graph is called *weighted*. Weighted graphs have *weighted spanning trees*. It is important for many applications to find the minimum spanning tree.

Example 55. *Consider a connected weighted graph G (Figure 17.3), where the weights are $w_1 = 2$, $w_2 = 5$, $w_3 = 1$, $w_4 = 3$. This graph has three different spanning trees shown in Figure 17.4. These trees have different weights, namely $W(T_1) = w_2 + w_3 + w_4 = 9$, $W(T_2) = w_1 + w_3 + w_4 = 6$, and $W(T_3) = w_1 + w_2 + w_4 = 10$; among them, the tree T_2 has the smallest weight – it is the minimum spanning tree of the graph G.*

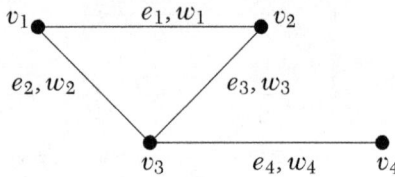

FIGURE 17.3 Graph G.

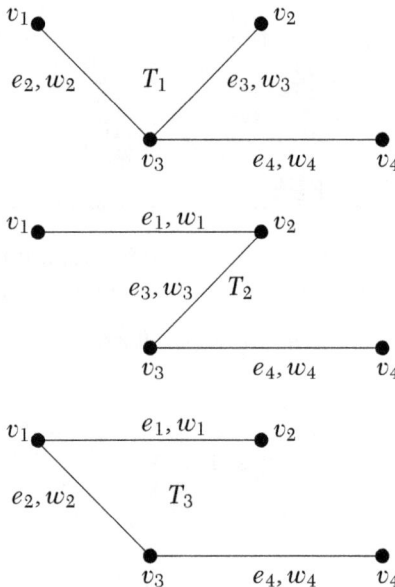

FIGURE 17.4 Spanning trees T_1, T_2, T_3.

There are several algorithms for finding a minimum spanning tree in a graph. We present the well-known algorithm of Kruskal. The connectedness of a graph is, obviously, a necessary condition for the existence of a spanning tree.

Kruskal's Algorithm for Finding a Minimum Spanning Tree

Given a connected weighted graph $G = G(V, E)$ with n vertices, find its minimum spanning tree. We assume that all weights are nonnegative numbers.

1. Select an edge e with the smallest weight. If the graph has several edges with the same minimum weight, we can choose any of them: the edge e and its end-vertices form the initial subgraph (subtree) T_1 of G.

2. Construct the spanning trees $T_2, \dots, T_m$. Specifically, for $m = 1, 2, \dots, n-2$, select an unused edge with the smallest weight, such that this edge does not make a cycle with the previously selected edges. In particular, we can use an edge with the same weight as the one in the previous step. Append the edge chosen and, if necessary, its end-vertices to the subgraph T_m generated at the previous step to built the next subgraph T_{m+1}.

3. Repeat Step 2 $n-2$ times. The subtree T_{n-1}, where n is the order of the given graph G, is a minimum spanning tree of G.

We demonstrate Kruskal's Algorithm using the next example, see Problem 397.

TABLE 17.1 The Weights of the Complete Graph K_8 in Problem 397.

	c_1	c_2	c_3	c_4	c_5	c_6	c_7	c_8
c_1	0	5	10	7	22	27	25	13
c_2	5	0	8	12	28	23	17	6
c_3	10	8	0	1	9	19	3	26
c_4	7	12	1	0	4	14	2	21
c_5	22	28	9	4	0	11	16	18
c_6	27	23	19	14	11	0	15	20
c_7	25	17	3	2	16	15	0	24
c_8	13	6	26	21	18	20	24	0

Remark 18. *Not every graph among $T_2, \dots, T_{n-2}$ is a tree, some of them can be forests, but T_{n-1} is a tree.*

Problem 395. *Modify the algorithm if some weights are negative.*

Problem 396. *Prove that Kruskal's algorithm generates a minimum spanning tree in any connected graph.*

Problem 397. *Find a minimum spanning tree for the complete graph K_8, where the weights of the edges are given in the symmetric Table 17.1.*

Solution. In this problem, the vertices are denoted as c_i, c_j, etc.; the (i, j) – entry of the table is the weight $w_{i,j} = w_{j,i}$ of the edge incident to the vertices c_i and c_j. The reader can notice that the weights are all the integer numbers from 1 through 28 inclusive.

The following figures exhibit all the consecutive steps of Kruskal's algorithm being applied to Problem 397. The smallest weight is $w_{3,4} = 1$; thus, we start with the graph with eight isolated vertices $c_1, \ldots, c_8$ (Figure 17.5) and first connect the vertices c_3 and c_4 by an edge of weight 1 (Figure 17.6). The second smallest weight is $d_{4,7} = 2$. Adding an edge of weight 2 connecting the vertices c_4 and c_7, we get the graph shown in Figure 17.7.

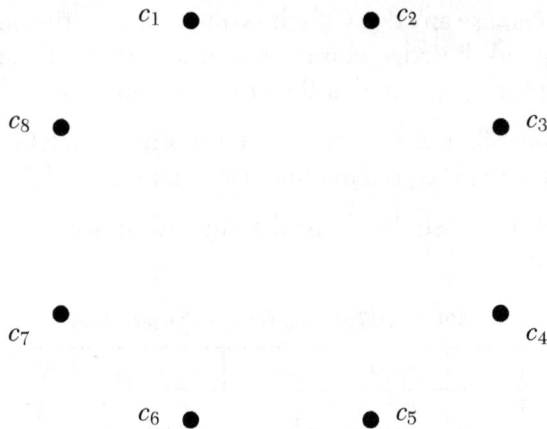

FIGURE 17.5 The initial graph without edges. All vertices are isolated.

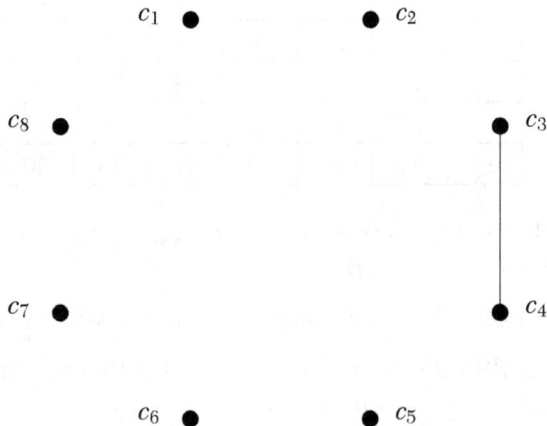

FIGURE 17.6 First step of Kruskal's algorithm.
The first (non-spanning) subgraph with only one edge is formed.

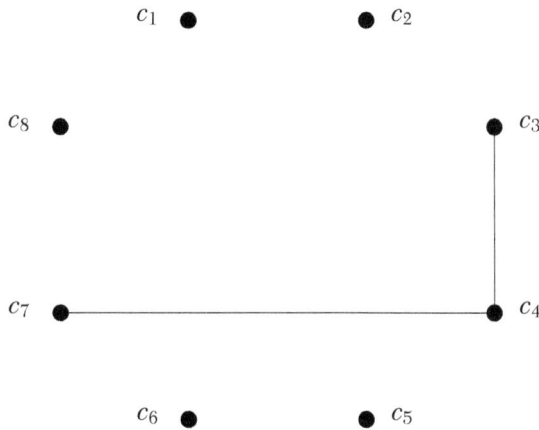

FIGURE 17.7 A subtree with two edges.

The next smallest weight is $d_{3,7} = 3$. However, we cannot connect the vertices c_3 and c_7, because such an edge would form a cycle with the two previously included edges (Figure 17.8), which is forbidden by Part 2 of the algorithm.

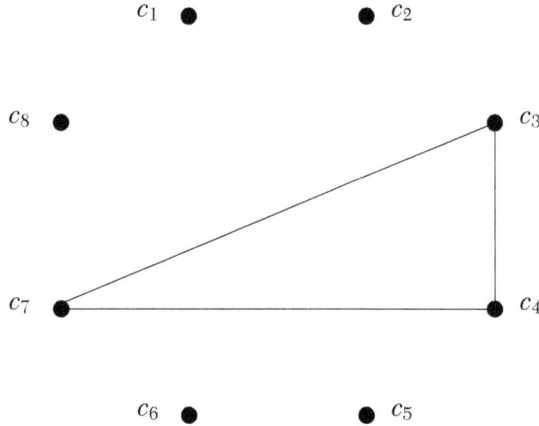

FIGURE 17.8 This subgraph with three edges is not a tree.

Thus, we look for the next smallest weight, $w_{4,5} = 4$, and at the next step, we connect the vertices c_4 and c_5 by the edge of weight 4 (Figure 17.9). Figures 17.10-17.13 show the sequel subgraphs leading to a minimum spanning tree of weight 27 (Figure 17.13).

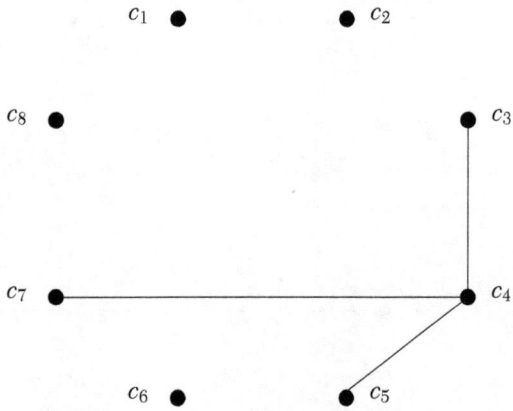

FIGURE 17.9 The subgraph with three edges.

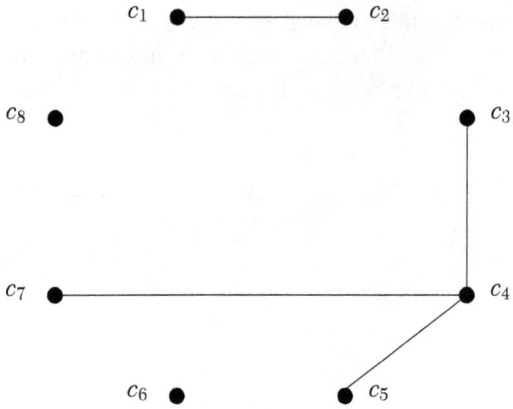

FIGURE 17.10 This subgraph is not a tree, since it is not connected.

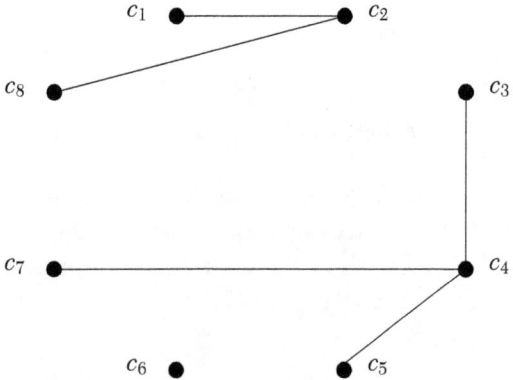

FIGURE 17.11 This subgraph with five edges also is a forest, not a tree.

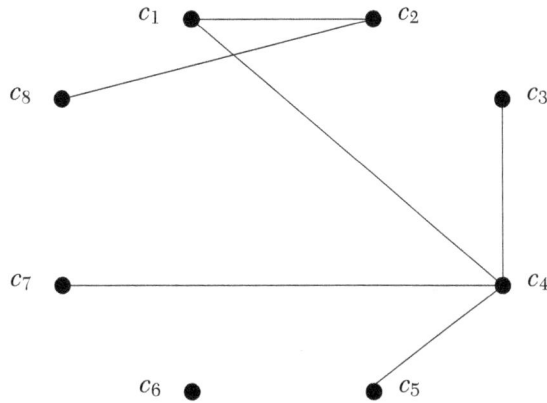

FIGURE 17.12 Second to the last step of the algorithm. Two subtrees merge into a tree with six edges. This subtree is not a spanning tree yet since the vertex c_6 is still isolated.

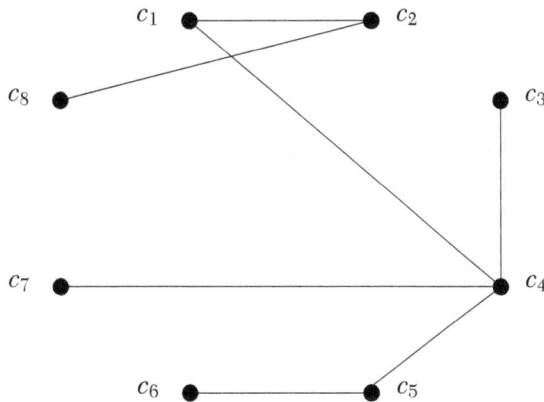

FIGURE 17.13 The minimum spanning tree of the initial graph in Problem 397, its weight is $w(T) = 36$.

Problem 398. *Prove that in any simple finite graph $G = (V, E, f)$*

$$2q \leq (p - cc(G))(p - cc(G) + 1),$$

where $cc(G)$ stands for the number of connected components of the graph G.

Problem 399. *Find a minimum spanning tree in the graph in Figure 17.14.*

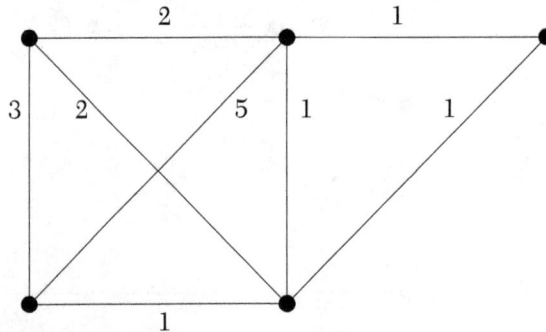

FIGURE 17.14 Find a minimum spanning tree in this graph.

We have considered so far the graph models, such that the two endvertices of any edge were symmetric in the sense none has any preference – this is similar to a two-way street. However, there are also one-way streets; corresponding mathematical models are represented by *directed graphs (digraphs)*.

Definition 100. *A digraph is a triple G = {V, E, F}, where the set of vertices V is any nonempty set, the set of edges E = $E_1 \vee E_2$ consists of two parts, where $E_1 \subset V$ and $E_2 \subset V \times V$, and the incidence function is a map $f : E \to V \cup V \times V$. An edge $e' \in E_1$, whose image contains only one vertex $v \in V$, i.e., $f(e') = \{v'\}$, is called a loop at the vertex v'. If the image of an edge contains two different vertices, $f(e') = (v_i, v_j), v_i \neq v_j$, then e' is a directed edge, v_i is called its initial vertex and v_j is its end vertex.*

The diagrams are convenient for human beings, but to represent graphs in the computer memory, it is more suitable to use their matrix representation.

Definition 101. *A square matrix A = (a_{ij}) of size $|V|$ is called the adjacency matrix of the digraph G, iff each its element a_{ij} is equal to the number of the directed edges from the initial vertex v_i to the final vertex v_j in G.*

The nonsymmetric matrix A in Problem 400 is the adjacency matrix for digraph G in Figure 17.15.

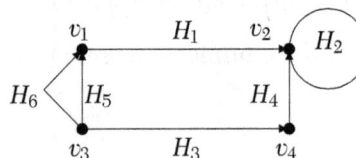

FIGURE 17.15 Digraph G; an edge H_2 is the directed loop at the vertex v_2. The incidence matrix B of this graph G is shown below.

Problem 400. *Describe, if any, acyclic paths in digraph G (Figure 17.15) from $v_3 \to v_2$ and from $v_2 \to v_3$. For the same digraph, give an example of paths of length 2, of length 3, and of length 4 from v_3 to v_2.*

$$A = \begin{pmatrix} 0 & 1 & 0 & 0 \\ 0 & 1 & 0 & 0 \\ 2 & 0 & 0 & 1 \\ 0 & 1 & 0 & 0 \end{pmatrix}.$$

If a digraph represents a binary relation ρ, where a directed edge (an arrow) from $v_i \to v_j$ indicates that the elements v_i and v_j are in this relation, i.e., $v_i \rho v_j$, then that matrix is the adjacency matrix of both the digraph and the relation.

Consider, for instance, a county with towns $v_1, \ldots, v_4$. Some of them are connected with highways H_i, $1 \le i \le 6$. These highways $H_1 - H_6$ are shown in Figure 17.15.

Problem 401. *Verify that the square $A^{[2]} = A \times A$ of the incidence matrix A with the elements $a_{i,j}^{[2]}$ is*

$$A^{[2]} = \begin{pmatrix} 0 & 1 & 0 & 0 \\ 0 & 1 & 0 & 0 \\ 0 & 3 & 0 & 0 \\ 0 & 1 & 0 & 0 \end{pmatrix}.$$

Its element $a_{3,2}^{[2]} = 3$ implicates that there are three highways of length 2 from v_3 to v_2, and indeed, those are the roads $H_5 - H_1$, $H_6 - H_1$, and $H_3 - H_4$.

Problem 402. *Compute the matrix $A^{[3]} = A^{[2]} \times A$ and check that its elements $a_{i,j}^{[3]}$ are equal to the number of directed ways of length 3 from v_i to v_j.*

Now we return to problems of graph enumeration and prove Cayley's formula on the number of labeled trees. We need the next definition.

Definition 102. *Abstract graphs $G_1 = (V_1, E_1, f_1)$ and $G_2 = (V_2, E_2, f_2)$ are called isomorphic, denoted as $G_1 \cong G_2$, iff there exist two bijective functions $\phi : V_1 \to V_2$ and $\psi : E_1 \to E_2$, compatible with the incidence functions f_1, f_2 in the sense that $f_1(\psi(e)) = \phi(f_2(e))$ for all edges $e \in E_2$. A geometric graph and an abstract graph or two geometric graphs are isomorphic if corresponding abstract graphs are isomorphic according to this definition. If $v_2 = \phi(v_1)$ or $e_2 = \psi(e_2)$, then the two vertices v_1 and v_2, or the two edges e_1 and e_2 are called corresponding vertices or corresponding edges.*

We repeat that an actual shape of an edge makes no difference in these definitions.

Problem 403. *Prove that if $G_1 = (V_1, E_1, f_1)$ and $G_2 = (V_2, E_2, f_2)$ are two isomorphic graphs, abstract or geometric, then $|V_1| = |V_2|, |E_1| = |E_2|$, and the degrees of the corresponding vertices are equal. Are these necessary conditions also sufficient for two graphs to be isomorphic?*

Problem 404. *Are the two graphs in Figure 17.16 isomorphic?*

FIGURE 17.16 Are these two graphs isomorphic?

It is useful in many problems to assign a piece of additional information, numerical or literal, to every edge. For example, it may be the length of a trip between two places, or possible flow through the pipe, or directions like *one-way* on streets, or any other additional information. Such an extra label is called a *weight* of an edge. In formal way, weight $w \in W$ is a map defined on the set of edges of the graph, such that the co-domain of W contains possible values of the weights. A graph with the weights provided is called a *weighted graph* $G = (V, E, f, w)$. The vertices, like edges, can also carry additional information, i.e., can be weighted.

Theorem 51. Cayley First Theorem. *There are p^{p-2} nonisomorphic labeled trees with $p \geq 1$ vertices.*

Proof. The following is an inductive proof over the number of vertices p. We label the vertices with first natural numbers. If $p = 1$, then $p^{p-2} = 1$, and the statement is obvious since the unique labeled tree with one vertex is this isolated vertex labeled by 1. Let be now $p \geq 2$ and T be a labeled tree of order p. Delete the end-vertex with the smallest label, record the label of the adjacent vertex, and repeat this step until only two vertices remain. This procedure generates a sequence of $p - 2$ natural numbers ranging from 1 through p with possible repetitions. The sequence is called the *Prüfer code* of the tree T. By the theorem on the cardinality of the power-set, there are p^{p-2} such sequences, and since there is an obvious one-to-one correspondence between such codes and the labeled trees of order p, the proof is done.

Example 56. *For instance, if* $p = 3$ *then* $p^{p-2} = 3$. *All the nonisomorphic labeled trees with three vertices and, hence, with two edges, are shown in Figure 17.17. The only distinction between them is the order of labels.*

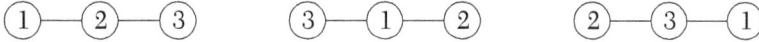

FIGURE 17.17 Nonisomorphic labeled trees with three vertices.

Corollary 12. *Let* $1 \le d_1 \le d_2 \le \ldots \le d_p$ *be the degree sequence of a tree of order* p. *The number of labeled trees of order* p *with this degree sequence is given by the multinomial coefficient (See Theorem 25),*

$$C(p-2; d_1 -1, d_2 -1, \ldots, d_p -1) = \frac{(p-2)!}{(d_1 -1)! \cdots (d_p -1)!}.$$

Proof. Indeed, $\sum_{i=1}^{p} d_i = 2p - 2$. If v_i is a pendant vertex then $d_i = 1$, and it is clear from the proof that this label does not appear in the Prüfer code at all. If $d_i > 2$, then together with the removal of this vertex we must remove $d_i - 1$ adjacent vertices, hence v_i appears in the Prüfer code $d_i - 1$ times, and the same is true for any other vertex, which proves the corollary.

Problem 405. *Is* $(1, 1, 1, 2, 2, 2, 3, 3, 3, 3)$ *the degree sequence of a graph? The same question about the sequence* $(1, 1, 1, 5, 5, 5)$? *Is it the degree sequence of a tree?*

Problem 406. *Label the tree in Figure 17.18 and compute its Prüfer code. Repeat this with another labeling of the same tree and compare the new Prüfer codes.*

Problem 407. *Restore a labeled tree if its Prüfer code is 133132.*

Solution. We do not distinguish vertices and their labels. From the proof of Theorem 51, we see that since the length of the code is 6, the tree must have $6 + 2 = 8$ vertices. The vertices (labels) 1, 2, and 3 are present in the code; thus, they were not removed at the first deletion step; hence the very first vertex removed was 4, and this vertex was connected to 1. The next smallest vertex was 5, and it was connected to 3. The next vertex, which is 6, was connected to 3 again. The vertex removed after that was 7, and it must have been connected to 1. The vertex 1 does not appear in the code after that, thus now it is the smallest, and we remove it, keeping in mind that it is adjacent to 3. At this stage, only three vertices, 2, 3, and 8 remain, but we cannot remove 2 now; hence, we have to remove 3, which is, apparently, adjacent with 2. Thus, vertices 8 and 2 are connected – see Figure 17.18.

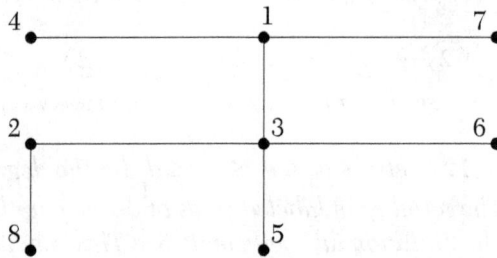

FIGURE 17.18 This tree has the prüfer code 133132, see Problem 407.

Problem 408. *Prove that there is a one-to-one correspondence between the nonisomorphic labeled trees and Prüfer codes.*

Problem 409. *A tree has p vertices. What is the largest possible number of its pendant vertices?*

Problem 410. *Prove that in any tree of order $p \geq 2$ there are at least two pendant vertices. Moreover, a stronger statement holds true – any acyclic graph of order $p \geq 2$ has at least two pendant vertices.*

Problem 411. *Prove that a graph is a forest if and only if for any two distinct vertices there is at most one path connecting them.*

Problem 412. *Generalize Part 4 of Theorem 50 to forests: If a forest of t trees has v vertices and d edges, then $v = d - t$.*

Problem 413. Cayley's Second Formula. *For $1 \leq k \leq n$, prove that there are $k(n + k - 1)^{n-2}$ labeled forests with $n + k - 1$ vertices and k connected components, such that k distinguished vertices belong to different connected components.*

Problem 414. *Prove that $1 \leq d_1 \leq d_2 \leq \ldots \leq d_p$ is the degree sequence of a tree of order p if and only if $\sum_{i=1}^{p} d_i = 2p - 2$.*

Problem 415. *Prove that every sequence of the integers $1 \leq d_1 \leq d_2 \leq \cdots \leq d_p$, such that $\sum_{i=1}^{p} d_i = 2p - 2k$, $k \geq 1$, is the degree sequence of a forest with k connected components.*

Problem 416. *Prove that the number $F(p)$ of forests of order p satisfies the recurrence relation.*

$$F(n) = \sum_{k=1}^{n} C(n-1, k-1)k^{k-2}F(n-k).$$

Problem 417. *Theorem 50 claims that a tree of order p has p – 1 edges. For non-acyclic graphs, this conclusion clearly fails. Nevertheless, prove that a connected graph of order p must have at least p – 1 edges.*

Problem 418. *How many edges are to be removed from a connected graph with 12 vertices and 15 edges to generate a spanning tree of the graph? Does this number depend upon the order in which the edges are being removed?*

Problem 419. *There are 300 cities in a state and 3000 highways connecting them, such that each city is connected with at least one other city. How many of the highways can simultaneously be closed for repair if no city should be completely isolated from the others?*

Problem 420. *Show that a graph is connected if and only if it has a spanning subtree.*

Definition 103. *A graph $G' = (V', E')$ is called a subgraph of a graph $G = (V, E)$, iff $\phi \neq V' \subset V$,*

$E' \subset E$ and the incidence relation in G' is the same as in G, i.e., f' is a restriction of f' onto E'.

Problem 421. *Find all the subgraphs and the paths of length 3 or less in the graph G (Figure 16.1).*

Problem 422. *A graph $G = (V, E)$ is called connected iff for every pair of its vertices u and v there exists a path connecting the vertices. Are there connected graphs among the graphs in this chapter?*

Problem 423. *A connected subgraph $G_1 = (V_1, E_1)$ of a graph $G = (V, E)$ is called a connected component of G iff either $G_1 = G$ or no vertex in the set-difference $V \setminus V_1$, that is, outside of V_1 is connected with any vertex in V_1. Find all connected components in the graphs above.*

Problem 424. *An edge e is called a cut-edge or a bridge iff its removal increases the number of connected components in the graph. Verify that in graph G_1 (Figure 16.1) only the edge e_5 is a cut-edge.*

The graph T is also a subgraph of graph G_3 – it contains all the vertices of G_3 and some of its edges. What is important for us now, is that T is a tree containing all the vertices of G_3. Such a tree is called a spanning tree of a graph. A graph may have several spanning trees.

Problem 425. *Draw a graph having only one spanning tree.*

Problem 426. *Draw all the spanning trees of graph G_1.*

Problem 427. *Give an example of a connected but not complete graph.*

Problem 428. *Does there exist a complete but not connected graph?*

17.2   BINARY SEARCH

Binary search is one of the most important applications of trees. A detailed exposition of searching and sorting algorithms belongs to a text in computer science, see, e.g., a treatise of D. Knuth. Here we only consider a binary search algorithm, which is closely connected with the weighted trees. Given an array of data and the element x, this algorithm looks for the location of x in this array, and either returns the location or a certain neutral item, say 0, meaning that x is not in this array. The elements of the array are supposed to be pair-wise comparable, all different, and placed in increasing order.

The algorithm is called *binary*, because at every step it splits the array into two equal or almost equal sub-arrays. If the array contains an odd number of elements, one of these sub-arrays is one element more than another. To be specific, we assume that the lower sub-array always contains at least as many elements as the upper one. The binary search algorithm is as follows. At the first step, we divide the array into two sub-arrays, as was just described. Let M be the largest element of the lower sub-array, and m be the smallest element of another one; since we assume that all the elements are different, $M < m$. Then we compare these elements with the given entity x. If $x \le M$, then x belongs to the lower sub-array, otherwise $x \ge m$ and it belongs to the upper sub-array. If $M < x < m$, we easily see that x is not in the array and must return 0.

Next, we apply the algorithm to the appropriate sub-array. After finitely many steps, we see that either $x = M$, or $x = m$, or we get 0.

If instead of a list, we have a weighted binary tree, i.e., a tree where every vertex has at most two descendants, then at every node of the tree, we must compare x with the weight of this node and move along the tree either to the left or to the right, correspondingly. We also mention that if every vertex has precisely two outgoing edges, the tree is called a full binary tree.

Problem 429. *Similarly, give definitions of the ternary, quaternary, etc., trees, and the definitions of full ternary, quaternary, etc., trees.*

Problem 430. *Sort the list* {5, 1, 3, 2, 5, 9, 8}.

Problem 431. *List the ordered pairs of elements of the set* {1, 2, 3, 4, 5} *in the lexicographic order.*

Problem 432. *Search for 45 in the array*

$$5, 12, 17, 22, 30, 39, 45, 60, 69, 79, 82, 101, 110.$$

Problem 433. *Prove that there are 504 3–arrangements without repetitions of the integers* {1, 2, 3, …, 9}. *Arrange them in the lexicographic order and find among them the triple* (2, 5, 7).

Problem 434. *Design the binary search tree for the words (sip, eat, hop, quack, attack, fasten, pat).*

In Theorem 45, we have proved a criterion for a graph to be an Eulerian graph, i.e., to have an Eulerian cycle. This result can be easily generalized to digraphs, but to do that, we have to give another definition.

Definition 104. *Given a vertex v of a digraph G, its out-degree v^+ is the number of the oriented edges, emanating from v, and its in-degree v^- is the number of the oriented edges, ending at v.*

Lemma 26. *Given an Eulerian digraph G, if we replace each directed edge with a (non-) directed edge connecting the same two vertices, the parity of every edge remains the same.*

Theorem 52. *A connected digraph is Eulerian iff, for each its vertex, its in-degree equals to its out-degree, $v^+ = v^+$.*

Problem 435. *The proof is similar to a proof of Theorem 45, and we leave it to the reader.*

17.3 EXERCISES

Exercise 17.1. *Prove that any tree of order $p \geq 2$ is bipartite.*

Exercise 17.2. *Draw all nonisomorphic trees with five vertices and those with six vertices.*

Exercise 17.3. *Consider graphs in Figure 6.2 (p. 99), 14.2 (p.192), 16.1 (p.214) – some of them are not connected. Find a minimum spanning tree for every connected component of these graphs.*

Exercise 17.4. *Forty-one points in the plane are connected by straight segments, such that any two points are connected by either a segment or a broken line, and for any two points, this broken line is unique. Prove that there are precisely 40 segments connecting the points.*

Exercise 17.5. *Find all spanning trees of the graph $K_{3,3}$ in Figure 16.4 and those of the tree T in Figure 17.2.*

Exercise 17.6. *Draw the labeled trees with the Prüfer codes 234, 3123, 4444, 7485553. Is there a labeled tree with the Prüfer code 126?*

Exercise 17.7. *A forest has 67 vertices and 35 edges. How many connected components does it have?*

Exercise 17.8. *Determine whether the graphs in Exercise 17.3 are Eulerian, semi-Eulerian, or neither.*

Exercise 17.9. *For graph G_3, find its minimum and maximum spanning trees.*

Exercise 17.10. *What graphs coincide with their spanning trees?*

Exercise 17.11. *Draw a diagram having only one spanning tree.*

Exercise 17.12. *How many nonisomorphic spanning trees does the bipartite graph $K_{3,3}$ have?*

Exercise 17.13. *n towns are connected by highways without intersections, such that a driver can reach every town from each other town, and there is the only route between any two towns. Prove that the number of highways is $n - 1$.*

Exercise 17.14. *How many are there nonisomorphic trees with n vertices if the degree of any vertex is no more than 2?*

Exercise 17.15. *Calculate the maximal number of vertices at level three in a full binary tree.*

Exercise 17.16. *Calculate the number of vertices in a full ternary tree with 40 vertices.*

COMPUTATIONS AND ALGORITHMS

An essential part of any problem in sciences may be computational: it is enough to remind that the solution of the classical fourcolors problem was accomplished by massive numerical calculations. Some calculations use analogous devices, but for the most part, current calculations are done by electronic computers,[1] which have discrete memory and operate in discrete time. Hence, we discuss here certain mathematical questions, which come from the discrete nature of computers and computations.

At the very end of 19th century, soon after G. Cantor created the set theory, mathematicians and philosophers discovered *paradoxes* at its very hurt. That was unimaginable to have antinomies at the foundation of *precise* science. This discovery was one of the sources of the crisis of the foundations of mathematics and attracted the close attention of famous researchers to the *foundations of mathematics*. During the 20th century, a lot of efforts were made, and even though there still is no solution, completely acceptable by everyone, the situation is much clearer now. Several major programs were developed, including *logicism, intuitionism*, and *constructivism*. The former was an attempt to reduce the whole mathematics to formal logic. It is considered mostly unsuccessful so that we do not discuss it here.

Another important attempt to develop the noncontradictory foundations of mathematics was made by intuitionists. The essential reason for the appearance of paradoxes in the foundations of mathematics was the use of *non-predicative*, that is, internally contradictory definitions – for instance, the Barber antinomy (see Problems 84-86 and the whole Lecture 3) is an example of such a definition. However, the intuitionists – Brauer and his followers believe that this is not the only one and even not the major

---

[1] However, a century back, when electronic computers did not exist, some important calculations were performed by the many hundreds people working simultaneously, and those people were called *computers*.

problem. Any mathematical reasoning consists of little, intuitively clear steps; these steps are really obvious. However, during the millennia our intuition was developed in the world of *finite* things, *finite sets*. Mathematicians and philosophers of the classical epoch, including Cantor, bravely applied the methods and schemes of reasoning, developed before, to the infinite sets, treated as "actual," that is, *completed infinity*, even though still two centuries back, the Prince of Mathematics Carl Friedrich Gauss warned against unlimited frivolous use of the infinite sets. Only Brauer clearly realized the dangers of such an approach; another great mathematician, German Weil [51], should be mentioned here.

In particular, Brauer and his school raise objections to the unlimited application of the law of excluded middle (the excluded third) $p \wedge \neg p \equiv 1$ to arbitrary infinite sets, and so that against the indirect proofs of existence "by contradiction," based on this law. Due to those efforts, many important parts of arithmetic, number theory, geometry, and analysis were developed anew by the constructivist methods, however, more complicated than in classical mathematics. Moreover, there are certain chapters of classical mathematics, which up to now remain beyond constructive mathematics.

Another attempt to resolve the crisis of foundations was made by David Hilbert (1862-1943), one of the greatest mathematicians of all time. His major idea was that to prove the noncontradictory character of the set theory or of any other mathematical theory, one must develop it as an *axiomatic theory*. To accomplish that goal, first of all, one must understand what a mathematical theory is without any regard to its specific content. Hilbert suggested using only *finitary methods*, so that it was obvious that the *actual infinity* has never been used in the reasoning.

Hilbert, his collaborators, and students have significantly developed their *formalization* program. With regards to a *formal* theory, let us remind the way we developed the algebra of propositions in Chapter 2. First of all, a primary notion of an alphabet was introduced. Combining various symbols, we introduced *propositions*. It is worth repeating that while building a formal theory, we are not interested in the meaning of specific symbols, rather we only observe how certain *strings* of symbols transform into another *finite* string of symbols after *finitely* many steps.

So that, what is a *formal mathematical theory*? What does distinguish it from a *substantive theory*? Again, let us remind the development of the propositional algebra in Chapter 2. We started with a primary notion of a proposition. However, it is impossible to decide whether an object is a proposition by making use only of the formal manipulations with symbols. To answer this question, one must know the content, the meaning of this particular symbol.

Theorems of the algebra of propositions also have a meaningful nature, and they are not about formal transformations but about the meaning of the results. So that the propositional algebra is not a formal theory. On the contrary, in a formal theory, we are not interested in the content, the meaning of the symbols under consideration[2] and only follow the way how some finite strings of the symbols, after finitely many steps, transform into other finite strings. As an example of a formal mathematical theory, we consider *propositional calculus*, which is a *formalization* of the algebra of propositions.

When we develop a formal theory, it is impossible to do only with the symbols and terms of the theory. We have to use a substantive *meta-language* outer with respect to the theory in hand. In our example, this is the language of usual informal logic. Thus, the implication $\rightarrow$ is a symbol of the formal language, while the symbol $\Rightarrow$ is a symbol of the metalanguage. We start by fixing the alphabet. Since we formalize the theory about propositions, first of all, we introduce the *formal* symbols for propositions, that is, the letters, maybe with subscripts, $p, p_1, p_2, \ldots, p_n, \ldots$. Besides the letters, the alphabet contains the symbols for connectives $\neg, \vee, \&, \rightarrow$ and two parentheses, the opening, left parenthesis "("and the closing, right parenthesis")."

Next, we describe the *formulas* of the theory. They are defined *inductively*, as in the algebra of propositions; however, rules for abbreviating the formulas do not exist in the formal theory; simplified formulas are objects of the metatheory. Hence,

(1) All the propositional letters being formulas
(2) If A and B are formulas, then the combinations

$$(A \vee B), (A \wedge B), (A \rightarrow B), \neg A$$

are formulas as well. No other formula exists in the theory. Here A and B are symbols of the metatheory.

The following 10 formulas are declared to be the *axioms* of the formal theory under construction.

A1. $(p_1 \rightarrow (p_2 \rightarrow p_1))$
A2. $((p_1 \rightarrow (p_2 \rightarrow p_3)) \rightarrow ((p_1 \rightarrow p_2) \rightarrow (p_1 \rightarrow p_3)))$
A3. $((p_1 \& p_2) \rightarrow p_1)$
A4. $((p_1 \& p_2) \rightarrow p_2)$
A5. $(p_1 \rightarrow (p_2 \rightarrow (p_1 \& p_2)))$
A6. $(p_1 \rightarrow (p_1 \vee p_2))$
A7. $(p_2 \rightarrow (p_1 \vee p_2))$

---

[2] See a famous claim ascribed to Hilbert [8, pp. 32–33].

A8. $((p_1 \to p_3) \to ((p_2 \to p_3) \to ((p_1 \vee p_2) \to p_3)))$
A9. $((p_1 \to p_2) \to ((p_1 \to \neg p_2) \to \neg p_1))$
A10. $(\neg\neg p \to p)$

The axioms of the formal theory are not claims, which are obvious for us and for anyone else without proof; there is no such thing as obviousness in a formal theory. Even though it is useful to check that the axioms A1–A10 are identically true in the algebra of propositions, which of course is not a random coincidence: this is how we have chosen the axioms. The axioms of a formal theory are just finite sequences of symbols, distinguished among the other formulas and playing a certain special role in transformations of the formulas.

A version of the formal theory that we are developing now also contains two *rules of inference*; one of them is *modus ponens* (MP)

$$(MP)\frac{A,\, A \to B}{B}.$$

Intuitively, (MP) is clear; its formal meaning is that if A and $A \to B$ are formulas of the formal theory, then the B is also a formula. Another rule is the substitution rule (S),

$$(S)\frac{A}{A(Q_1, Q_2, ..., Q_m \parallel B_1, B_2, ..., B_m)}.$$

Here $A, B_1, ..., B_m$ are formulas of the formal theory, $Q_1, ..., Q_m$ are propositional letters, $A(Q_1, Q_2, ..., Q_m \parallel B_1, B_2, ..., B_m)$ is the result of a simultaneous substitution of each of the B_j instead of the corresponding $Q_j, j = 1, ..., m$. If a certain $B_i = Q_i$, then this symbol does not change, and if Q_i does not appear explicitly in A, then this substitution is not performed, it is vacuous.

Definition 105. *Derivation in a formal theory is any finite sequence of the formulas $A_1, A_2, ..., A_n$, such that every A_i is either an axiom or was derived from the preceding terms with the help of either (MP) or (S). If a formula A has a derivation, then A is called a derivable formula or a theorem of formal theory and is designated as $\vdash A$.*

In formal expositions, the formulas are usually parenthesized, but we try to avoid unnecessary parentheses if the meaning is clear.

Example 57. *As an example, we prove that the formula $(A \to A)$ is a theorem of our formal theory.*

Proof. We start with the axiom A2. Applying the substitution rule (S) as

$$(p_1, p_2, p_3 \,\|\, A, (A \to A), A),$$

we deduce $\vdash (A \to ((A \to A) \to A)) \to ((A \to A)) \to (A \to A)$.

In the axiom A1, we do a substitution $(p_1, p_2 \,\|\, A, (A \to A))$, whence

$$\vdash (A \to ((A \to A) \to A)).$$

The rule (MP), applied to the two derived formulas, infers

$$\vdash (A \to (A \to A)) \to (A \to A).$$

Now we make a substitution $((p_1, p_2) \,\|\, A, A)$ into the axiom A1, thus getting $\vdash (A \to (A \to A))$. To finish the deduction of the formula $(A \to A)$, we apply the rule (MP) to the two last derived formulas. Let us demonstrate the entire deduction:

$$D_1 : ((p_1 \to (p_2 \to p_3)) \to (p_1 \to p_2) \to (p_1 \to p_3));$$
$$D_2 : ((A \to ((A \to A) \to A)) \to ((A \to (A \to A)) \to (A \to A)));$$
$$D_3 : (p_1 \to (p_2 \to p_1));$$
$$D_4 : ((A \to (A \to A) \to A));$$
$$D_5 : ((A \to (A \to A)) \to (A \to A));$$
$$D_6 : (p_1 \to (p_2 \to p_1));$$
$$D_7 : (A \to (A \to A));$$
$$D_8 : (A \to A).$$

We see that the derivation is relatively long; it consists of 8 formulas, even though D_6 can be omitted. However, it is not a major problem, especially for computers, and we return to that issue soon. First, we discuss again the meaning of formalizations of substantive theories, using as an example the propositional calculus.

When we develop a formal analog of the propositional algebra, we can expect that the theorems of the formal theory are relevant to identically true formulas of the algebra of propositions. It is easily proved that every inferred formula is *identically true* as a formula of propositional algebra.

Moreover, one can check straightforwardly that the axioms A1–A10 are true formulas if we interpret them in the algebra of propositions. And similarly, both rules of inference do not violate identical truth of formulas. Therefore, every deduced formula is identically true in the algebra of propositions. What is more, this statement is convertible, that is, every identically true formula of the propositional algebra is a theorem of the propositional calculus. In this sense, the latter is a *complete and noncontradictory theory*[3]. The converse statement requires relatively simple though relatively long reasoning, see, for example, [30, Chap. 38 and ff.] Hilbert suggested a formalization of a theory as a method of proof of its consistency.

Definition 106. *A formal theory is called inconsistent, if a certain formula and its negation are both simultaneously inferred in theory.*

Theorem 53. *The calculus of propositions is consistent.*

Proof. Indeed, if formula A is deduced, then it is identically true. The same is true for $\neg A$ – we mean here the propositional calculus formulas. On the other hand, if we consider the same formulas in the *algebra* of propositions, we immediately observe that if one of the formulas A or $\neg A$ is a tautology, then the other must be identically false. But due to the completeness of the calculus of propositions, both A and $\neg A$ cannot be inferred simultaneously, which proves that the propositional calculus must be consistent.

Before the Hilbert program was created, there was only one method to prove the inconsistency of a certain formal theory, namely, to find its *interpretation* as a certain substantive theory. This means that we somehow attach a certain content to every symbol of a formal theory. After that, the identical truth of the axioms and of the rules of inference can be verified at a substantial level. If a formal theory allows a meaningful interpretation, it is clearly, consistent. Just recently, in this book, we had used an interpretation of a propositional calculus in the algebra of propositions when propositional letters were interpreted as true-false propositions and the symbols of logical operations as the logical functions.

We now prove that the complete theory, which we have just developed, is *complete* as well in the following sense: if one extends a system of axioms by attaching to it any formula, which is not deduced in this theory, that expanded theory becomes *contradictory*.

Theorem 54. *The system of axioms A1–A10 is complete in the above sense.*

Proof. Let a formula $F(p_1, \ldots, p_n)$ is not deducible in this calculus. Hence, F is not identically true, that is, there is a Boolean vector $(a(\epsilon_1, \ldots, \epsilon_n) \in \mathbb{B}^n$ such that $F(\epsilon_1, \ldots, \epsilon_n) = 0$. Fix an arbitrary formula c and n formulas $g_1 - g_n$,

$$g_i = \begin{cases} c \vee \neg c, & \text{if } \epsilon_i = 1 \\ c \,\&\, \neg c, & \text{if } \epsilon_i = 0 \end{cases}.$$ Then the formula $F(G_1, \ldots, G_n)$ is identically

---

[3] Keep in mind that this term may have different meanings in different contents – see, e.g., Theorem 53 below.

false, while $\neg F(G_1, ..., G_n)$ is identically true, hence, it is an inferred formula. But if $\neg F$ is inferred, then it remains inferred in any extended theory. Now we add to the system of axioms $A1 - A10$ the formula $F(p_1, ..., p_{10})$ as an axiom. In the extended system of axioms, the formula F is an axiom, so that, it is inferred. Therefore, after the same substitution, one gets the inferred formula $F(G_1, ..., G_n)$, and as we have noticed above, the inferred formula $\neg F(G_1, ..., G_n)$. We proved that both formulas F and $\neg F$ are inferred; hence, the extended system of axioms is contradictory.

Similarly to the algebra of propositions, it is possible to formalize the algebra of predicates, and this formal theory will also be complete and consistent. However, these theories, unlike say, arithmetic, are relatively simple. Already in 1930-1931, K. Gödel proved that formal arithmetic is incomplete in the sense that it is possible to formulate true statements, which are not deducible in this formal theory. To prove them, one must introduce in the system new, more powerful methods. And this hierarchy goes forever. To cite Norbert Wiener (1894–1964) [52, p.150], "Thus logic has had to pull in its horns."

When we use the two-millennia old Euclidean algorithm to compute the Greatest Common Divisor of two natural numbers, anyone has no doubts, that after a certain time, we will get an answer. However, during centuries, the computations and algorithms have become more and more involved, laborious, and people derived a natural idea to make these calculations automatic. Leaving aside the history of these developments (see, e.g., the Wikipedia article "Mechanical Calculator," achieved on Jan. 05, 2020, and the references therein), we mention only that these devices inevitably forced the researchers to ask the questions "What is computation?" and "What can be computed?"

Several independent answers were given by A. Church (1903-1995) – Church-Turing thesis, by A. A. Markov Jr. (1903-1979) – normal algorithms, and even before them, by A. Turing (1912-1954) – Turing machine. It turns out that these three solutions, which look absolutely different, are actually all equivalent in some precise sense; they describe the same class of computational processes. Any process formalizes certain types of computations, and so far any calculation, we can think of is within this description. So that, we currently believe that any reasonable computation can be within this scheme, and nothing else does exist. For a more detailed exposition of these reasonable computational processes, we address the reader to more advanced books, for example, [47].

During those years, the researches developed other models of the computations, like finite states machines, push-down automata, etc. In the following chapter we give a brief introduction into the finite automata. This computational model, initiated by S. C. Kleene (1909-1994), is more powerful than Combinatorial Circuits, developed in Chapter 14, but less powerful than the general Turing machine, see Chapter 23. Here "less powerful" means that there is an algorithm, which can be realized as a Turing's machine but cannot be realized as a finite automaton.

18.1   EXERCISE

Exercise 18.1 *Prove that the formula* $p \to (\neg\neg p)$ *is a theorem of our formal theory.*

CHAPTER 19

FINITE AUTOMATA

An (abstract) automaton is a mathematical model of some discrete processes. In this text, we concentrate on certain models of computing realized by electronic computers. As usual, we start with a simple model example.

19.1 SIMPLE MODEL

Example 58. *We want to develop a mathematical model of a device, which is "listening" a text in English (at a certain radio frequency) and turns on the alarm after getting the signal SØS.*

Solution. This device is called an *automaton, automata* is its plural. The automaton must have a part, which "listens" the arriving signals, we call it an *input* or *input channel*. Since we design a mathematical model, no physical characteristics of an automaton are of any interest to us. Therefore, we assume that the input signals are the 26 small letters of the Latin alphabet and the 10 digits. These input symbols make *input words* of arbitrary finite length, separated from one another by the special symbol ⊔.

Unlike the Boolean functions and combinatorial circuits above, the automata can have an internal structure, they can be in various internal *states*. In other words, our automata *must have memory*. The automaton with finitely many internal states is called *finite*; otherwise, it is an infinite automaton. In this section, we consider only finite automata. The current state and the current output at a certain instant τ depend upon the previous state and the previous input. Depending upon these two parameters, the automaton generates its output and transfers into the next state. One sees from this description that from the formal point of view, this model is nothing but a few sets and the maps between these sets and their Cartesian products, and we have to specialize these objects. Adding more features to this model, we arrive at the Turing machine, considered in more detail in Chapter 23.

Now, we design the internal structure of an automaton. Suppose that the behavior of the automaton depends on the previous inputs; otherwise, this is a simple contact scheme without memory, studied earlier. Let the current input be the letter $\varnothing$. If this input follows the pair $\sqcup S$, the automaton must be in the alert state since, after the break $\sqcup$, it can get the pair $S\varnothing$, which may start the alarm triple $S\varnothing S$, and the alarm must follow; denote this output as 1. Otherwise, if the following symbol is anything but S, the reaction must be 0, thus, denote this output as 0, and the automaton must wait for the next input.

If the input is the letter S, the automaton must recall three preceding inputs, and if that was the triple $S\varnothing S$, then the alarm must go on. Thus, again the automaton must have memory, which is formalized by making use of the internal states. The state, in which the device was before we start it on, is called the initial state s_1, and the automaton's reaction in this state on any input must be 0. If the automaton is in the initial state, and the input is anything but $\sqcup$, the state must be the same, since, until the next alarming input $\sqcup$, the pair $S\varnothing$ cannot come. However, if the automaton is in the state s_1 and gets the input $\sqcup$, then s_1 must be changed by another state; we call this state "The expectation of the start" and denote it as s_2.

Let the automaton be in state s_2 and get any input, except for either S or $\sqcup$, then s_2 must return to s_1. However, if the input, in this state, is the character S, then s_2 must be changed by a new state; we denote it s_3 and call it "waiting for $\varnothing$." The input $\sqcup$ does not change s_2. The output for any input at the state s_2 must be 0.

Let the automaton be now in state s_3. If the input is any symbol, except for $\varnothing$, then the state s_3 must transfer to s_1, but if the input is $\varnothing$, then s_3 must transfer to a new state s_4, "alertness." The automaton's reaction in this state is always 0. Finally, the state s_4 must transfer to s_1 independently upon the input. What is more, the reaction of the automaton on the input S must be 1, while for any other input it must be 0. Our simple model of an automaton is ready. Let us denote the automaton as $\mathbb{M}_1$ and analyze some of its features.

The device may have several input channels, but we can always make their Cartesian product and consider one vector - input channel; so that we always consider automata with only one input, and quite similarly, with only one output.

The inputs can arrive at arbitrary moments, but we assume that the time between two consecutive inputs is always the same. Moreover, we assume that the automaton *instantly* transforms the input into the output, and at the following moment, the automaton is at the next internal state. This means that our automaton can be described as a triple of sets, the input alphabet $X = \{x_1, \dots, x_l\}$, the output alphabet $Y = \{y_1, \dots, y_m\}$, and the internal alphabet or the alphabet of states $S = \{s_1, \dots, s_n\}$, with the two maps attached, the transfer function $f : X \times S \to S$ and the output function $g : X \times S \to Y$. The state $s' = f(x, s)$ is called a successor (state) of the state s with respect to the input x.

Definition 107. *A quintuple of symbols* $\mathbb{M} = \langle X, Y, S, f, g \rangle$ *is called an abstract automaton, or Mealy (1927-2010) machine, or a sequential machine.*

Let us denote the state, input, and output of the automaton at the time instance r_k as $s^{(k)}$, $x^{(k)}$, $y^{(k)}$, respectively; thus,

$$s^{(k+1)} = f(s^{(k)}, x^{(k)}), y^{(k+1)} = g(x^{(k)}, s^{(k)});$$

these are called *characteristic* or *canonical equations* of the automaton.

We always assume that the sets X and Y are finite. If S is also finite, the automaton is called finite; otherwise, it is infinite. Automata defined as above, are called *deterministic*. We do not consider here non-deterministic, or stochastic automata. Our automata are *completely determined*, and any element of the set X can be an input at any state $s \in S$. This scheme can be used in many problems; for instance, it can describe the functioning of electronic computers (without glitches).

Problem 436. *Design an automaton, which takes a finite sequence of digits* 0 *and* 1 *and counts the amount of* 1s *modulo* 3.

Problem 437. *Design an automaton, which takes input* x *and returns output* y, *where* $y^{(k)} = x^{(k)} \oplus y^{(k-1)} \oplus w^{(k-1)}$ *and* $w^{(k)} = x^{(k)} \oplus w^{(k-1)}$, x, y, w *are Boolean variables and* $\oplus$ *is the addition modulo* 2.

19.2 TABLES AND GRAPHS OF AUTOMATA

We use the example of the "listening" automaton $\mathcal{M}_1$ above to show how to find the table of an automaton and draw its graph. The table of an automaton consists of two parts, the table of outputs and the table of the internal states or transfer table; it is actually a representation of the canonical equations in the table form. The internal alphabet consists of 4 states, and the input alphabet in this example, by coincidence, also has 4 elements, thus, we derive the Table 19.1 of the automaton $\mathcal{M}_1$.

TABLE 19.1 Table of the Automaton $\mathcal{M}_1$. Here α is Any Symbol of the Input Alphabet, Except for S and $\varnothing$.

$\mathcal{M}_1$	S				Y			
$S \backslash X$	$\sqcup$	S	$\varnothing$	α	$\sqcup$	S	$\varnothing$	α
s_1	s_2	s_1	s_1	s_1	0	0	0	0
s_2	s_2	s_3	s_1	s_1	0	0	0	0
s_3	s_1	s_1	s_4	s_1	0	0	0	0
s_4	s_1	s_1	s_1	s_1	0	1	0	0

The tables of automata allow us to easily compute the cardinality of certain classes of automata.

Problem 438. *How many automata have the parameters $|X| = l, |Y| = m$, and $|S| = n$?*

Solution. The result immediately follows from the theorems on the cardinality of the power-set and the Cartesian product. Since automaton functions are $f: X \times S \to S$ and $g: X \times S \to Y$, we have $|X \times S \to S| = n^{ln}$ and $|X \times S \to Y| = m^{ln}$, hence there are $(mn)^{ln}$ such automata.

In some problems it is more convenient to use graphical representation of the automata. As an example, we draw the graph of the same automaton $\mathcal{M}_1$. It has as many vertices as the automaton has internal states. This is an ordered graph (digraph), where an input goes from state s_i to a state s_j iff there is an input a, which transfers s_i to s_j. What is more, this input must be complemented by the corresponding output.

Problem 439. *The weights of the digraph $\mathcal{M}_1$ are shown in Figure 19.1:*
$w_{1,1} = (S,0) \vee \varnothing, 0 \vee \alpha, 0); \quad w_{1,2} = (\sqcup, 0); \quad w_{2,1} = (\varnothing, 0) \vee (\alpha, 0); \quad w_{2,2} = (\sqcup, 0);$
$w_{3,1} = (\sqcup, 0) \vee (S,0) \vee (\alpha, 0); w_{3,4} = (\varphi, 0); w_{4,1} = (\sqcup, 0) \vee (S,1) \vee (\varphi, 0) \vee (\alpha, 0).$

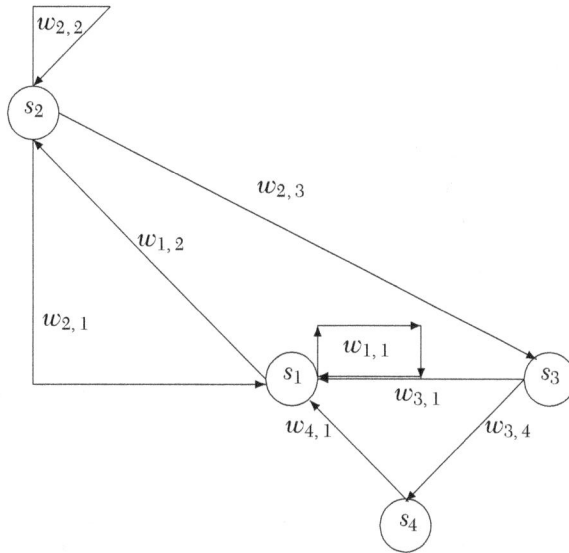

FIGURE 19.1 Digraph of the automaton $\mathcal{M}_1$ in Problem 439; w_{ij} stands for the weight of the directed edge from s_i to s_j.

19.3 CLASSIFICATION: EQUIVALENT STATES AND EQUIVALENT AUTOMATA

Consider a finite automaton $\mathbb{M} = \langle X, Y, S, f, g \rangle$. If $|S| = 1$, that is, the machine is always in the same state, then there is no reason to consider internal states of the automaton; this automaton moves any input x into an output $y = y(x)$ depending on the input, but not on the time moment. In this case, we have an automaton without memory, or a logic (switching) circuit. If $|X| = 1$, the input is irrelevant, it is always the same; this is an autonomous automaton, or the automaton without input. If $|Y| = 1$, the device is called automaton without output.

Definition 108. *A finite automaton* $\mathbb{M} = \langle X, Y, S, f, g \rangle$ *is called a Moore (1925-2003) automaton, iff the output function* $g : X \times S \to Y$ *does not essentially depends upon the first argument, that is,* $g : S \to Y$ *and there exists a map* $g_1 : X \times S \to Y$ *such that for any* $s \in S$ *and* $X = \{x_1, \dots, x_l\}$

$$g(s) = g_1(x_1, f(x_1, s)) = \cdots = g_1(x_l, f(x_l, s)).$$

Definition 109. *The automata* $\mathbb{M} = \langle X, Y, S, f, g \rangle$ *and* $\mathbb{M}_1 = \langle X_1, Y_1, S_1, f_1, g_1 \rangle$ *are called comparable, iff* $X = X_1$ *and* $Y = Y_1$.

Definition 110. *Let* $\mathbb{M}$ *and* $\mathbb{M}_1$ *be comparable automata; in particular, they can coincide. The states* $s \in S$ *and* $\sigma \in S_1$ *are called equivalent,* $S \sim \sigma$, *iff the reactions* $\mathbb{M}(s)$ *in the state* s *and* $\mathbb{M}_1(\sigma)$ *in the state* σ *for every sequence of the inputs are the same.*

Definition 111. *Two comparable automata are called equivalent,* $M \sim M_1$, *iff for every state* $s \in S$ *there exists an equivalent state* $\sigma \in S_1$, *such that* $S \sim \sigma$, *and vice versa.*

Problem 440. *Consider the automata* $\mathbb{M}_2$ *and* $\mathbb{M}_3$, *see Tables 19.2 and 19.3, where* $X = X_2 = X_3 = \{\alpha, \beta\}$, $Y = Y_2 = Y_3 = \{0, 1\}$, $S_1 = \{s_1, s_2\}$, $S_2 = \{s_3, s_4, s_5\}$, *and the automata are defined by the tables*

TABLE 19.2 Table of the Automaton $\mathcal{M}_2$.

$\mathcal{M}_2$	S		Y	
$S \backslash X$	α	β	α	β
s_1	s_1	s_2	0	1
s_2	s_2	s_2	0	0

TABLE 19.3 Table of the Automaton $\mathcal{M}_3$.

$S \backslash X$	α	β	α	β
s_3	s_3	s_4	0	1
s_4	s_4	s_5	0	0
s_5	s_4	s_4	0	0

Prove that $M_2 \sim M_3$. *We can assume that* $s_1 \sim s_3$ *and* $s_2 \sim s_4 \sim s_5$.

Every Moore automaton is, clearly, a Mealy machine. The converse statement is also true as the following theorem states.

Theorem 55. *For every Mealy automaton there exists an equivalent Moore automaton.*

Proof of Theorem 55. Given a Mealy machine $\mathbb{M}$, we want to build an equivalent Moore automaton $\widetilde{\mathbb{M}}$. Let us set $\tilde{S}$ and define the maps $\tilde{f} : X \times \tilde{S} \to \tilde{S}$ and $g_1 : X \times \tilde{S} \to Y$ as $\tilde{f}(x, \tilde{s}) = (g(x, s), f(x, s))$ and $g_1(x, \tilde{s}) = g(x, s)$, $\forall y \in Y$.

Also set $\tilde{g}(\tilde{s}) = g_1(x, \tilde{s}')$. The end of the proof is clear, and we leave it to the reader.

Problem 441. *Prove that the output $\tilde{g}(\tilde{s})$ does not depend on the preceding internal state $\tilde{s}'$, nor upon the input $x : \tilde{s}' \to \tilde{s}$.*

Thus, we get the Moore automaton $\widetilde{\mathbb{M}}\langle X, Y, \tilde{S}, \tilde{f}, \tilde{g}\rangle$, we sought for.

19.3.1 Minimization of Automata

We observed in Problem 440 that two equivalent automata could have different internal spaces with unequal numbers of states. In many problems, it is important to find an automaton with the same behavior but with the smallest possible number of states, that is, to *minimize an automaton*. To this end, we need the next definition, which refines the notion of the equivalent states.

Definition 112. *Two states $s \in S$ and $\sigma \in S'$ of the two comparable automata $\mathbb{M}$ and $\mathbb{M}'$ are called k–equivalent, $k = 1, 2, \ldots$, if the automata $\mathbb{M}(s)$ and $\mathbb{M}'(\sigma)$ have the same reaction, that is, have the same output sequence for any input sequence of the length k; this is denoted as $S \overset{k}{\approx} \sigma$. Two states, which are not k–equivalent, are called k–distinguishable.*

Two states are equivalent iff they are k–equivalent for any natural k. If two states are k–equivalent, then they are l–equivalent for any $1 \leq l \leq k$, and if two states are k–distinguishable, they are l–distinguishable for every $l \geq k$.

Now we show a minimization algorithm[1]. Let us remark that the binary relations of the equivalency and of k–equivalency are the equivalence relations on the set of states S of an automaton. The factor-sets of S with respect to these relations are denoted as S_∞ and S_k, $k = 1, 2, \ldots$, respectively. Introduce also the equivalence classes $S_k = \{G_1^k, \ldots, G_{p_k}^k\}$ and $S_\infty = \{G_1, G_2, \ldots\}$.

Lemma 27. *(1) When k is increasing, the equivalence classes G_i^k do not collide, but can only split, that is, if two states $s, s' \in G_i^k$, then for every $l, 1 \leq l \leq k$, there exists an index i_l such that $s, s' \in G_{i_l}^l$.*

(2) If $S_{k_0} = S_{k_0+1}$ with a certain k_0, then
$$S_{k_0} = S_{k_0+1} = S_{k_0+2} = \cdots = S(\infty).$$
(3) If $|S| = n < \infty$, then $S_{n-1} = S_\infty$.

---

[1] There exist different minimization algorithms, see, e.g., [5].

Proof. Statement (1) is just a rephrased remark made before the Lemma 27, so that we turn to statement (2). Let s and σ be (k_0+1)-equivalent states, $s \overset{k_0+1}{\sim} \sigma$, and prove that in this case $S \overset{k_0+2}{\cong} \sigma$ as well. Consider an arbitrary input $x \in X$ and the following states $f(x, s)$ and $f(x, \sigma)$. Since $S \overset{k_0+1}{\cong} \sigma$, it follows that $fx, s \sim k0 + 1fx, \sigma$. But $S_{k_0} = S_{k_0+1}$, hence $f(x, s) \overset{k_0+1}{\cong} f(x, \sigma)$ for every $x \in X$. So that, $S \overset{k_0+2}{\cong} \sigma$. Now we conclude by mathematical induction that $S_{k_0} = S_{k_0+1} = S_{k_0+2} = \cdots$, and by the preceding remark, $S_{k_0} = \cdots = S_\infty$.

Take up the statement (3). We consecutively build the factor-sets $S_1, S_2, \ldots$. If $S_i \neq S_{i+1}$, then due to the part (1), at least one equivalence class in S_i splits in two or more classes in S_{i+1}. Even if just one state separates at each step, each the equivalence class in S_n will contain one state, so that further split is impossible. In this case, no two equivalent states do exist, and $S = S_\infty$.

Finally, we describe the construction of the factor-sets $S_1, S_2, \ldots$. Let s_i and s_j be q–equivalent. Then $g(x, s_i) = g(x, s_j)$ for every $x \in X$, that is, the ith and jth rows of the Y-table are identical. Therefore, the set S_1 is derived from the Y-table by grouping the states, which have the same reactions (outputs) for every input.

Problem 442. *Minimize the automaton M_4, given by Table 19.4, where the alphabets are $X = \{\alpha, \beta, \gamma\}$, $Y = \{0, 1\}$.*

TABLE 19.4 Table of the Automaton M_4.

M_4	X			Y		
$S \backslash X$	α	β	γ	α	β	γ
s_1	s_2	s_2	s_5	1	0	0
s_2	s_1	s_4	s_4	0	1	1
s_3	s_2	s_2	s_5	1	0	0
s_4	s_3	s_2	s_2	0	1	1
s_5	s_6	s_4	s_3	1	0	0
s_6	s_8	s_2	s_6	0	1	1
s_7	s_6	s_2	s_8	1	0	0
s_8	s_4	s_4	s_7	1	0	0
s_9	s_7	s_2	s_7	0	1	1

Problem 443. *Find the outputs of the automata* $\mathbb{M}_4(s_2)$ *and* $\mathbb{M}_4(s_7)$ *on the sequence* $\gamma\alpha$.

Solution of Problem 442. The exit table of $\mathcal{M}_4$ shows that its internal states are divided into two classes of 1-equivalence and mark them by s_a and S_b as follows; $S = S_a \cup S_b$, where $S_a = \{s_1, s_3, s_5, s_7, s_8\}$ and $S_b = \{s_2, s_4, s_6, s_9\}$, thus, it is convenient to rearrange the S-table of $\mathcal{M}_4$.

TABLE 19.5 Rearranged Table S of the Automaton $\mathcal{M}_4$.

$S \setminus X$	α	β	γ
s_1	s_2	s_2	s_5
s_3	s_2	s_2	s_5
s_5	s_6	s_4	s_3
s_7	s_6	s_2	s_8
s_8	s_4	s_4	s_7
s_2	s_1	s_4	s_4
s_4	s_3	s_2	s_2
s_6	s_8	s_2	s_6
s_9	s_7	s_2	s_7

To find the 2-equivalence classes of the internal states of $\mathcal{M}_4$, it is convenient to show the new 1-equivalence classes in the subscripts of the states as, for example, $s_7 = s_{7a}$ or $s_2 = s_{2b}$, and to construct Table 19.5.

The second subscripts show that the input α transfers all the states of the class a to the state of the 1–equivalence class b. But the latter is 1-equivalent. So that, all the states of the class a identically react onto any two-symbol input sequence beginning with α. Moreover, the subscripts in the second and the third columns and in the first five rows are also coincide. This shows that the states of class a are not only 1-equivalent, but also 2-equivalent.

Let us look next at the b–states, and their reactions at the input γ. One can see that the states s_2, s_4, s_6 go to class b, but s_9 transfers to class a. Since class a and class b are 1–distinguishable, the states s_2, s_4, s_6, on the one hand, and the sate s_9, on the other hand, are 2–distinguishable; moreover, we can observe that the input sequence of length 2, which distinguishes them, must begin with γ.

Now it is clear that at the next step of the algorithm, one is to split the class b into two classes of 2–equivalence, when the class S_a remains the same;

however, the class S_b splits into two 2–equivalence classes, $\{s_2, s_4, s_6\}$ and $\{s_9\}$. The set of 2-equivalent states consists of three subsets,

$$S_a = \{s_1, s_3, s_5, s_7, s_8\}, S_{bb} = \{s_2, s_4, s_6\} \text{ and } S_{bc} = \{s_9\}.$$

Similar reasoning shows that the states s_2 and s_4 are not only 2, but also 3-equivalent; however, they are not 3-equivalent to s_6. Thus, the set S_8 remains unchanged yet, and also, S_{bc} cannot change. However, the class S_{bb} divides, $S_{bb} = S_{bbb} \cup S_{bbc}$, where $S_{bbb} = \{s_2\}$ and $S_{bbc} = \{s_4\}$. And only at the next step, while developing 4-equivalent states, the class S_a splits as

$$S_a = S_{aaaa} \cup S_{aaab} = \{s_1, s_3, s_8\} \cup \{s_5, s_7\}.$$

Since the indices show that none of these classes can split into smaller classes, that is, $S_4 = S_5$, we stop here. By Lemma 27, $S_\infty = S_4$.

Finally, we show the minimal automaton $\widetilde{\mathbb{M}}_4$, which is equivalent to $\mathbb{M}_4$. The alphabets X and Y are the same; the internal alphabet consists of five states only; we denote them $\tilde{S} = \{a, b, c, d, e\}$, where, of course, e is not the base of natural logarithms. Our reasoning gives the following table,

TABLE 19.6 Table of States S of the Automaton $\mathcal{M}_4$ with Double Subscripts.

$\mathcal{M}_4$	S		
$S \backslash X$	α	β	γ
s_1	s_{2b}	s_{2b}	s_{5a}
s_3	s_{2b}	s_{2b}	s_{5a}
s_5	s_{6b}	s_{4b}	s_{3a}
s_7	s_{6b}	s_{2b}	s_{8a}
s_8	s_{4b}	s_{4b}	s_{7a}
s_2	s_{1a}	s_{4b}	s_{4b}
s_4	s_{3a}	s_{2b}	s_{2b}
s_6	s_{8a}	s_{2b}	s_{6b}
s_9	s_{7a}	s_{2b}	s_{7a}

Problem 444. *Find the minimal forms of the automata* $\mathbb{M}_1, \mathbb{M}_2, \mathbb{M}_3$

Problem 445. *Construct the graph of* $\widetilde{\mathcal{M}}_4$.

TABLE 19.7 Table of the Automaton $\widetilde{\mathcal{M}}_1$.

S \ X	α	β	γ	α	β	γ
a	b	b	e	1	0	1
b	a	b	b	0	1	1
c	e	c	e	0	1	1
d	a	c	d	0	1	1
e	d	b	a	1	0	0

19.3.2 Automaton Operators

An automaton receives some information, written in the input alphabet, transforms it, and sends it out in the output alphabet. One of the main questions which we are addressing now is what information, what algorithms can be realized in this model?

Let X be an arbitrary finite set, called from now on an *alphabet*, and its elements are called the *letters*. Any finite sequence of (maybe repeating) letters is called a *word w*; the number of characters in the word is called its *length $|w|$*. The collection of all the words with the letters from X is called the *vocabulary* on X and is denoted as $\Omega(X)$.

If $\mathbb{M} = \langle X, Y, S, f, g \rangle$ is any automaton, then $\Omega(X)$ and $\Omega(Y)$ are called the input alphabet and output alphabet, respectively.

Any map $T : \Omega(X) \to \Omega(Y)$ is called a lexicographic (vocabulary) operator with the domain X and the co-domain Y.

Consider an initial operator $\mathbb{M}(s)$ and send to it all the input words $w \in \Omega(X)$; as the output we will get various words $v \in \Omega(Y)$. Therefore, we put a lexicographic operator

$$T = T_{\mathbb{M}(s)} : \Omega(X) \to \Omega(Y), v = Tw$$

into the correspondence to the initial operator $\mathbb{M}(s)$. Thus, the operator T is acceptable by the automaton $\mathbb{M}(s)$, or the initial automaton $\mathbb{M}(s)$ accepts the operator T.

Definition 113. *A lexicographic operator $T : \Omega(X) \to \Omega(Y)$ is called an automaton operator or a deterministic function iff there exists an initial automaton $\mathbb{M}(s)$, $s \in S$, realizing T. If $\mathbb{M}$ is finite, then the operator T is called finite automaton operator, or boundedly deterministic function.*

Our goal is to derive the criterion of a finite automaton operator, that is, to determine which automaton operators are finite automaton operators.

Let a lexicographic operator $T : \Omega(X) \to \Omega(Y)$ be accepted by an initial automaton $\mathbb{M}(s)$, not necessarily finite. It is clear from the definition that the automaton T transforms one-letter words into one-letter ones, two-letter words into two-letter, etc., that is, the length of an input word s equal to the length of the output word.

Definition 114. *A lexicographic operator $T : \Omega(X) \to \Omega(Y)$ is called synchronous iff it preserves the length of a word, that is, $|Tw| = |w|$ for every word $w \in \Omega(X)$.*

Before we discuss this definition, we need an auxiliary claim.

Lemma 28. *An automaton operator is synchronous.*

Proof. Consider again an automaton operator $T_{\mathbb{M}(s)}$, fix an input word $w \in \Omega(X)$, and feed all the possible input words starting with w, that is, the words ww' with $w' \in \Omega(X)$, to the initial automaton $\mathbb{M}(s)$. It should be mentioned that when a word $w = abc...$ is an input word, then the inputs are started on the left, that is, the first input symbol is a, the second is b, etc. Since a synchronous operator processes the output words letter-wise, the corresponding output is $T(ww') = T(w)u$, where the first part $T(w)$ does not depend on w'. The second part depends on both w and w', thus we denote it as $u = T_w(w')$ and rewrite that equation as

$$T(ww') = T(w)T_w(w'). \tag{19.1}$$

Definition 115. *A lexicographic operator T is called causal iff it satisfies (19.1) for any words $w, w' \in \Omega(X)$.*

Keeping in mind Lemma 28, we have obviously proved the next claim.

Lemma 29. *Any automaton operator is causal.*

Definition 116. *A synchronous and causal lexicographic operator is called an operator without anticipation.*

Example 59. *Let the input and output alphabets are $X = Y = \{0, 1\}$ and a lexicographic operator $T_1 : \Omega(X) \to \Omega(Y)$ is defined as*

$$T_1(x^1 x^2 \cdots x^k) = y^1 y^2 \cdots y^k$$

where $y^i = \begin{cases} 0 & \text{if a word } x^1 x^2 \cdots x^i \text{contains an even number of } 1 \\ 1 & \text{otherwise.} \end{cases}$

We treat 0 *as an even number. It is clear that* T_1 *is an operator without anticipation.*

Example 60. *Let the input and output alphabets are* $X = Y = \{0, 1\}$ *and a lexicographic operator* $T_2 : \Omega(X) \to \Omega(Y)$ *is defined as*

$$T_2(x^1 x^2 \cdots x^k) = y^1 y^2 \cdots y^k$$

where $y^i = \begin{cases} 0 & \text{if a word } x^1 x^2 \cdots x^i \text{contains more 1s than 0s} \\ 1 & \text{otherwise;} \end{cases}$

T_2 *is also an operator without anticipation.*

In Lemmas 28–29, we have proved that any automaton operator is an operator without anticipation. The converse statement is also valid.

Theorem 56. *A lexicographic operator* $T : \Omega(X) \to \Omega(Y)$ *is an automaton operator (deterministic function) iff it is an operator without anticipation.*

To finish the proof of the theorem, given an operator T without anticipation, we have to construct an automaton $\mathbb{M}(s)$, which is realizing T. We will draw a weighted tree of the automaton $\mathbb{M}(s)$. The next example clearly demonstrates the universal algorithm of constructing the weighted trees of automata. The tree of the automaton T_1 in Example 59 is shown in Figure 19.2.

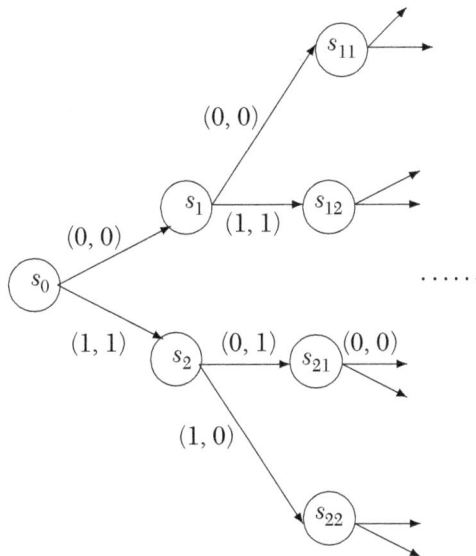

FIGURE 19.2 Infinite tree of the automaton T_1 in Example 59.

Problem 446. *Construct an automaton that accepts the operator T_2 in Example 60.*

However, not every automaton is a finite automaton – we just saw that the initial tree in the previous example was infinite. Now we tackle the question of how to determine when an automaton is finite. First, we remark that together with the empty set $\varnothing$ it is convenient to consider an empty word Λ, that is, the word with no character, hence $|\Lambda| = 0$. For every word w, it is $w\Lambda = \Lambda w = w$.

Fix a word w in equation (19.1), and let the word w' run through the whole vocabulary $\Omega(X)$. Thus, the word $T(w)$ is constant, while the remainder is varying depending upon the word w'. So that, to every word $w' \in \Omega(X)$ there corresponds a word $T_w(w') \in \Omega(Y)$. Hence, we get another lexicographic operator $T_w : \Omega(X) \to \Omega(Y)$. This new operator is called the residual operator of the operator T, generated by the word w.

Problem 447. *Let $n(w)$ be the difference between the number of ones and the number of zeros in the word w. Prove that if T_2 is the operator in Example 60, then $T_{2,w} = T_{2,u}$ iff $n(w) = n(u)$.*

Lemma 30. *Prove that if T is a lexicographic operator without anticipation, then every residual operator is also an operator without anticipation.*

Proof. By uv we, as always, mean the concatenation of the two words u and v in this order, that is, first we write the word u, and then, on the right of u, we write the word v. For example, if $u = ab$ and $v = ababc$, then $uv = abababc$. Hence, $|uv| = |u| + |v|$. Moreover, due to the synchronous property of T,

$$|w| + |w'| = |ww'| = |T(ww')| = |T(w)T_w(w')| = |T(w)| + |T_w(w')|.$$

Since $|T(w)| = |w|$, it follows that $|T_w(w')| = |w'|$, hence T_w is a synchronous operator.

To prove the causality of the operator T, we consider the equation

$$T(ww'w'') = T(ww')T_{ww'}(w'') = T(w)T_w(w')T_{ww'}(w'').$$

At the same time,

$$T(ww'w'') = T(w)T_w(w'w''),$$

which implies

$$T_w(w'w'') = T_w(w')T_{ww'}(w'').$$

This proves the causality of T.

Problem 448. *Prove that concatenation is noncommutative, associative operation. What is its neutral element?*

Definition 117. *Let T be a lexicographic operator without anticipation. The total number of all its residual operators, including $T_\Lambda = T$ is called the weight of T.*

Problem 449. *Find the weight of Operators T_1 and T_2, defined in Examples 59–60.*

Solution. As before, let the input and output alphabets be $X = Y = \{0, 1\}$. Introduce the lexicographic operator $\overline{T}_1 : \Omega(X) \to \Omega(Y)$ as $\overline{T}_1(x^1 x^2 \cdots x^k) = (y^1 y^2 \cdots y^k)$, here

$$y^i = \begin{cases} 1 & \text{if a word } x^1 x^2 \cdots x^i \text{ contains an even number of 1s} \\ 0 & \text{otherwise.} \end{cases}$$

Consider a word $0w$, where $w \in \Omega(X)$. It is obvious that there are exactly as many 1s among the first k symbols of this word, as the 1s among the first $k-1$ symbols of the word w, therefore, $T_1(0w) = 0T_1(w)$. At the same time, the number of 1s in the word $1w$ is one more than in the word w, therefore, $T_1(1w) = 1\overline{T}_1(w)$. It is clear that in the general case, the conclusion is the same: if the word w has an even number of 1s, then $T_{1w} = T_1$, and if the number of 1s is odd, then $T_{1w} = \overline{T}_1$. Hence, the weight of T_1 is 2. However, since the quantity $n(w)$ can take on infinitely many values, the weight of T_2 is infinite.

Now we can state the major claim about the lexicographic operators.

Theorem 57. *A lexicographic operator T is a finite automaton operator (or a boundedly deterministic function) iff T is an operator without anticipation with finite weight. Moreover, the number of states of a minimal automaton realizing T ie equal to the weight of T.*

A proof of Theorem 57 is given at the end of this chapter.

Example 61. *Apply this algorithm to the tree of the automaton in Figure 19.2.*

Solution. We see that operator T_1 corresponds to the state s_0, which also corresponds to the states s_1 and s_{22}, while to the states s_2 and s_{21} there corresponds the operator $\overline{T}_1$. Now the algorithm in the proof generates the graph in Figure 19.3.

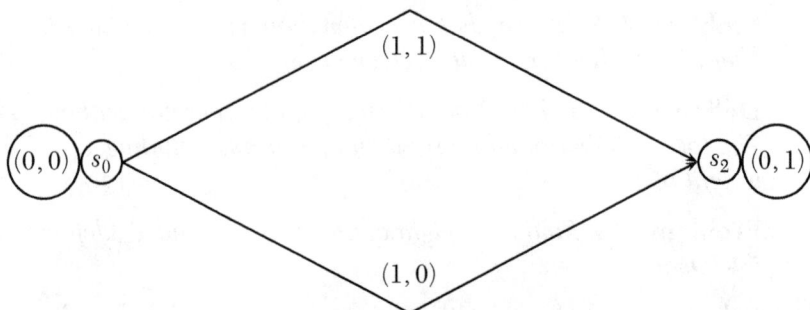

FIGURE 19.3 The finite graph, which is equivalent to the infinite tree of the automaton T_1 in Example 59.

19.3.3 Regular Grammars

As we discussed above, every automaton represents a lexicographic operator, transforming input words into output words according to the specified rules. Below we consider alternative approach to finite automata, based on the grammars (events); this approach was initiated by S. C. Kleene. It can be used for the automata without output, thus in this lecture, without any loss of generality, we consider the Moore automata.

Definition 118. *Given an alphabet X, any ordered, maybe empty, string w of symbols of x, is called a word in X. A set of words on X is called a vocabulary in X. The empty word is denoted as* Λ.

A subset, maybe empty, of the vocabulary $\Omega(X)$ is called an event over X.

Lemma 31. *Let X be an arbitrary finite alphabet and* $Y = \{0, 1\}$. *There is a one-to-one correspondence between the set of all the events over X and the set of all the lexicographic operators without the anticipation* $T : \Omega(X) \to \Omega(Y)$.

Proof. Given the operator T, let us build the grammar $E(T) \in \Omega(X)$, which consists of the words $w \in \Omega(X)$, whose images $T(w)$ have the last symbol of !. Since T is an operator without anticipation, it is straightforward to show that this correspondence is also one-to-one.

Problem 450. *(1) Prove that this is a one-to-one correspondence. (2) Prove that to the empty grammar there corresponds the operator, which moves any word w into the word, consisting of* $|w|$ *zeros.*

Definition 119. *We say that an event is represented in an automaton, iff the corresponding, by Lemma 31, lexicographic operator is an automaton* *operator. An event is represented in a finite automaton, iff the corresponding, by Lemma 31, lexicographic operator is a finite automaton operator.*

Let E be a finite automaton event on X, which is represented by the finite initial Moore automaton $\mathbb{M}(s)$, $\mathbb{M} = \langle X, Y, S, f, g \rangle$, $s \in S$. Let us feed all possible words $w \in \Omega(X)$ to the $\mathbb{M}(s)$. By Lemma 31, if $w \in E$, then the reaction of $\mathbb{M}(s)$ to the input w will end with the symbol 1, otherwise, its last symbol is 0. But $\mathbb{M}$ is a Moore automaton, its reaction depends solely upon its internal state. Therefore, the set of all the internal states is a partition $S = S_E \cup S_{\bar{E}}$, $S_E \cap S_{\bar{E}} = \varnothing$, such that $s \in S_E$ iff the output, corresponding to s, is the 1.

Definition 120. *It is said that a family of states S_E represents an event E in the automaton $\mathbb{M}$.*

Now, since the reactions of the automaton are related to its internal states, we can neglect the outputs and assume that an event is represented in an automaton by those and only those states, to where the automaton transfers in as a result of input words belonging to the event.

Our goal in this section is to derive a criterion of the representability of events in an automaton. First, we define certain operations over events. Remind that an event is just a set of words. As always, *the empty set* is denoted as $\varnothing$. It is also convenient to use here another symbol, ε for the *empty string* or the *empty word*.

Definition 121. *Disjunction, or union $E_1 \vee E_2$ of the events E_1 and E_2 is their set-theory union.*

Definition 122. *Product of the events E_1 and E_2, in this order, is the event*

$$E = E_1 E_2 = \{ w = w_1 w_2, w_1 \in E_1, w_2 \in E_2 \}.$$

It is clear that $E_1 E_2 \neq E_2 E_1$, in general, that is, the concatenation of words and the product of the events are not commutative.

Problem 451. *Design an example of two commutative events.*

Problem 452. *Prove the distributive laws for the events $E, F, G \subset \Omega(X)$,*

$$E(F \vee G) = EF \vee EG$$

$$(F \vee G)E = FE \vee GE$$

Remark 19. *Such algebraic systems have a special name; thus, the algebra of events is called an associative monoid over its alphabet.*

Definition 123. *The event* $\{E\} = E \vee E^2 \vee E^3 \vee \cdots$, *where* $E^2 = EE$, $E^3 = EEE, \ldots$, *is called the iteration (or the closure) of the event E. The event* $\{E\}^* = \{\varepsilon\} \vee \{E\} = \{\varepsilon\} \vee E \vee E^2 \vee E^3 \vee \cdots$ *is called the (Kleene) star of the event E.*

Problem 453. *Consider the languages* $L_1 = \{\alpha\}$, $L_2 = \{\alpha, \beta\}$, $L_3 = \{\alpha\beta\}$, $L_4 = \{\varepsilon\}$. *Compute the unions, concatenations, and stars of these languages.*

Consider an alphabet $X = \{x_1, x_2, \ldots, x_l\}$. The letters $x_1, x_2, \ldots, x_l$ are one-symbol words, therefore, one-word events. These l events, and only these events, are called the *elementary events* over the alphabet X.

Definition 124. *An event is called regular, iff it can be derived from the elementary events with the help of disjunction, concatenation, and iteration, being used finitely many times in any order.*

Example 62. *Let the alphabet* $X = \{x, y, z\}$. *The event*

$$E = \{x \vee (xy)\}\{x\{z\}\} \vee y \qquad (19.2)$$

is regular.

Formulas (19.2) and alike are called the regular expressions of the event E (over the alphabet X).

Definition 125. *An event E is called regular up to the empty word* Λ, *iff it can be written as* $E = F \vee \Lambda$, *where F is a regular event (in the sense of Definition 124).*

Theorem 58. *(Kleene). An event is finitely automaton iff it is regular up to the empty word.*

To prove the theorem, we are to study linear equations and systems of linear equations in the algebra of events. The reader will observe some similarity between the systems of events under consideration and the systems of linear equations in elementary algebra.

Lemma 32. *Let the events E and F be given. The equation* $X = XE \vee F$ *with respect to an event X has the unique solution* $X_0 = F \vee F\{E\}$ *in the algebra of events, such that if E is regular and F is regular up to the empty word, then* X_0 *is also regular up to the empty word.*

Proof of Theorem 58. Given an operator T without anticipation, we have to construct an automaton $\mathbb{M}(s)$, which is realizing T. Suppose that an input alphabet consists of l symbols $X = \{x_1, \ldots, x_l\}$, $|X| = l$. Fix a point s_0 in the plane, – this is the root of the future tree, and choose l more vertices

$\{s_1, s_2, \ldots, s_l\}$ of the graph, corresponding to the input characters in X. Draw the edges from s_0 to each of the vertices $\{s_1, \ldots, s_l\}$, these edges together with s_0 make the zeroth level of the tree. Next, we pick l^2 points, representing the vertices of the second level and mark them $s_{11}, s_{12}, \ldots, s_{1l}, s_{21}, \ldots, s_{2l}, s_{31}, \ldots, s_{ll}$. The first level of the tree consists of l vertices $\{s_1, \ldots, s_l\}$ and l^2 edges from them to the vertices $s_{11}, s_{12}, \ldots, s_{1l}, s_{21}, \ldots, s_{2l}, s_{31}, \ldots, s_{ll}$. Continuing the process, we get the infinite tree rooted at s_0.

To finish the construction, we must choose the weights of edges, so that the weighted tree represents the automaton $\mathbb{M}(s_0)$, realizing the automaton operator T. To each edge from s_0 to s_i, $1 \leq i \leq l$, we put into a correspondence a pair (x_i, Tx_i). To each edge from s_i to s_{ij}, $1 \leq j \leq l$, we put into a correspondence a pair (x_j, y_{ii}), where y_{ij} is computed from the equation $T(i, j) = T(x_i)y_{ij}$. All the next steps are completely similar and single-valued, since T is an operator without anticipation. It is obvious that this infinite weighted tree is the graph of the given initial automaton realizing the given operator T.

Proof of Theorem 57. Let the operator $T : \Omega(X) \rightarrow \Omega(Y)$ can be accepted by a finite initial automaton $\mathbb{M}(s)$, $\mathbb{M} = \langle X, Y, S, f, g \rangle$, $s \in S$. If $\mathbb{M}$ is not minimal, then its minimal form $\widetilde{M}$, with the number of states at most as the number of states of $\mathbb{M}$. Hence, we will consider $\mathbb{M}$ to be a minimized automaton.

It is important to mention a relationship between the residual operators of T and the states of the initial automaton $\mathbb{M}$. Let an input word w transfers the initial automaton $\mathbb{M}(s)$ into an initial automaton $\mathbb{M}(s')$. Then the latter automaton is realizing the residual operator T_w, since if the automaton $\mathbb{M}$ had equivalent states, they would correspond to the same residual operator. So that, it is impossible to have more residual operators than the number of states of the automaton, which accepts them; therefore, there is only a finite number of these states. What is more, if an automaton is minimal, then the weight of the operator is *equal* to the number of states.

Now, let T be a lexicographic operator without anticipation and with a finite weight. As in theorem 40, we construct an infinite tree, realizing the operator. We have already established that to each vertex there corresponds a residual operator, generated by the input word, which leads to this vertex from the root, and vice versa. Let us look consecutively through the vertices of the tree, layer by layer, starting at the root, and write down the corresponding residual operators $T_\Lambda = T, T_{x_1}, \ldots, T_{x_l}, T_{x_1 x_2}, \ldots$. Since T has a finite weight, we eventually arrive at the layer such that every its operator has appeared in a previous layer.

Now we can transform the tree as follows. Let to a vertex s_α there corresponds the same residual operator as to a vertex s_β, which occurred earlier, and such that there exists a edge going from s_γ to s_α, and this edge has a weight (x, y). Now we remove the whole this brunch starting at the edge from s_γ, and replace it with an edge from s_γ to s_β with the same weight (x, y). When we do these "trimmings" with all the vertices of that layer, instead of an infinite tree we get a finite graph, corresponding to the finite initial operator $\mathbb{M}(s_0)$ – this graph accepts the operator T. To end the proof, if it is necessary, one has to minimize this automaton.

Theorem 59. *(V. Bodnarchuk) Consider in the algebra of events, a system of equations with respect to the events* X_i,

$$X_i = X_1 E_{j1} \vee X_2 E_{j2} \vee \cdots \vee X_n E_{jn} \vee F_j, 1 \le j \le n,$$

where all the non-empty events E_{ij} are regular, and all the nonempty events F_j are regular up to the empty word. Then the system has the unique solution, which is also regular up to the empty word.

Proof. We solve the system by the method of consecutive elimination with the help of Lemma 23. After every iteration, the coefficients of the unknowns are derived from E_{ij} by making use of the three basic operations, concatenation, union, and iteration, and the free terms from E_{ij} and F_j, that is, at every iteration, all our assumptions regarding the coefficients are preserved. Thus, Theorem 59 follows from Lemma 32.

Proof of Theorem 58. Necessity. Let the states $s_{i_1}, s_{i_2}, \ldots, s_{i_p}$ represent an event E in the finite initial automaton $\mathbb{M}(s_0)$. Consider the events E_i consisting of the words w that move the automaton $\mathbb{M}(s_0)$ into the state $s_i, 1 \le i \le n$. In particular, $E_i \supset \Lambda$. It is obvious that $E = E_{i_1} \vee \cdots \vee E_{i_p}$ and since the disjunction of regular events is regular, it is sufficient to prove that every event E_i is regular up to the empty word.

Next we introduce the sets M_s of all the states directly preceding the state s, namely,

$$M_s = \{\sigma \in S : \exists x \in X, \text{ such that } f(x, \sigma) = s\}.$$

Considering the graphs of automata, we derive the following equations, for $s_i \ne s_0$,

$$E_i = \vee_{\sigma \in M_s} (\vee_{x \in X : f(x, \sigma) = s_i} E_\sigma x).$$

If $i = 0$, the equation has a free term,

$$E_{s_0} = \bigvee_{\sigma \in M_{s_0}} \left(\bigvee_{x \in X: f(x, \sigma) = s_0} E_\sigma x \right) \vee \Lambda.$$

Now Theorem 59 directly implies that the event E is regular up to the empty word. Thus, we have proved the necessity of Theorem 58.

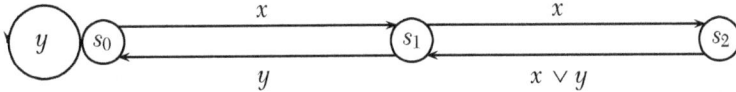

FIGURE 19.4 Graph of the automaton in Example 62.

Sufficiency. We use the graph to compose a system of equations

$$\begin{cases} E_0 = E_0 y \vee E_1 y \vee \Lambda \\ E_1 = E_0 x \vee E_2 (x \vee y) \\ E_2 = E_1 x. \end{cases}$$

Next, we use the third equation to eliminate E_2,

$$\begin{cases} E_0 = E_0 y \vee E_1 y \vee \Lambda \\ E_1 = E_0 x \vee E_1 (x^2 \vee xy). \end{cases}$$

Now, the first equation implies by Lemma 32,

$$E_0 = (E_1 y \vee \Lambda)\{y\} \vee E_1 y \vee \Lambda = E_1 \{y\} \vee \{y\} \vee \Lambda.$$

Insert this expression in the second equation,

$$E_1 = E_1 (x^2 \vee xy \vee \{y\}x) \vee (\{y\}x \vee x),$$

hence

$$E_1 = (\{y\}x \vee x)\{x^2 \vee xy \vee \{y\}x\} \vee \{y\}x \vee x.$$

Finally, from here we get the regular expressions of the events, we sought for, $E_2 = E_1 x$ and

$$E = E_1 \vee E_2. \tag{19.3}$$

We have to establish the sufficiency of the conditions of Theorem 58. That is, given a regular expression of an event, one must find a finite automaton, accepting this event. The following is an algorithm that was derived by McNaughton – Yamada and independently by Glushkov, which builds this automaton. We expose the algorithm by an example, but it will be obvious that the construction has a general nature, and gives a proof of the sufficincy of Theorem 58.

Example 63. *Construct a finite automaton corresponding to a regular expression $E = x\{xyx \vee y\} \vee x\{y\}x$ over the alphabet $X = \{x, y\}$.*

Solution. We start by numbering all the letters in the expression starting from $x = 1$ and $y = 1$ independently; thus, it becomes $E = x_1\{x_2y_1x_3 \vee y_2\} \vee x_4\{y_3\}x_5$ over the alphabet $X = \{x, y\}$. Mark the characters that can be initial letters of the words in E; those are x_1 and x_4. Choose the vertices x_1, x_4 and an additional vertex O, and draw edges from O to x_1 and x_4; this is a seed of the graph of E:

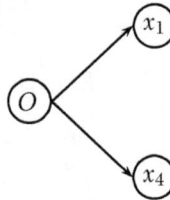

FIGURE 19.5 Graph of the automaton in Example 63. Initial step.

Next, we notice that in words from E, the letter x_1 must be followed either x_2 or y_2; thus, we draw the following layer of the graph:

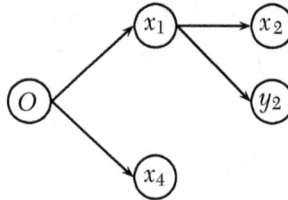

FIGURE 19.6 Graph of the automaton in Example 63. The second step.

After x_4, in the words from E, there may be only y_3:

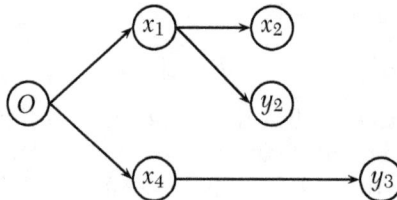

FIGURE 19.7 Graph of the automaton in Example 63. The third step.

We continue in a similar way and get *the graph of a regular expression E*, see Figure 19.8.

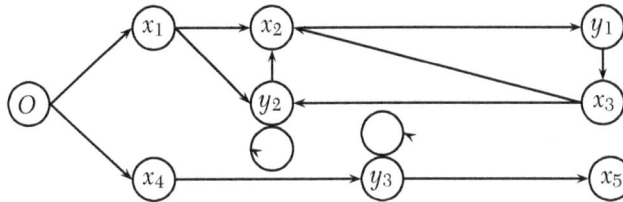

FIGURE 19.8 Graph of the regular expression E in Example 63.

Now we mark the edges, arriving at the vertices carrying the symbols x or y, with the same symbol x or y, see Figure 19.9.

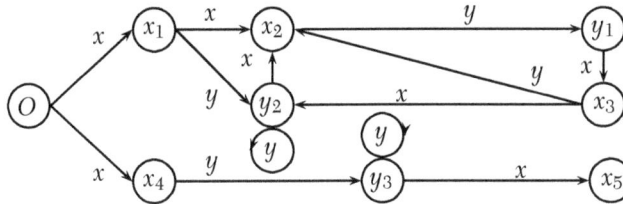

FIGURE 19.9 Marked graph in Example 63.

However, this graph does not have to be a graph of a deterministic well-defined finite automaton. For example, from the root O of this graph emanate two x–arrows and no y-arrow. So that, to finish the construction, we must do *optimal determination*, when certain subsets of the vertices of the developed graph, see Figure 19.9, will be the states of the future automaton. It is convenient to do that process as a series of tables.

We start with the following table, where the upper row contains only the state O, the left-most column contains (except for the upper cell) the entrance alphabet X, and the second column contains, at beneath the O, the states – its successors.

	O			...
x	$\{x_1, x_4\}$			...
y	$\varnothing$			...

The second row here shows that the inputs x move the state O to either $\{x_1\}$ or $\{x_4\}$, while there is no y–input leading from O. Therefore, we define two new states, $\{x_1, x_4\}$ and $\varnothing$.

	O	$\{x_1, x_4\}$	$\varnothing$	...
x	$\{x_1, x_4\}$			...
y	$\varnothing$			...

Now, x–arrows lead from x_1 or x_4 to x_2, and y–arrows to y_2 and y_3; hence, there appear two more new states, $\{x_2\}$ and $\{y_2, y_3\}$. The table extends to

	O	$\{x_1, x_4\}$	$\varnothing$	$\{x_2\}$	$\{y_2, y_3\}$	...
x	$\{x_1, x_4\}$	$\{x_2\}$	$\varnothing$	$\varnothing$	$\{x_2, x_5\}$	...
y	$\varnothing$	$\{y_2, y_3\}$	$\varnothing$	$\{y_1\}$	$\{y_3\}$	...

The states $\varnothing$, y_1, and y_3 have already occurred in the table, while the states x_3 and x_5 have just appeared, thus, the next table is as follows:

	O	$\{x_1, x_4\}$	$\varnothing$	$\{x_2\}$	$\{y_2, y_3\}$	$\{x_2, x_5\}$	$\{y_1\}$	$\{y_3\}$	...
x	$\{x_1, x_4\}$	$\{x_2\}$	$\varnothing$	$\varnothing$	$\{x_2, x_5\}$	$\varnothing$	$\{x_3\}$	$\{x_5\}$	...
y	$\varnothing$	$\{y_2, y_3\}$	$\varnothing$	$\{y_1\}$	$\{y_3\}$	$\{y_1\}$	$\varnothing$	$\{y_3\}$	...

There are two new states, x_3 and x_5, while the states $\varnothing$, y_1 and y_3 are already present in the table. The only new state y_2. At the next step no new state has appeared, hence, the determination process is over. The new notations for these states are introduced in the last row of the table.

	O	$\{x_1, x_4\}$	$\varnothing$	$\{x_2\}$	$\{y_2, y_3\}$	$\{x_2, x_5\}$	$\{y_1\}$	$\{y_3\}$	$\{x_3\}$	$\{x_5\}$	$\{y_2\}$
x	$\{x_1, x_4\}$	$\{x_2\}$	$\varnothing$	$\varnothing$	$\{x_2, x_5\}$	$\varnothing$	$\{x_3\}$	$\{x_5\}$	$\{x_2\}$	$\varnothing$	$\{x_2\}$
y	$\varnothing$	$\{y_2, y_3\}$	$\varnothing$	$\{y_1\}$	$\{y_3\}$	$\{y_1\}$	$\varnothing$	$\{y_3\}$	$\{y_2\}$	$\varnothing$	$\{y_2\}$
	s_0	s_1	s_2	s_3	s_4	s_5	s_6	s_7	s_8	s_9	s_{10}

Problem 454. *Draw the graph of the new automaton.*

It may be of interest to know what states of the built automaton represent the given event E. To do this, let us notice that the last symbols of the words from E can be only x_3, y_2, and x_5. Since these three letters appear only in the states s_4, s_5, s_8, s_9, and s_{10}, these and only these events represent the event E.

Problem 455. *Verify the last statement, that is, prove that the event E is represented by the states s_4, s_5, s_8, s_9, and s_{10}.*

This completes the proof of Kleene Theorem 58.

Proof of Lemma 32. To prove that X_0 satisfies equation $X = XE \vee F$, it is enough to insert X_0 into the equation and employ Problem 125. The regularity of this solution is obvious, and we only have to prove its uniqueness.

To do that, we iterate the equation, thus

$$X = XE \vee F = (XE \vee F)E \vee F = XE^2 \vee FE \vee F = \cdots$$
$$\cdots = XE^{n+1} \vee (FE^n \vee FE^{n-1} \vee \cdots FE \vee F).$$

If a word $w \in X_0 = F \vee F\{E\} = F \vee FE \vee FE^2 \vee \ldots$, then at some k, $w \in FE^k$. Now for $n \geq k$ we see that $w \in X$, hence $X_0 \subset X$. On the other hand, let $w \in X$. Every word from XE^{n+1} contains at least $n+2$ letters. So that, if $|w| < n+2$, then $w \in FE^n \vee FE^{n-1} \vee \cdots \vee F$, that is, $w \in X_0$. Therefore, we derive the converse embedding $X \subset X_0$, and the uniqueness follows.

Example 64. *To end this section, we present an example of a nonregular event, by following* [21].

Solution. Let $X = \{x, y\}$ and an event E consists of those and only those words that contain equal amounts of the symbols x and y. Suppose that the event E is represented in a finite initial automaton $\mathbb{M}(s_0)$ with the set of states S_E. Consider the states

$$s_0, s_1 = f(x, s_0), s_2 = f(x, s_1), \ldots, s_n = f(x, s_{n-1}), \ldots.$$

We show that for every n, all these states must be pairwise different. Indeed, otherwise, one can find the words w_k and w_m consisting only of the symbol x, such that $|w_k| = k, |w_m| = m$, $k \neq m$, and such that the states $s_k = s_m$. Let the word v consists of k symbols y. By their definitions, $w_k y \in E$ and $w_m y \neq E$. Then if we apply the word v to the states s_k and s_m, the following states are different, since one belongs to S_E, and another does not. This, however, contradicts the equation $s_k = s_m$. Hence, the event E cannot be presented in a finite automaton, and by the Kleene theorem, E is not regular.

19.4 EXERCISES

Exercise 19.1. *Construct the tables of the automata in Problems 446, 447, and 449.*

Exercise 19.2. *Let $X = \{x, y\}$ and an automaton without output is given by the graph in Figure 19.4. Find a regular expression of the event given by the states s_1 and s_2 in this diagram.*

20

INTRODUCTION TO GAME THEORY

The name "Game Theory" (GT) as it was used in the groundbreaking treatise of J. von Neumann, 1903–1957, and O. Morgenstern, 1902–1977, [38], and stays since then forever, is slightly unfortunate, since "games" often refer to entertainment, while the applications of the GT cover areas such as Operation Research, Decision Making, in particular Business Solutions and Military Actions, designing Efficient Auctions, and many other quite serious endeavors. The GT provides a mathematical framework for studying *conflicts* and *cooperation* between individuals and organizations of various complexity, up to blocks of states. After [38], there appear many good textbooks of the GT, for example, [41]. Leaving aside any discussion of the discrete nature of the GT, we give here the very succinct introduction that follows, to some extent, the excellent introductory text [48].

Consider, for instance, a well-known game, (which is just a game, really!) "Tic-Tac-Toe," also called "Noughts and Crosses," or "Xs and 0s," where two parties P_1 and P_2 (One – Crosses, Another – Noughts) in turn put their symbols into the available cells of the 3×3 chart. A player who first puts her symbols into three cells of any line (that is, a row, a column, or any big diagonal) wins the round of a game. We see that there are at most 9 moves in each round.

The kids quickly learn that if both sides play correctly (*"rationally"*), the game has no winner; the outcome is always a draw. We use this example to introduce some terminology of the GT, and then consider more examples. In this short foray, we consider only games with two players, *two-person games*. After a round is over, every player gets some *payoff*, also called the *utility* of this player; she either wins, or loses, or gets a tie; these results can be quantified as 1 for a win, 0 for a draw, and −1 for a loss. If players can make errors, then one player may win; hence her payoff is 1, and the other player loses with the payoff of −1.

Since there are only two players, the total *payoff* for both players at each round is $1 + (−1) = 0$. In the case of a tie, the total payoff is also $0 + 0 = 0$, so that, the "Noughts and Crosses" is a *Zero-Sum Two-Person Game*. It is not difficult to imagine a military conflict, such that in the end both sides destroy one another - no winner, or to imagine two companies producing similar products, that share the market, and both survive and get some profit – hence, no loser. Zero-sum games are a partial case of the games with a *constant sum*, but actually are equivalent to these games; however, in this short introduction we consider only zero-sum games.

It must be decided who starts the round, that is, puts her symbol first. To set this example in the set-theory language, consider a map BEG: $\{1, 2\}$ → $\{P_1, P_2\}$. Now the game is the two-element set $P = \{P_1, P_2\}$, where Pi stands for the ith Participant, or Player, or Party, and a map BEG, such that $BEG(1)$ makes the first move and initiates the round, that is, puts her symbol (0 or X) into any of 9 available cells – they are all empty yet. Now, the second player $BEG(2)$ follows, etc., until the round ends with either somebody's win or a tie. Numbering the cells of the upper row of the whole 3×3 square as 1, 2, 3, the cells of the second row as 4, 5, 6, and the cells of the lowest row as 7, 8, 9, inserting the zeros and crosses into the corresponding cells, and separating the zeros and crosses with commas, we can write down the result of any round as a sequence of at most 9 zeros and crosses. The next example completely describes one round of the game. This is called a *descriptive* presentation of the game.

Example 65. *For instance, if the first player draws crosses and the second one puts zeros, then the next 9 sequences:* $\{,,,,0,,,,\}$, $\{X,,,,0,,,,\}$, $\{X,,,,0,0,,,\}$, $\{X,,,X,0,0,,,\}$, $\{X,,,X,0,0,0,,\}$, $\{X,,X,X,0,0,0,,\}$, $\{X,0,X,X,0,0,0,,\}$, $\{X,0,X,X,0,0,0,X,\}$, *and* $\{X,0,X,X,0,0,0,X,0\}$ represent a typical round. There is no need to write the last sequence or to fill out the right-lower corner in the last chart, since the result is, obviously, a tie. The same round can be depicted as the next series of the 3×3 charts, starting with the empty chart.

	0	

X		
	0	

X		
	0	0

X		
X	0	0

X		
X	0	0
0		

X		X
X	0	0
0		

X	0	X
X	0	0
0		

X	0	X
X	0	0
0	X	

In the round of "Tic-Tac-Toe," shown above, the very first symbol occurs at the center of the field. However, the first player has two more options, namely, to place this symbol in the corner or in the middle of a side. These various options are called *strategies* of the player.

Problem 456. *Play the game using both these options, such that the first symbol is either in the corner or at the side. Do there exist the winning strategies in these cases?*

Problem 457. *In response, the second player also has several options. Are there among them winning strategies, or "Tic-Tac-Toe," if rationally played, is always the tie game?*

So that, we assume that every player has at least two (and, in general, a finite set of) strategies to choose from. These strategies are called *pure strategies*, and they prescribe what must be the next step of the player. Every pure strategy is the player's algorithm how to proceed in these circumstances, given that the player possesses all the *necessary information*. These games are called the *games with perfect information*, and below, we study only such games. In many cases, however, the players do have *imperfect information* and should *average several pure strategies*, where the weights of pure strategies are their probabilities – the results are called the *mixed strategies*. To simplify our exposition, we suppose that every player has exactly *two strategies*.

Let Σ stand for the set of strategies. A game can be described as the triple $G = \{P, BEG, \Sigma\}$. Under our assumptions,

$$\sigma = \{\sigma_1, \sigma_2\}; \text{ where } \sigma_1 = (\sigma_{11}, \sigma_{12}), \sigma_2 = (\sigma_{21}, \sigma_{22}).$$

In the two problems above, the reader should have constructed certain strategies for the "Tic-Tac-Toe" game.

The description of a game in Example 65 above, involving the charts, is called an *extensive form* or a *game tree*. We do not see the tree of the game here explicitly, but the above sequence of the charts can be easily converted into an equivalent tree whose leaves are the charts above.

The descriptive form is a very detailed, complete depiction of the game, prescribing each step of every player. A shorter and often more convenient description can be done if we represent a game in *matrix form*. Again, we discuss only two-person games with the zero-sum outcome. We also emphasize that since all the games here are two-person ones, no cooperation between the players is considered, and we study hereafter only *non-cooperative games*.

A shorter and often more convenient description of non-cooperative games is their representation as *matrices*, called the *strategic* or *normal form*. This form lists the available strategies for every player and the resulting *payoffs*, also called the *utilities* of this outcome for the player. Repeat that we consider only games whose matrices are symmetric with respect to the main diagonal.

TABLE 20.1 An Example of a Game in the Matrix Form.

I\II	$\sigma_{2,1}$	$\sigma_{2,2}$	Row minima
$\sigma_{1,1}$	**−2\3**	**1\0**	−2\0
$\sigma_{1,2}$	0\1	2\1	0\1
column maxima	0\3	2\1	0\1

The matrix representation of a two-person game is given in Table 20.1. The players here are denoted as *I* and *II*; the symbol $a \setminus b$ shows the outcome a for player I and the outcome b for player II. For instance, if I play the strategy $\sigma_{1,2}$ and II plays $\sigma_{2,2}$, then the payoff of I is 2, and that of II is 1.

Remark 20. *This matrix does not represent a game with a constant sum.*

The left-most column shows, in its two middle cells, the two strategies that player I has, namely, $\sigma_{1,1}$ and $\sigma_{1,2}$. The middle cells of the upper row show the two strategies, $\sigma_{2,1}$ and $\sigma_{2,2}$, available to player II. The 2×2 sub-matrix in the middle contains the **bold-faced** payoffs of the game. Since it is a non-cooperative game, we suppose that both players make their moves simultaneously and independently upon one another, without knowing the other player's choices. And of course, since they are rational players, they are choosing their moves to increase or, at the least, not to decrease their payoffs. The next definitions imply that the set of strategies equipped with a binary relation of (weak) dominance has a structure of the partially ordered space, see Lecture 6.

Definition 126. *If a player P has two or more strategies, then a strategy $\hat{\sigma}$ is called weakly dominant (w. d.) for P iff the payoff of the player P, using this strategy, is at least as good or better at least once than her payoff when she uses any other strategy σ, independently upon what strategy is used by any other participant Q of the game; weak domination is a binary relation on the set of strategies and is denoted as $\hat{\sigma} \succeq \sigma$.*

A strategy $\hat{\sigma}$ is called dominant for player P, or dominating a strategy σ, which is called a dominated strategy, iff the payoff of the player P using this strategy is better than her payoffs when she uses any other strategy σ independently upon what strategies are used by any other participant Q of the game. Denote the binary relation of domination as $\hat{\sigma} \succ \sigma$.

To demonstrate this binary relation in terms of the payoff matrix, we assume that P has only two strategies, $\hat{\sigma}$ and σ. The strategy $\hat{\sigma}$ weakly dominates a strategy σ, iff the P payoff in the cell $(\hat{\sigma}, \tau)$, is no worse or better than her payoff in any cell (σ, τ) of every τ-column.

TABLE 20.2 Strategy $\hat{\sigma}$ Weakly Dominates a Strategy σ iff $p \geq p'$ for all the Strategies τ of the Player Q, and $p' > p$ at Least in One Column of the Table.

P\Q	$\cdots$	τ	$\cdots$
$\hat{\sigma}$	$\cdots$	(p, q)	$\cdots$
σ	$\cdots$	(p', q')	$\cdots$

For instance, consider the payoffs for player I, Table 20.1 above. If she plays the strategy $\sigma_{1,1}$ and simultaneously player II plays the strategy $\sigma_{2,1}$, then we read out at the crossing of the intersection of the corresponding column and row, that player I gets the payoff **–2** and player II gets the payoff **3**. But if I play the strategy $\sigma_{1,2}$, then the payoff of I is now bigger than the previous one. Hence, the strategy $\sigma_{1,1}$ is not dominant. On the other hand, we can easily check that with $\sigma_{1,2}$, the maximal payoff of I is **2**, which cannot be increased, whichever strategy II has chosen. Thus, $\sigma_{1,2}$ is a dominant strategy for I.

We remark in passing that not every game has a dominant strategy, and a strategy, which is not dominant, nor is it dominated, is called *intransitive*. Thus, $\sigma_{2,1}$ is an intransitive strategy.

Problem 458. *Prove that the family of strategies with the (weak) dominance binary relation is a partially ordered set, see Chapter 6. What are its minimal, maximal, least, largest elements? Is it a linearly ordered set?*

The (weak) dominance leads to crucial notions in the game theory. Any game is a competition, where every side wants to improve its payoff, by making use of all the available strategies. If each player uses the best strategy, which gives her the largest possible payoff, we say that the game is in *equilibrium*. That combination of the strategies represents a *solution* and is the goal of the game.

20.1 CLASSIFICATION OF STRATEGIES

Dominant strategies, if one can find them, may be useful in solving games, since a game can be simplified by eliminating the dominated strategies. If by eliminating iteratively rows and columns, the game matrix can be reduced to just one row and one column, the cell at their intersection represents a solution, which does not have to be unique, that is, the winning equilibrium strategy for the game.

Problem 459. *Solve the following game, Table 20.3, by eliminating, one after another, the rows and columns. By changing the order of elimination, find two different solutions of the game.*

TABLE 20.3 The Matrix for Problem 459.

P\Q	$\sigma_{2,1}$	$\sigma_{2,2}$	$\sigma_{2,3}$
$\sigma_{1,1}$	3\7	2\4	2\7
$\sigma_{1,2}$	1\7	1\4	2\5
$\sigma_{1,3}$	1\7	2\4	1\8

Returning to Table 20.3 and repeating some reasoning, let us suppose that the second player, II, has chosen the strategy $\sigma_{2,1}$. Then the payoffs for the player I are either **−2** or **0**. Since $0 > -2$, the strategy $\sigma_{1,2}$ is better than $\sigma_{1,1}$ for player I, and must be chosen by I, assuming her *rational* play. If II plays another strategy, say $\sigma_{2,2}$, then the strategy $\sigma_{1,2}$ is also better for I, since her payoff in this case is also better, **2 > 1**. So that, the strategy $\sigma_{1,2}$ is weakly dominant for player I in the sense of Definition 126, since it gives at least as good or better payoffs for the player I, independently upon the choice of the strategy the other payer does. According to Definition 126, the strategy $\sigma_{1,2}$ is the *dominant strategy* for player I; $\sigma_{1,1}$ is the dominated strategy for I.

Look now at the payoffs of player II. If she uses the strategy $\sigma_{2,1}$, and I plays $\sigma_{1,1}$, then the output of II is **3 > 0**, but if II plays $\sigma_{2,2}$, then her output is worse, **0 < 1**. So that, player II does not have a dominant strategy.

In the real life, we have to make decisions every day, and in most cases, we do that intuitively. But sometimes, when we have to make important decisions, the game theory can help. If we can explicitly figure out our decisions and quantify the dominate and dominated strategies among them, then we can *eliminate* the latter out of consideration, and even predict, to some extent, the opponents' strategies, thus, staying with the narrower circle of possible choices.

Since a payoff can be any real number, there are infinitely many strategies, even if we consider only integer payoffs and two-person games. However, some typical transformations can essentially change the numerical outputs but preserve the character of the game. That is why there are few typical games under consideration, those are, for example, "Matching Pennies," "Rock, Paper, Scissors," "Prisoner Dilemma," "Stag Hunt," which provide game-theory models for many real-life situations. For instance, in the United States, and in many other countries, we drive on the *right*, while in some other countries (Japan, Great Britain), people drive on the left. This situation is well described as a *coordination game*. We finish this chapter and the book by considering in more detail the well-known game "Matching Pennies," which under different names and in various disguises often appears in the literature.

This is a two-person game, where we denote the players Odd and Even. In the beginning, each player has a penny. They flip them and simultaneously demonstrate the outcomes, Heads (H) or Tails (T). If both players have the same picture, then Even gets both coins, thus getting 2 cents, or 2 thousand of something, or what else. If both coins have different pictures, then Odd gets both coins, thus getting 2 cents. If both players wish, the game repeats, etc.

The following is a matrix representation of one round of this game. Here, column headers "Heads" or "Tails" mean claims of the Odd; similarly, the row headers mean the claims of Even. The symbol 1\–1 in a cell means that the Even receives 1 point and the Odd looses 1 point or "wins" –1 point, etc.

TABLE 20.4 Matching Pennies Game in Matrix Form.

Even\Odd	Heads	Tails
Heads	1\–1	–1\1
Tales	–1\1	1\–1

Problem 460. *Check that this is a zero-sum game.*

Every player has two strategies, to claim "Heads" or "Tails." Let us check whether either of these strategies is dominant or dominated. They were supposed the Odd plays the Heads strategy. Now, if Even plays the Tails, Odd's payoff is 1, which cannot be increased and is bigger than Odd's previous payoff. Thus, "Heads" is not a dominant strategy for the Odd. Similarly, "Tails" is not a dominant strategy either. Due to the symmetry, no player in this game has a dominant strategy, the best approach for everyone is to choose either "Heads" or "Tails" at random.

If only one player, say player I, has a dominant strategy, but the other does not, then a solution is obvious, I must steadily play the dominant strategy. If none has a dominant strategy, they both should play any strategy at random. What if both have dominant strategies? In this situation, no player has any incentive to change their strategy; thus the game came to an equilibrium, called the Nash equilibrium, and we leave that topic for deeper courses in Game Theory. In the end of the chapter, we study an ancient game, the game of "Nim."

Two players have in front of them three heaps of stones, or beads, or matches, or something similar. The player, who starts the game, chooses the heap and takes any positive number of stones from the heap, maybe even the whole heap. Then another player does the same, etc. The player, who takes the last stone, wins the round.[1] "NIM" is a typical strategy game; its analysis can be found, for example, in [18, Sect. 9]. The next example clearly demonstrates a possible solution to the game; it can be straightforwardly converted into the formal description of the strategies of the players.

Example 66. *Suppose that the 1st heap contains 11 items, the 2nd heap contains 5, and the 3rd heap contains 7 items. Solve the game, that is, determine the winner and find the winning strategy.*

---

[1] This is called the "NIM" in normal form. Keep in mind that there are versions of the game with the opposite condition: the player, who must take the last stone on the board, looses the game; this form is called the misère "NIM."

Solution. It is convenient to change these decimal integers into binary numbers and expand them as the sums of the powers of 2. The results are shown in the following Table 43, where all the sums in the right column are made of equal, largest length by inserting necessary quantities of 0s on the left. Thus, in Table 20.5, the longest line in the right-most column is the upper one,

$$1 \cdot 2^3 + 0 \cdot 2^2 + 1 \cdot 2^1 + 1 \cdot 2^0$$

and has the length of 4, so that some powers of 2 with zero coefficients appeared in the second and third lines,

TABLE 20.5 The Table for Example 66.

A (decimal) number Of items in the heap	Its binary Representation	A binary expansion Of this binary integer
The heap 1: 11 =	1011 =	$1 \cdot 2^3 + 0 \cdot 2^2 + 1 \cdot 2^1 + 1 \cdot 2^0$
The heap 2: 5 =	0101 =	$0 \cdot 2^3 + 1 \cdot 2^2 + 0 \cdot 2^1 + 1 \cdot 2^0$
The heap 3: 7 =	0111 =	$0 \cdot 2^3 + 1 \cdot 2^2 + 1 \cdot 2^1 + 1 \cdot 2^0$

An important role in the solution is played by the coefficients of the powers of 2^k, the right column of the table. These coefficients appear in the table as vertical binary three-vectors with $0 - 1$ components, which in more usual way can be written as $(1, 0, 0)$ - the coefficients of 2^3, $(0, 1, 1)$ - the coefficients of 2^2, $(1, 0, 1)$ - the coefficients of 2^1, and $(1, 1, 1)$ - the coefficients of 2^0. The components of these binary vectors are written from top-down and are either 0 or 1. Since there are three components in every such vector, their *decimal* sum can have only 4 values, 0, 1, 2, or 3. If the sum is 1 or 3, we call this vector an *odd vector*; otherwise, we call it an *even vector*; we obviously count 0 as an even number.

In "NIM" problems, we have to distinguish two kinds of data: class E contains the problems with all these three-vectors being even vectors; and class O containing the problems with at least one odd vector. The strategy depends about this parity. If a "NIM" problem belongs to the O class, then the starting player haas a winning strategy and wins (if she plays correctly independent upon the moves of the second player). Otherwise, that is, if all the initial three-vectors are even, then the second player wins if she plays correctly, independently upon the initial and all the other moves of he first player.

First, we consider the class O; for instance, the game in Example 66 belongs to that class. Here two vectors, $(1, 0, 0)$ of coefficients of 2^3, and $(1, 1, 1)$ of coefficients of 2^0 are odd vectors. In this case, the winning strategy is based on the following observation: If among the three-vectors of the coefficients there there is at least one *odd* vector, then the first player has a winning strategy, that is, the first player has a winning move. More specifically, there is a move, after which all the vectors become *even vectors*. After that, any move of the second player takes an even vector, and she has no choice but to change it to the odd vector again. Iterating these pairs of moves, we see, that the hame ends when the first player takes the last item on the board. Above describes the dominant strategy for the first player, if the initial configuration is odd.

Let us apply the verbal description above to our example. The left-most binary vector of the coefficients is $(1, 0, 0)$, where the 1 appears in the column of 2^3. Hence, if we take away $8 = 2^3$ items, that is, leave $11 - 8 = 3$ items in the first heap, the last, right-most vector is still odd. So that, proceeding this way, let us take away $8 + 1 = 9$ items, arriving at the next three heaps, where all the three vectors are even.

TABLE 20.6 The Table for Example 66 After the First Move.

A (decimal) number Of items in the heap	Its binary Representation	A binary expansion Of this binary integer
The heap 1: 2 =	0010 =	$0 \cdot 2^3 + 0 \cdot 2^2 + 1 \cdot 2^1 + 0 \cdot 2^0$
The heap 2: 5 =	0101 =	$0 \cdot 2^3 + 1 \cdot 2^2 + 0 \cdot 2^1 + 1 \cdot 2^0$
The heap 3: 7 =	0111 =	$0 \cdot 2^3 + 1 \cdot 2^2 + 1 \cdot 2^1 + 1 \cdot 2^0$

Suppose that now the second player takes away the whole third heap, and the next configuration is as in Table 20.7.

TABLE 20.7 The Table for Example 66 After the First Move of the Second Player.

A (decimal) number Of items in the heap	Its binary Representation	A binary expansion Of this binary integer
The heap 1: 2 =	0010 =	$0 \cdot 2^3 + 0 \cdot 2^2 + 1 \cdot 2^1 + 0 \cdot 2^0$
The heap 2: 5 =	0101 =	$0 \cdot 2^3 + 1 \cdot 2^2 + 0 \cdot 2^1 + 1 \cdot 2^0$
The heap 3: 0 =	0000 =	$0 \cdot 2^3 + 0 \cdot 2^2 + 0 \cdot 2^1 + 0 \cdot 2^0$

Now the vectors are $(0, 0, 0)$, $(0, 1, 0)$, $(1, 0, 0)$, $(0, 1, 0)$, and we again have at least one odd vector. The first player can remove three items from the second heap, again returning to all even vectors.

TABLE 20.8 The Table for Example 66 After the Second Move of the First Player.

A (decimal) number Of items in the heap	Its binary Representation	A binary expansion Of this binary integer
The heap 1: 2 =	0010 =	$0 \cdot 2^3 + 0 \cdot 2^2 + 1 \cdot 2^1 + 0 \cdot 2^0$
The heap 2: 2 =	0010 =	$0 \cdot 2^3 + 0 \cdot 2^2 + 1 \cdot 2^1 + 0 \cdot 2^0$
The heap 3: 0 =	0000 =	$0 \cdot 2^3 + 0 \cdot 2^2 + 0 \cdot 2^1 + 0 \cdot 2^0$

The end of this game is obvious. If the initial configuration is even, then as we remarked above, any move of the first player changes it to odd, and now the second player can apply the winning strategy, described above for the first player. To describe this game in matrix form, we must determine the *payoffs* of these strategies precisely, One way to do that is to count the total number of the removed items separately for each player.

Problem 461. *What are other possible ways to define payoffs in this problem?*

20.2 EXERCISES

Exercise 20.1. *Draw the tree of the game for Example 66*

Exercise 20.2. *Draw the Hasse diagram for strategies in this example.*

CHAPTER 21

INFORMATION THEORY AND CODING

People are searching for information, people are encrypting information, people are hiding information... The word "information" has been used even in this book several times already, however, without any explanation. As most of the people, we do not define what information is. For us, information is a primary, undefined concept, like a function, or a set, or time. We live in time but fail to give a definition. However, we can define time intervals, like an hour, or a year, or a minute.

When we read a book or listen to the radio, we get some information. When we talk with our friends or instructors, we exchange the information. To exchange the information, we must code it. The information coding is an important part of discrete mathematics, and we introduce it in this chapter.

If we do not know where to turn at a road fork, and a stranger (or GPS) says "Make left," we thus resolve a "yes-no" question, so that we got a unit piece of information, the *one bit*.[1]

The crucially important fact is that the information must be transferred, sometimes, very far away, so that it is physically impossible to use sound, acoustical waves. Thus, the information must be transformed and carried by electromagnetic oscillations – we don't know now yet another kind of waves, which can propagate far enough. Hence, the information must be represented as the sequence of ups and downs, and eventually must be *digitized*, that is, must be written as the sequence of *zeroes* and *ones*. This change of representation of the information is called *coding*.

---

[1] *A bit* is a contraction of "binary unit," however, our usage here is far from the literary meaning of these words.

So that, the information (for us) is a sequence of *binary words – messages*, written in the *binary alphabet* {0, 1}. Any transmission of information is a transformation of these binary words, which is called *coding*.

The words "Code," "Coding" are overloaded. Most computer programmers mean by the "coding" writing a computer program in any algorithmic language, like Python. However, many other people would think about coding as representing a certain text in a different alphabet, for instance, translating from English to Chinese, or vice versa. For example, the ASCII (American Standard Code for Information Interchange) encodes English characters, digits, etc., so that they can be stored in computers.

Problem 462. *The words in the Morse code consist of two symbols only:* 0 *and* 1. *Does there exist an alphabet consisting of just one symbol?*

We use, in this chapter, the binary alphabet $\mathbb{B} = \{0, 1\}$ for both the input and the output alphabets. This is possible since any decimal digit, and moreover, any finite set of characters, can be written in the binary system. Any text can be represented by using just these two Boolean symbols. The need for using this alphabet occurs very often. Suppose you talk to your friend in a classroom. Your speech is carried on with the oscillations of the surrounding air. These oscillations are only heard at very short distances, and if you want to talk to your friend in another state or even in another building on the same campus, you are to use the telephone, that is, it is necessary to transfer low-frequency acoustical oscillations into high-frequency electromagnetic ones. Thus, we *must* digitize our messages, and then to transmit them further through a certain channel.

More specifically, to derive a formula for the amount of information $I = I(k)$, where k is the length of a message, consider an example. Let our road contain not one, but $k \geq 1$ forks, and at every of them, we must resolve the same "yes-no" question. Therefore, the total answer is a sequence of k partial answers, which together have 2^k possible answers. Moreover, every particular answer decreases our uncertainty, indeterminacy, the lack of information exactly in half. Denoting $2^k = E$, we get $k = \log_2 E$, where $\log_2 x$ stands for the binary, that is, base 2, logarithm.

The logarithmic function inevitably occurs here because the following three assumptions are made, maybe implicitly.

I. If the road fork has more existing ways, in other words, if we have more choices, and we got the correct answer, we obtain more information; thus, the *quantity information function I(k)*, we are designing, must be a *monotonically increasing function*, that is, $I(k) \geq I(l)$ for $k \geq l$. Of course, we assume that all the outcomes, that is, all the ways out, are equally likely. The probability terms "equally likely," etc., are discussed in the next chapter.

II. If we make two independent experiments, the total amount of information is additive, that is, $I(k \cdot l) = I(k) + I(l)$. The product $k \cdot l$ here stands for the concatenation of two independent experiments of the lengths k and l, respectively.

III. If the sample space of a random experiment has only one element, that is, the outcome is certain and the experiment cannot give any information, we are to normalize this function as $I(k = 1) = 0$.

Moreover, if we add a non-essential smoothness assumption, such as the continuity of the information function, then up to a multiplicative factor, there is only one function, satisfying all of these properties, namely, the *binary logarithm*,

$$I(k) = m\log_2 k.$$

Due to Lemma 1 in Lecture 1 (the Change of Base formula) one can change the base of logarithms here for any convenient value, which results only in changing the constant coefficient m; we prefer the binary logarithms and $m = 1$, which simplifies some formulas and gives the amount of information $I = \log_2 k$ in bites. In the "fork"-example above, if all the ways are equally likely and $p = 1/k$, then

$$I = -\log_2 p.$$

It is worth mentioning that p here is a probability of a certain event, hence, $0 \le p \le 1$ and $-\log_2 p \ge 0$.

If the outcomes are not equally likely, let us assume that the outcome k_i occurs with the probability p_i, hence the logarithms $\log_2 p_i$ must appear in the final formula for the quantity of information. It is necessary to average partial amounts and we arrive at the well-known formula for an averaged posteriori amount of information,

$$\overline{I} = -\sum_{k=1}^{n} p_k \log_2 p_k. \tag{21.1}$$

The following problem or alike often occur in various subjects.

Problem 463. *Suppose that you have 10 coins, and one among them is false, lighter than a good coin. Given a scale without weights, how many weighings are necessary to find the false coin?*

Solution. According to the formulas above, the message "A randomly chosen coin is false" gives the amount of information

$$I_1 = -\log_2(1/10) = \log_2 10 = \ln 10 / \ln 2 \approx 3.32$$

bits (of information), where $\ln x$ is the natural logarithm. Any experiment with the scale, that is, a weighting, has three outcomes; therefore, one weighting gives $I_2 = -\log_2(1/3) \approx 1.58$ bits of information. Since, $3.32/1.58 \approx 2.10$, the number of weightings must be at least $[2.10] + 1 = 3$, and indeed, you can find the false coin in three weighting. For example, it can be done as follows. First, pick any two groups of three coins among the given 10, and compare them on the scale. If they have the same weight, the false coin is among the four coins that initially were left out, and we must compare them by using two comparisons, that is, two weightings. Otherwise, pick the three coins, which are lighter, and compare two of them, leaving one outside.

Problem 464. *Solve the same problem, if we do not know whether the false coin is lighter or heavier than a fair one.*

Problem 465. *Let there be scales without weights, k fair coins, and one counterfeit coin, which weighs less than a fair coin. What is the smallest number of weightings, which is necessary to find the wrong coin?*

Problem 466. *An experiment has two outcomes with the probabilities of 0.35 and 0.65. Find the exact and average amounts of information given by this experiment.*

Solution. Using the formulas above, the exact probabilities of these outcomes are $I_1 = -\log_2 .35 \approx 1.51$ and $I_2 = -\log_2 .65 \approx 0.62$, while the average amount of information is $1.51 \cdot 0.35 + 0.62 \cdot 2.65 \approx 0.93$ bits.

21.1 MEASURE OF INFORMATION

So far we considered the quantity of information, I. However, in many problems it is useful to consider the negation of this quantity. Suppose that we have a random experiment with finitely many outcomes. Before the experiment, each outcome X has a certain *a priori* average amount of indefiniteness. This average amount $H(X)$ is called the (a priori) entropy of the outcome X. It is clear that the formula for the entropy is the same as the formula for the negation $\overline{I}$ above. We leave any discussion of the entropy for the more advanced courses,[2] just mention its occurrence in the cryptography book [13], which, of course, is not a random coincidence.

---

[2] It is worth mentioning, that the first modern development of the coding theory was made by Claude Shannon, 1916-2001, [46], based in part, on the ideas of Norbert Wiener.

21.2 CHANNELS

The word "channel" is the crucial concept here. Centuries ago people knew that any road has a certain specified maximum *capacity*; however, the capacity of telephone wires seemed to be endless. Now we understand well that any channel has certain maximum capacity.

However, *no physical channel* is *noiseless*. The noise can corrupt the signal, that is, introduce some errors. The errors here mean that sometimes instead of 0 we receive 1, or vice versa. So that, the code, used to digitize the signal, should *detect some errors* in the transmitted signal and, even better, should *automatically* correct them. Such codes are studied in the algebraic coding theory, which is introduced in this chapter. First, we remind a few definitions, see, for example, [4].

Any transmitted message consists of words $w \in \mathbb{B}^m$, that is, a message is a finite sequence of m 0's and 1's. The system transforms this word into an output word of n 0's and 1's, which is a word in $\mathbb{B}^n$; this mapping is denoted as $c : \mathbb{B}^m \to \mathbb{B}^n$ ("c" is for "code"), and is called an *encoding function*. The $c(w)$ is called the *code-word*. If $n < m$, the system will inevitably merge, lose some symbols, that is, certain information, and we never recover the initial word. So that, if we want the code to be able to detect errors and, even better, automatically correct them, the system must have certain redundancy, thus it must be $n > m$. Suppose for example, $m = 3$ and $n = 5$. Then $|\mathbb{B}^3| = 2^3 = 8$ and $|\mathbb{B}^5| = 32$. Hence, every word-preimage has $2^5 \div 2^3 = 4$ possible corresponding words-images. Very roughly, these words create some neighborhood around the word $c(w)$, where the images "live" without interfering with one another.

The simplest way to detect finitely many errors is to use the *repetition code*. Suppose we are certain that the code-word contains at most *one error*. This means that if we received the word $w_r = (0101)$, we are certain that the sent word was either the word w_r itself, or maybe, one of the four words $w_1 = (1101)$, or $w_2 = (0001)$, or $w_3 = (0111)$, or $w_4 = (0100)$. To detect this error, let us transmit this word not once, but *three* times. Since we have at most one error, at least two among these three received words must coincide; hence, we accept the repeated word as the true transmitted word, and correct the received information correspondingly.

An essential drawback of the repetition code is very large increase in the lengths of transmissions, which is not always acceptable, for instance, at the communications in space. Another problem occurs with the number of possible corrupted words. If we consider 4-character words, there are

$4 = C(4, 1)$ possible corruptions. If two errors are possible, then there possible $C(4, 2) = 6$ corruptions, and this amount grows very quickly. Hence, it is necessary to look for more economical ways to detect and correct errors.

Problem 467. *Suppose there are c characters, and every code-word consists of s symbols. How many corrupted words is it possible to receive (and check), if a channel can make at most r errors? How many times must we repeat each word to be able to detect an error? To be able to correct the error?*

As was remarked above, an encoding function must be injective, otherwise, the decoding will fail. The simple example above shows that we can have certain empty space around the image $c(w)$ for each word $w \in \mathbb{B}^m$; namely, in the example there are 4 images and in general there are 2^{n-m} words-images. That allows the images to be "separated" and allows us to detect errors, since if any image gets into the "empty space," that indicates a transmission error.

The number of 1's in a word w is called the *weight* of w, and is denoted as $|w|$; for example, the weight of $|(1,0,0,1,1)| = 3$. We see that the weight depends only upon the number of 1's in a word, but not on the dimension of the ambient space. The following definition belongs to R. Hamming, 1915-1998.

Definition 127. *The sum modulo 2 of the Boolean vectors $a = (a_1, a_2, ..., a_n)$ and $b = (b_1, b_2, ..., b_n)$ is defined component-wise, as*

$$\rho(a, b) = |a \oplus b|,$$

where $a \oplus b$ is calculated component-wise, and is called the Hamming distance.

For example, $\rho((1, 0, 1, 0, 1), (0, 0, 0, 1, 1)) = 3$, $\rho((1, 0, 1, 0, 1), (1, 1, 1, 1, 1)) = 2$.

Since $0 \oplus 0 = 1 \oplus 1 = 0$ and $0 \oplus 1 = 1$, only different components of the Boolean vectors a and b give nonzero input into this sum, hence, equivalently, the Hamming distance between two Boolean vectors a and b can be defined as the total number of different components in a and b, which can be also written as $|a \oplus b|$.

Lemma 33. *The Hamming distance $\rho(a,b)$ satisfies the following three standard axioms of the metric in Hilbert space,*

$\rho(a, b) \geq 0$, and $\rho(a, b) = 0$ iff $a = b$
$\rho(a, b) = \rho(b, a)$, $\forall a, b \in \mathbb{B}^n$
$\rho(a, b) \leq \rho(a, c) + \rho(c, b)$, $\forall a, b, c \in \mathbb{B}^n$.

Proof. The first two properties are obvious. The third one, called the *triangle inequality*, follows directly from the triangle inequality in $\mathbb{B}^m$, that is, from the inequality $|a \oplus b| \le |a| + |b|$. The latter is evident, since any component, such that a and b are the same, contributes 0 to the left-hand side, while at any component, where the left-hand side is 1, at least one of a and b must have 1, thus, contributing 1 to the right-hand side.

The next feature shows how well the different code-words of a given encoding function are separated from each other.

Definition 128. *Given an encoding function $c : \mathbb{B}^m \to \mathbb{B}^n$, the smallest among the distances between the images $c(a)$ and $c(b)$ over all the pairs of the code-words a and b, is called the minimum distance of the encoding function c.*

Example 67. *Let $m = 3$, $n = 3$, and the encoding function c is given by the Table 21.1.*

TABLE 21.1 Example 67.

$\mathbb{B}^3$	$c(\mathbb{B}^3)$
(0,0,0)	(0,0,0,0,0,1)
(0,0,1)	(0,0,0,0,1,0)
(0,1,0)	(1,0,0,1,0.0)
(0,1,1)	(1,1,1,0,0,0)
(1,0,0)	(1,1,1,1,0,0)
(1,0,1)	(0,0,0,1,1,1)
(1,1,0)	(1,1,1,1,1,1)
(1,1,1)	(0,1,0,0,1,1)

Here, the minimum distance is $\rho((1,1,1,0,0,0),(1,1,1,1,0,0)) = 1$.

This quantity shows the amount of free space available about the corrupted code-words. That space can be used to determine the corrupted code-words and, maybe, to correct them.

Definition 129. *The code-word $x \in \mathbb{B}^n$, whose transmission image is denoted as x_{trans}, can be transmitted with no more than κ errors iff x and x_{trans} have at most κ distinct symbols. The encoding function c detects at most κ errors, iff, when the code-word $c(w)$ was transmitted with at most κ errors, it is not a code-word; hence, it is detected as an error.*

In Chapter 10, we discussed the parity check-digits in ISBN. These extra digits allow the user to control, only partially though, the correctness of the book codes. The algebraic codes, we are discussing now, are more powerful means to verify whether a transmission was correct.

Theorem 60. *An encoding function* $c : \mathbb{B}^m \to \mathbb{B}^n$ *can detect at most κ errors iff its minimum distance is no less than $k + 1$.*

Proof. Let the minimum distance between two code-words is at least $\kappa + 1$. Take a word $b \in \mathbb{B}^m$, let $x = c(b) \in \mathbb{B}^n$ be the corresponding code-word, and its image under translation was received as x_t. If the code-word $x_t \neq x$, the distance between x and x_t must be $\rho(x, x_t) \geq \kappa + 1$, so that x would be transmitted with at least $\kappa + 1$ errors. Hence, if x were transmitted with at most κ errors, then the x_t cannot be a code-word. Whence, the encoding function c can detect at most κ errors.

Conversely, let x and y be code-words, such that the minimum distance is $\rho(x, y) = r$, where $r \leq \kappa$. If x was transmitted and was received, by error, as y, then $r \leq \kappa$ errors were made but were not detected. Hence, the encoding function cannot, in general, detect κ errors with certainty. In particular, the (7, 4)-Hamming code can correct at most one error.

To extend the possibilities of a coding function, we can try to enlarge the co-domain by adding extra dimensions.

Example 68. *Determine, how many errors can detect the encoding function* $c : \mathbb{B}^3 \to \mathbb{B}^5$ *defined by the equations*

$$c((0,0,0)) = (0,0,0,1,1)$$
$$c((0,0,1)) = (0,0,1,0,1)$$
$$c((0,1,0)) = (1,0,0,0,1)$$
$$c((0,1,1)) = (1,1,0,0,1)$$
$$c((1,0,0)) = (0,1,0,0,1)$$
$$c((1,0,1)) = (0,1,1,1,1)$$
$$c((1,1,0)) = (1,1,1,1,1)$$
$$c((1,1,1)) = (0,1,0,1,1).$$

Solution. The minimum distance here is

$$\rho((1,0,0,0,1),(0,1,0,0,1)) = 1 = \kappa + 1,$$

therefore, $\kappa = 0$, and this code in general, cannot detect any error.

We can detect no error, but we can try to modify the code-words.

Example 69. *Find how many errors can detect the encoding function* $c : \mathbb{B}^3 \to \mathbb{B}^6$ *given as*

$$c((0,0,0)) = (0,0,0,1,0,1)$$
$$c((0,0,1)) = (0,0,1,0,1,0)$$
$$c((0,1,0)) = (1,0,0,0,0,1)$$
$$c((0,1,1)) = (1,1,0,0,1,0)$$
$$c((1,0,0)) = (0,1,0,0,0,1)$$
$$c((1,0,1)) = (0,1,1,1,1,0)$$
$$c((1,1,0)) = (1,1,1,1,0,1)$$
$$c((1,1,1)) = (0,1,0,1,0,0).$$

Solution. Now the minimum distance is $\rho((0,1,1,1,1,0),(0,1,0,1,0,0)) = 2 = \kappa + 1$, therefore, $\kappa = 1$, and this code can detect 1 error.

Problem 468. *Consider code-functions* $c : \mathbb{B}^3 \to \mathbb{B}^5$; *there are* $5^3 = 125$ *of them. Can you construct a code that detects 1 error?*

Problem 469. *By increasing the dimension n of the codomain* $c : \mathbb{B}^3 \to \mathbb{B}^n$, *construct codes which detect 1,2, 3… errors.*

Problem 470. *Determine how many errors can detect and how many correct the code with the characteristic function (1)* $f(x) = x_1 \oplus x_2 \oplus \cdots \oplus x_n$ *(2)* $f(x) = x_1 x_2 \cdots x_n \vee \overline{x_1} \, \overline{x_2} \cdots \overline{x_n}$.

Problem 471. *When the words of the binary code* $C = \{0110, 1010, 1110\}$ *are transmitted through a communication channel, at most one error can occur. What words can be received after the code C is transmitted?*

It is also possible to *decrease* the dimension of the domain of a code-function, but in most cases, this is unacceptable since we can lose important information.

In Theorem 60, a criterion was developed for finding the amount of the errors committed. A similar but more involved criterion for correcting the errors can be found, for example, in [32].

21.3 EXERCISES

Exercise 21.1. *Solve Problem 466 given that the experiment has three outcomes with the probabilities 0.2, 0.4, 0.4.*

Exercise 21.2. *Among eight coins one is false. It is known that the false coin is lighter than a good one. Given a scale without weights, how many trials are necessary to find the false coin?*

Exercise 21.3. *Determine how many errors can detect the encoding function*

$$c : \mathbb{B}^3 \to \mathbb{B}^4,$$

$$c((0,0,0)) = (0,0,0,1)$$
$$c((0,0,1)) = (0,0,1,0)$$
$$c((0,1,0)) = (1,0,0,0)$$
$$c((0,1,1)) = (1,1,0,0)$$
$$c((1,0,0)) = (0,1,0,0)$$
$$c((1,0,1)) = (0,1,1,1)$$
$$c((1,1,0)) = (1,1,1,1)$$
$$c((1,1,1)) = (0,1,0,1)$$

Solution. The minimum Hamming distance in the example is $p((0,0,0,1),(0,1,0,1)) = 1 = \kappa + 1$, therefore, $\kappa = 0$ and this code, in general, cannot detect any error with certainty.

CHAPTER 22

PROBABILITY THEORY WITH A FINITE SAMPLE SPACE AND THE BIRTHDAY PROBLEM

Any operation, procedure, experiment, such that its results cannot be predicted in advance, like tossing a coin, is called a *random experiment*. This is not a definition, we just introduce a *primary notion* like the concepts of a set or a function introduced in Chapter 3. If an experiment has the only possible result, the outcome is certainly known in advance, and this experiment is not random.

22.1 RANDOM EXPERIMENTS

Definition 130. *All the possible results of a random experiment are called its outcomes. The totality of all possible outcomes is called the sample space S of the experiment. The elements of the sample space, that is, the possible outcomes of a random experiment, are also called elementary events. Any set E of outcomes, that is, any subset $E \subset S$, is called an event. The empty event $E = \varnothing$ is also called impossible or improbable, the universal event $E = S$ is called certain.*

The outcomes, belonging to a given event, are called *favorable* outcomes for this event. The sample space depends upon the problem. For instance, when we throw a coin, then in addition to two typical outcomes, heads and tails, a coin might rest on the edge, even though this phenomenon is not easy to observe, or it can roll away and disappear, but the latter two possibilities are practically improbable, they are negligible. Thus, while discussing experiments with flipping a coin, we always consider the sample space consisting of only two points, a head H and a tail T, $S = \{H, T\}$. If we roll a die (a six-faced cube) with faces marked by the digits 1, 2, 3, 4, 5, 6, or by several dots, then the sample space of this random experiment is $S = \{1, 2, ..., 6\}$. However, if a die is marked by $\{1, 2, 3, 4, 5, 5\}$, then the sample space of the experiment is $S = \{1, 2, 3, 4, 5\}$.

Example 70. *Consider a lottery with prizes of $1, $5, $100, and $10 000. The drawing is a random experiment and if we have only one ticket, the sample space consists of five points, $S = \{\$0, \$1, \$5, \$100, \$10\,000\}$. However, in some cases, we may only be interested in the very fact of winning (W) or loosing (L) the game and can choose another sample space, say $S_1 = \{W, L\}$. Depending on the issue we are interested in, there may also be other possible choices for the sample space in a problem.*

To correctly solve a probabilistic or combinatorial problem, one must explicitly specify the sample space of the problem; otherwise, different people can read the same words in different ways and arrive at different conclusions. Hereafter we consider only random experiments with *finite* sample spaces.

In many problems, it is necessary to consider composite events, consisting of simple ones, and combine simple sample spaces in more complex spaces.

Problem 472. *Describe the sample space, if two distinguishable coins are rolled simultaneously.*

Solution. Since the pair $\{H, T\}$ and $\{T, H\}$ are indistinguishable, the sample space consists of three elements $\{H, H\}$, $\{T, T\}$, $\{H, T\}$, where the first two can be identified with $\{H\}$ and $\{T\}$, respectively. However, if the coins are distinguishable, then the sets $(H, T) \neq (T, H)$ are different, and by the Product Rule, the new sample space contains 4 points. If three different fair dice are rolled simultaneously, then the sample space consists of $6^3 = 216$ ordered triples, $S = \{(1, 1, 1), (1, 1, 2), ..., (6, 6, 6)\}$.

22.2 PROBABILITY DISTRIBUTIONS

Up to this point, we have discussed only sample spaces. The probability theory originates when a certain specific number $p(s)$, called the probability of the outcome s[1], is assigned to each point $s \in S$ of the sample space. The

---

[1] One can often hear in everyday talk, "It's probable" or "That's unlikely." Based on such individual judgment, some people play lotteries while the others do not, because they do not believe that there are reasonable chances to win. Any discussion of such subjective probabilities is beyond the scope of this book.

set of these values is called a *probability distribution* on the sample space S, because we distribute a certain given "supply" of probability among the points of S. These values cannot be assigned arbitrarily, though, they must satisfy certain assumptions, axioms of the probability theory; for more on that see, for example, [17]. We consider the following system of axioms:

PA1) $p(s) \geq 0$ for any point $s \in S$
PA2) if $E = \{s_1, s_2, \ldots, s_k\} \subset S$, then $p(E) = p(s_1) + p(s_2) + \cdots + p(s_k)$
PA3) $p(S) = 1$.

Therefore, we have assumed that probability values are nonnegative, the probability of any event E is the sum of the probabilities of elementary events composing E, and the total probability is 1. These axioms immediately imply that if $E_1, \ldots, E_k$ are any pairwise disjoint events, that is, $E_1, \ldots, E_k \subset S$ and $E_i \cap E_j = \varnothing, 1 \leq i, j \leq k$, then $p(E_1 \cup \cdots \cup E_k) = p(E_1) + \cdots + p(E_k)$, that is, the probability is *finitely additive*. Moreover, for any event E we have $p(E) = p(E \cup \varnothing) = p(E) + p(\varnothing)$, thus, $p(\varnothing) = 0$, the empty event must have zero probability.

In some cases, we can conduct a random experiment in reality; for instance, we can toss a coin many times and record the numbers of heads, $n(H)$, and tails, $n(T)$, occurred. If the experiment was repeated n times, the favorable outcomes for an event E were observed $k(E)$ times among the n outcomes, and the sequence $\{k(E)\}$ is stabilizing (has the limit) $f(E)$ when $n \to \infty$, then we can claim that the event E has the experimental or frequency probability. The ratio $f(E) = \dfrac{k(E)}{n}$ is called the *experimental* or *frequency probability* of the event E. Clearly, the frequency $f(E)$ depends, among other things, on the length n of the experiment. If with n increasing, $f(E)$ is stabilizing to a number $p(E)$, we can use $f(E)$ as an estimation of the probability $p(E)$ of the event E, but this is only a plausible approximation.

Indeed, there is nothing unusual to get two heads in a row, thus in this series of $n = 2$, the experimental probabilities are $p(H) = 1 / 1 = 1$ and $p(T) = 0$. However, if we use this very short sequence to estimate the probability of getting a tail, we have $p(T) = f(T) / 2 = 0$, which obviously makes no sense. Advanced courses in probability theory treat this issue in more detail – namely, it is discussed what is the appropriate length of an experiment.

Any set of positive numbers, satisfying axioms PA1)-PA3), can be used as a probability distribution. For example, experimenting with a coin and choosing the sample space $S = \{H, T\}$, we can assign $p(H) = 1 / 3$ and $p(T) = 2 / 3$. However, unless we have a specifically tailored (biased) coin, the results of our physical experiments will likely be essentially different from the results predicted by the mathematical model. So that, to assign a probability distribution, we use either some previous experience (the results of real experiments) or a theory, if it exists.

It is physically impossible (probably) to make a perfect coin; however, real experiments have confirmed that if a coin was chosen at random, then as the first approximation, it is quite realistic to assign the probabilities $p(H) = p(T) = 1/2$. On the other hand, the same experiments show that no real coin satisfies this probability distribution precisely but exhibits slight deviations from the theoretical probability 1/2. Nevertheless, it is customary in theoretical studies to accept the hypothesis of *equally likely probabilities* or equal chances[2], that is, to assign equal probabilities to each point of the sample space.

Definition 131. *It is said that the assumption of* equally likely probabilities *is valid for a given problem with the sample space* $S = \{s_1, s_2, ..., s_n\}$ *iff the probability distribution on S is given by n equations*

$$p(s_1) = p(s_2) = \cdots = p(s_n) = \frac{1}{n}.$$

Whether or not this assumption holds true in any particular case, should be verified by comparing our calculations with experiments. The following well-known example is illuminative. Our intuition might tell us that the number of girls born must on average be the same as the number of boys, and many computations using the equal probabilities 1 / 2 as the first approximation, give satisfactory results. However, the many-year observations have shown that in reality, the probability for a newborn baby to be a boy is slightly bigger, namely 0.51, versus 0.49[3] for a girl.

Hereafter, we suppose the hypothesis of equally likely probabilities to be valid, unless the opposite is explicitly stated.

The goal of this section is to show applications of the developed combinatorial results to the probability theory. First, we translate a few basic set-theory notions to probabilistic language. Remind that an event is just a subset of the basic (universal) set S, which is called hereafter the *sample space*. All the events under consideration are subsets of some fixed sample space S. Therefore, we can define the following operations with events through their set-theory counterparts.

---

[2] The terms *probability* or *chance* should not be confused with the term *odds*. The expression "odds in favor of an event E" means the ratio $\frac{p(E)}{p(\overline{E})}$, while the expression "odds against an event E" means the reciprocal ratio $\frac{p(\overline{E})}{p(E)}$.

[3] Some data indicate that this gap (maybe!) is shrinking.

Definition 132. *(1) The events E and $\overline{E} = S \setminus E$ are called complementary.*

(2) Two events are called disjoint or mutually exclusive iff their set-theory intersection is empty, that is, iff they have no common favorable outcomes. Thus, if E_1 and E_2 are disjoint events, then $p(E_1 \cap E_2) = 0$.

(3) A system of events $\{E_1, \dots, E_k\}$ is called exhaustive iff $\bigcup_{i=1}^{k} E_i = S$.

Example 71. *Let us toss a coin and choose the sample space $\{H, T\}$. Then the events "To get an H" and "To get a T" are disjoint, complementary, and together exhaust the sample space. The events "To get an odd number" and "To get a number less than 3" in one rolling of a die are not mutually exclusive. The event, complementary to "To get a number less than 3" is "To get a number greater than or equal to 3." Any event and its complement one make up an exhaustive system.*

The following properties are simple consequences of the definitions, of the axioms PA1–PA3, and of the elementary properties of sets, such as the Inclusion–Exclusion Principle, see Chapter 3. Remind that we consider only finite sample spaces; thus, all probability distributions are finitely additive.

Theorem 61. *(1) For any event E,*

$$p(\overline{E}) = 1 - p(E). \tag{22.1}$$

(2) For any events E_1 and E_2,

$$p(E_1 \cup E_2) = p(E_1) + p(E_2) - p(E_1 \cap E_2). \tag{22.2}$$

In particular, if E_1 and E_2 are disjoint, that is, $E_1 \cap E_2 = \varnothing$, then $p(E_1 \cap E_2) = 0$, and $p(E_1 \cup E_2) = p(E_1) + p(E_2)$.

(3) If $E \rightarrow E_1$, that is, event E implies event E_1, then $p(E) \leq p(E_1)$.

Remark 21. *Of course, case (2) is the Sum Rule and the Inclusion–Exclusion Principle translated into probabilistic terms.*

Problem 473. *A green and a blue fair right tetrahedra with faces marked 1 through 4, were tossed. We record the numbers on the faces they landed.*

(1) What is the probability that the sum of these numbers on the two pyramids is 7?

(2) What is the probability that the sum of these numbers is greater than or equal to 7?

(3) What is the probability that the sum of these numbers is greater than 7? Is greater or equal to 2? To 9?

Solution. (1) In this and similar problems, "fair" means that we accept the hypothesis of equally likely outcomes. Since the tetrahedrons are different, the sample space consists of $4 \times 4 = 16$ ordered pairs, $S = \{(1, 1), (1, 2), \ldots, (4, 4)\}$, where the pairs $(1, 2)$ and $(2, 1)$ are different. Hence, we must accept that the probability of any outcome is $1/16$, and consider the frequency probabilities. The sum of 7 can occur as either $3 + 4$ or $4 + 3$, and these outcomes are disjoint, since at the same throwing of a tetrahedron we cannot observe both a 3 and a 4. Hence, the answer to part (1) is $\dfrac{1}{16} + \dfrac{1}{16} = \dfrac{1}{8}$.

(2) Since the largest possible outcome in this part is a 4, the sum of 7 or more means either 7 or 8; therefore, comparing with part (1) of the problem, there is one more favorable outcome, the pair $(4, 4)$, which is also disjoint with the preceding ones, and the answer to part (2) is $3 \times \dfrac{1}{16} = \dfrac{3}{16}$.

(3) In part (2), the only favorable outcome is the pair $(4, 4)$; thus, the answer is $p((4, 4)) = 1 / 16$. The events in parts (1) and (2) are disjoint and their union is the event in part (2), that is why the answer in part (2) is the sum of those in parts (1) and (3).

We leave solutions to other questions to the reader; the answers are 1 and 0, respectively.

Problem 474. *Let E, F, G be three events with the same sample space. Prove that*

$$p(E \cup F \cup G) = p(E) + p(F) + p(G) - p(E \cup F) - p(E \cup G) - p(F \cup G) + p(E \cup F \cup G).$$

Problem 475. *Given all 2^n n–arrangements with repetitions from the set $A = \{0, 1\}$, we choose at random one of them, assuming that every arrangement has equal chances to occur. What is the probability of picking an arrangement containing an even number of 0s?*

Solution. The sample space consists of 2^n arrangements, 2^{n-1} of them (exactly half of the sample space), are favorable outcomes for our problem. Therefore, the probability we sought for, is $p = 2^{n-1} / 2^n = 1 / 2$.

Analyzing the solution, we observe an important feature of all similar problems:

To solve a probabilistic problem with the finite sample space and assuming the equally likely outcomes, we have to solve two enumerative combinatorial problems.

Problem 476. *All permutations of the letters of word MISSISSIPPI are written on balls, and one of these balls is chosen at random. What is the probability that we pick up the ball with the word MISSISSIPPI?*

Solution. The sample space consists of all permutations with repetition of the letters of word MISSISSIPPI, and by Theorem 18 it contains $C(11; 1, 4, 4, 2)$. elements. Among them there is only one favorable outcome, thus, the probability is $1 / C(11; 1, 4, 4, 2) = \dfrac{4!4!2!}{11!} \approx 0.000029$.

Problem 477. *Consider a regular deck of 52 cards. What is the probability that a 5–card hand contains exactly two clubs? At most two clubs? At least two clubs?*

Problem 478. *Among all the permutations with repetition of the letters of word DAD, one is chosen at random. What is the probability to find the chosen combination of letters in an English dictionary?*

Solution. The sample space consists of 3! / 2! = 3 permutations with repetition, ADD, DAD, DDA, but only the first two strings are meaningful English words, that is, are favorable outcomes in our problem. Therefore, the probability we sought for, is $p = 2/3$.

Any probability distribution on a sample space S puts a number $p(s) \in [0, 1]$ into a correspondence to a point s of the sample space, therefore, this distribution constitutes a function $f : S \to \mathbf{R}$. Since the domain of this function consists of the outcomes of a random experiment, the values of the function are also random. Such functions are called *random variables*. An initial probability distribution is also a random variable. In many problems it may be advantageous to change the values of a given probability distribution, as long as we preserve axioms PA1)-PA3).

Definition 133. *Given a random experiment with a sample space S, any real-valued function*

$$f : S \to \mathbf{R}$$

with the domain S is called a random variable *or a* random function *whenever it satisfies the three properties similar to the probabilistic axioms PA1)-PA3):*

RA1) $f(s) \geq 0$ for any point $s \in S$
RA2) If $E = \{s_1, s_2, \ldots, s_k\} \subset S$, then $f(E) = f(s_1) + f(s_2) + \cdots + f(s_k)$
RA3) $f(S) = 1$.
In particular, any probability distribution is a random function.

Problem 479. *Consider a sample space $S = \{1, 2, \ldots, n\}$, where n is a given natural number. Let f be a linear function, $f(s) = cs$, c being a real constant. Find the coefficient c so that the function f is a random variable on S.*

Solution. We must verify the properties RA1)-RA3); RA1) is clear if $c \geq 0$, RA2) is a rule of computing $f(E)$ through the values $f(s), \forall s \in S$, and we only have to compute the normalization constant c by making use of RA3). We have

$$1 = f(S) = f(1) + \cdots + f(n) = c \cdot 1 + c \cdot 2 + \cdots + c \cdot n = c \frac{n(n+1)}{2}$$

by the formula for the sum of the squares of n natural numbers, thus f is a random variable iff $c = \dfrac{2}{n(n+1)}$.

The equation $p(E_1 \cup E_2) = p(E_1) + p(E_2)$ tells us that two events are mutually exclusive. Another important mutual characteristic of a pair of events E_1, E_2 is their (stochastic) dependence or independence. It turns out that this property is connected with the equation $p(E_1 \cap E_2) = p(E_1) \cdot p(E_2)$, which is not always valid. Intuitively,

> Two events are independent,
> iff occurrence or non-occurrence of either of them   (22.3)
> does not affect the probability of another event to occur.

To define the dependence/independence of events, it is convenient to relate the notion of dependence with another important concept, namely, with the conditional probability of an event. First, we again model this notion by using an example.

Example 72. *Based on many years' statistic, the probability for a freshman to graduate from The Extremely Liberal College in at most four years is 0.85, while for the freshman majoring in sciences this probability is only 0.50. The sample space in this problem consists of all students ever graduated from the college. In the problem we have two probabilities – for all students the probability is 0.85, while for the science majors it is 0.5. The second number is different from the first one, because in computing it we have used some additional information on the students' majors, actually we reduced the sample space by removing all non-science majors. Since the second probability was computed under an extra condition, it is called the conditional probability.*

To arrive at a definition, we sketch a computation of the conditional probability of an event E, given another event (a condition) C, in terms of favorable outcomes. Computing the probability $p(E)$, we have to take into account all outcomes favorable to E and relate them to the whole sample space S. However, when computing the conditional probability, we certainly know that the event C has occurred, thus, now we consider only those favorable outcomes of E, which are favorable to C as well. Moreover, we must relate them not to the entire original sample space S, but only to the subset of outcomes favorable to C, hence, we must reduce the original sample space. If we express all these quantities in terms of the size $|S|$ of the sample space and of the probabilities $p(C)$ and $p(E \cap C)$, we derive the formula (22.4). It is convenient then to reverse this reasoning and use equation (22.4) as a definition of the conditional probability.

Definition 134. *Consider a random experiment with the sample space S, a generic event E, and a singled out event (condition) C, such that $p(C) > 0$. The conditional probability $p(E|C)$ of an event E, given the event C, is defined as*

$$p(E|C) = \frac{p(E \cap C)}{p(C)}. \tag{22.4}$$

It is often convenient to rewrite this formula as

$$p(E|C)p(C) = p(E \cap C).$$

Problem 480. *What is the probability to get a number 3 in one roll of a die given that the outcome is odd?*

Solution. Introduce the event $E_3 = \{x = 3\}$ and the condition $C = \{x \text{ is odd}\}$; we know that $p(E_3) = 1/6$ and $p(C) = 1/2$. The intersection of these events is $E_3 \cap C = E_3$, thus, $p(E_3 \cap C) = 1/6$. By (22.4), the conditional probability is $p(E_3|C) = (1/6)/(3/6) = 1/3$.

Problem 481. *What is the probability to get a 2 in one roll of a die given that the outcome is odd?*

Solution. It is clear without computations that if the outcome is odd, it cannot be 2, but let us formally compute the result. Let $E_2 = \{x = 2\}$ and $C = \{x \text{ is odd}\}$, $p(C) = 1/2$. The intersection of the two events is empty, $E_2 \cap C = \{2\} \cap \{1, 3, 5,\} = \emptyset$, thus, $p(E_2 \cap C) = 0$ and the conditional probability is $p(E_2|C) = 0/(1/2) = 0$.

Problem 482. *The 8 letters of the word "discrete" are written on 8 different cards and shuffled. Then we get the cards one at a time without return. What is the probability to get the word "discreet"?*

Solution. By the Multinomial Theorem, there are $\dfrac{8!}{2!}$ ways to arrange 8 characters with the two "t", thus, the probability is the reciprocal of this fraction. It is instructive, however, to consider the conditional probability. The probability to get the letter "d" at the first attempt is $1/8$. After that, the probability to get an "i" is not $1/8$, it is *conditional*, since the first selection has changed the sample number, and now it contains only 7 elements. Thus, from now and up to the selection "r" inclusive, the probabilities are $1/7, 1/6, 1/5, 1/4$, but the probability to get an "e" is $p(E \mid C = 2/3)$. These are conditional probabilities, since we take into consideration the *shrinking sample space*.

Problem 483. *Compute these numbers by making use of the conditional probability formula (22.4).*

Next, the probabilities to get the "e" and then "t" are $1/2$ and 1, respectively. Hence, the final probability is

$$1/8 \cdot 1/7 \cdot 1/6 \cdot 1/5 \cdot 1/4 \cdot 2/3 \cdot 1/2 \cdot 1 = \frac{2}{8!}.$$

Now we define the independence of two events in terms of conditional probability.

Definition 135. *Two events E and C are called (stochastically) independent iff*

$$p(E \mid C) = p(E), \tag{22.5}$$

otherwise, the events are called dependent.

Comparing (22.5) with (22.4), we see that two events are independent iff

$$p(E \cap C) = p(E)p(C), \tag{22.6}$$

thus, equation (22.6) formalizes our "intuitive" definition (22.3). It is worth noting that the independence is a symmetric property, which is obvious from (22.6), but not from (22.5).

Problem 484. *A card is drawn at random from a regular deck containing 52 cards. Are the events A – "To pick an ace" and C – "To pick a club" dependent or independent?*

Solution. The regular deck of cards contains 4 aces, so that $p(A) = 4/52 = 1/13$. Calculate the conditional probability $p(A \mid C)$. Obviously, $p(A \cap C) = 1/52$ and $p(C) = 13/52$, therefore $p(A \mid C) = \dfrac{p(A \cap C)}{p(C)} = 1/13$. Since $p(A \mid C) = p(A)$, we conclude that these events are independent.

Problem 485. *To win the jackpot in the New York Lottery Mega Millions game, one must guess correctly 5 numbers among 1, ..., 56 and one more number from 1, ..., 46. What is the probability to win the jackpot if you have one ticket?*

Solution. Since the order is not important, there are $C(56, 5) = 3\ 819\ 816$ ways to choose five numbers and $C(46, 1) = 46$ ways to select the Mega Ball number. For the last choice is independent from the first five numbers, (Why?) there are $C(56, 5) \cdot C(46, 1) = 175\ 711\ 536$ different tickets, and this is the cardinality of the sample space. Therefore, the probability we look for, is $1/175\ 711\ 536 \approx 5.69 \times 10^{-9}$.

Thus, if we have enough funds and time to buy $175\ 711\ 536$ one-dollar tickets, we definitely get the jackpot. Considering the appropriate taxes, not to mention a slight possibility that someone else has a winning ticket, we can estimate how large the jackpot is to be to pay off such an expense.

22.3 EXPECTATION: BAYES FORMULA

If we occasionally buy a lottery ticket, we cannot predict the future – the chances are slim, but who knows... Sometimes people win the jackpot. However, if we play any game of chance systematically, we may want to estimate our chances in the long run. The mathematical instrument for such estimations is called the mathematical expectation or the expected value of a random variable.

Definition 136. *Consider a probability distribution $p(s)$ on the finite sample space S and a random function $f(s), s \in S$. The mathematical expectation or the expected value of the random variable f is the sum*

$$E(f) = \sum_{s \in S} p(s) f(s). \tag{22.7}$$

Problem 486. *Find the expected value of the net gain in the preceding Problem 485, if the jackpot was $10\ 000\ 000$.*

Solution. The sample space has only two points, $s_1 = W$ with the probability $p(s_1) = 1/175\ 711\ 536$ and $s_2 = L$ with $p(s_2) = 1 - 1/175\ 711\ 536$. Corresponding gains are $f(s_1) = \$10\ 000\ 000 - \$1$ and $f(s_2) = -\$1$. Therefore,

$$E(f) = \$9\ 999\ 999 \cdot 1/175\ 711\ 536 - \$1 \cdot (1 - 1/175\ 711\ 536) \approx -\$0.94,$$

and in the long run we should expect to loose about 94 cents from each dollar spent. The house always wins.

Consider again the definition of the conditional probability (22.4). Since the operation of intersection of two sets is commutative, it implies the following property

$$p(E\,|\,C)p(C) = p(C\,|\,E)p(E). \tag{22.8}$$

Thus, we can express the conditional probability of two events through their conditional probability in reversed order, that is, as

$$p(C\,|\,E) = \frac{p(E\,|\,C)p(C)}{p(E)}$$

if $p(E) > 0$. This property can easily be extended to the case of several conditions.

Theorem 62. *Let events* C_1, C_2, ..., C_k *make a partition of the sample space S, that is, all C_j are non-empty, pairwise disjoint, and exhaust the sample space S. Then the following equation, called Bayes's formula, is valid for any event E with $p(E) > 0$ and every j, $1 \le j \le k$,*

$$p(C_j\,|\,E) = \frac{p(E\,|\,C_j)p(C_j)}{\sum\limits_{j=1}^{k} p(E \cap C_j)p(C_j)}. \tag{22.9}$$

Proof. We have

$$E = E \cap S = E \cap \left(\cup_{j=1}^{k} C_j\right) = \cup_{j=1}^{k} (E \cap C_j).$$

The intersections $E \cap C_j$, $1 \le j \le k$, are also mutually exclusive, thus $p(E) = \sum_{j=1}^{k} p(E \cap C_j)$. Combining the latter with the formula for the conditional probability $p(C_j\,|\,E) = p(E \cap C_j)/p(E)$, we deduce (22.9).

Problem 487. *Let us note that all $p(C_j) \ne 0$, since we have assumed $C_j \ne \varnothing$. Is it possible that the denominator in formula (22.9) is zero?*

Problem 488. *A die was randomly selected among a set, containing 999 999 regular dice and one die with all faces marked by 1, and was rolled 10 times. What is the probability that the fake die was chosen, given that a 1 was observed in all 10 trials?*

Solution. Consider three events,

C = {Observe a 1 in 10 consecutive trials}
D = {Choose a fair die with $p(D) = 1 - 10^{-6}$}
F = {Choose a fake die with $p(F) = 10^{-6}$}

By Bayes's formula,

$$p(F\,|\,C) = \frac{p(C\,|\,F)p(F)}{p(C\,|\,D)p(D) + p(C\,|\,F)p(F)}$$

$$= \frac{1 \times 10^{-6}}{(1/6)^{10} \times (1 - 10^{-6}) + 1 \times 10^{-6}} \approx 0.94.$$

The result is so close to 1 that it may look counterintuitive, and it is useful to compare it with the negligible probability to observe 10 consecutive 1's in rolling a fair die, which is $(1/6)^{10} \approx 1.65 \cdot 10^{-8}$.

Problem 489. *Describe the sample space if we simultaneously toss two indistinguishable coins, that is, the outcomes (H, T) and (T, H) must be identified. What is the sample space if we roll simultaneously 3 identical dice?*

Problem 490. *A woman can give birth to a girl, a boy, two girls, a girl and a boy, two boys, etc. Consider this as a random experiment with outcomes to be the number of children and the gender composition of the children born. Describe the sample space if the order at birth is important, that is, we consider the pairs "boy-girl" and "girl-boy" as different. What is the sample space if the order does not count?*

Problem 491. *Assuming that a newborn baby has equal chances to be a girl or a boy, what are the probabilities to have no boy, one boy, two boys, three boys, ..., n boys in a family with n children?*

Problem 492. *A die is rolled once. Find the complementary event to the following combined events.*

(1) "To get an odd number AND To get a number less than 3".
(2) "To get an odd number OR To get a number less than 3".

Problem 493. *Let E be any event, $\overline{E}$ its complement, and $p = p(E)$. What are the events $E \cap \overline{E}, E \cup \overline{E}, \overline{E \cap \overline{E}}, \overline{E \cup \overline{E}}$, and what are their probabilities, in terms of p?*

Problem 494. *Six faces of a regular die are marked by letters A, B, A, C, U, S. Find the probability that at six rolls of the die, the letters shown can be rearranged to spell "ABACUS."*

Problem 495. *Two friends have agreed to meet between 8:00 and 8:20 pm. The person, who comes first, is waiting for the other for 10 minutes and then leaving, unless the friend shows. Their arrivals are independent and equally distributed within that 20 min period. What is the probability of their meeting?*

Solution. Consider a square $\{(x, y) : 0 \le x \le T, 0 \le y \le T$, where x, y are the times of the friends' arrival, the points of this square make the sample space. Hence, the sample space as is, is infinite, but in this problem about the *geometric probability* we can define the probability as the ratio of the favorable area to the area of the entire square, that is, to T^2. For a point (x, y) represent a favorable point, its coordinates must satisfy $|x - y| < t$, that is, lie in the square between the lines $y = x \pm t$. The favorable area is, therefore, $T^2 - (T - t)^2$; hence the geometric probability is

$$\frac{T^2 - (T-t)^2}{T^2} = 1 - (1 - t/T)^2.$$

Problem 496. *The hunter shoots the dragon with the probability of 2/5, while an apprentice shoots the dragon with the probability of 1/4. Both the hunter and the apprentice shoot simultaneously. What is the probability that the dragon is shot at least once?*

22.4 BERNOULLI TRIALS AND THE BIRTHDAY PROBLEM

Problem 497. *Find the probability that a fair coin shows a heads twice in 3 consecutive tossing.*

Solution. The probability of getting a heads or a tails in a single tossing is 1/2, and we certainly accept that these 3, or 33, or... n trials are independent. Each trial has two outcomes, and the sample space for the whole experiment consists of $2^3 = 8$ elements. In combinatorial terms, this is the number of arrangements with repetitions. So that, since we accept the axiom of the equally likely outcomes, the probability of getting H in any triple is $\frac{1}{8}$. The favorable outcomes for the problem contain $2H$, and therefore, just one T, which can be either the first, or the second, or the third. Thus, there are 3 favorable triples, and finally, the probability is $\frac{3}{8}$.

Problem 497 gives an example of a *binomial distribution*, that is, a random experiment with exactly two outcomes, usually called *a success* and *a failure*, whose probabilities do not change in time. We follow the tradition and denote the probability of success as $p = Prob(success)$ and the probability of failure as $q = Prob(failure)$, thus, $0 \le p, q \le 1$. and $p + q = 1$. If we repeat a binomial experiment n times, assuming all the outcomes being independent (such series is called Bernoulli's trials), then a typical problem is to compute the probability of getting r successes in these n trials. Since there are $C(n, r)$ ways to select r "successful" trials among n trials, the probability of getting r successes is, by the Product Rule,

$$p(r, n) = C(n, r)p^r q^{n-r}. \qquad (22.10)$$

We used (22.10) in the solution of Problem 486 above with $n = 1$, $r = 1$, $p = 1/175\ 711\ 536$ and $q = 1 - 1/175\ 711\ 536$.

In combinatorial terms, the answer to Problem 497 can be written as $C(2, 3)/2^3$; keep in mind that $C(n, k) = C(n, n - k)$.

Problem 498. *A coin is tossed 7 times. Denote X the probability of getting T, that is, "tales" X times. Find the probabilities* $P(X = 2)$, $P(X \le 2)$, $P(X \le 3)$, $P(2 < X \le 5)$, $P(X < 10)$.

Problem 499. *There are six lamps in the classroom. The probability that any of them to be down is 0.25. The classroom cannot be used if less than four lamps work. Compute the distribution of probability, that after the turning all the lamps on, the room cannot be used.*

Solution. Since after we turn the lamp on, it is either on or off, this is a typical problem about the binomial distribution, with the probability of success $p = 0.75$ and the probability of failure $q = 0.25$. Lectures in the room can go if there are at least four lamps working. So that, by formula (22.10), the probability that the classroom is in the working condition, is

$$\sum_{k=4}^{6} C(6, k)p^k q^{6-k} = C(6, 4)0.75^4 0.25^2 + C(6, 5)0.75^5 0.25 + C(6, 6)0.75^6 \approx 0.83.$$

Problem 500. *Compute the average and the expectation of the binomial probability (22.10).*

The following is the classical birthday problem.

Problem 501. *What is the probability that among s members of the Discreet Club at least two have the same birthday?*

Solution. To simplify computations, we consider a non-leap year with 365 days and assume that for a randomly chosen person for any day of the year the probability, that this person was born exactly this day, is the same; thus, this probability is 1/365. Moreover, as we always do in problems involving people, we suppose that the birthdays of all the people involved are independent; in particular, any day of a year has equal probability to be someone's birthday.

It is easier in this problem to compute the probability of the complementary event, that is, the probability that no two members have the same birthday. First of all we notice that for any member there are 365 options to fix the birthday; hence the sample space contains 365^s points. To find the number of favorable outcomes, we choose s days from the 365 available for the s birthdays–this can be done in $C(365, s)$ ways. However, when we distribute members' birthdays among these s days, we can permute them in $s!$ ways, generating different favorable outcomes. Hence, there are $s! \cdot C(365, s) = P(365, s)$ favorable outcomes, so that the probability of the complementary event is $P(365, s)/365^s$, and the probability that at least two members have the same birthday is $1 - P(365, s)/365^s$. We see that if $s \geq 366$, then this probability is 1, which is obvious. An easy numerical experiment shows that this probability is increasing with s and becomes bigger than 1/2 for $s = 23$.

Remark 22. *It is instructive to rephrase this problem in terms of placing balls in urns.*

Problem 502. *There are d dolphins in the ocean. d_1 of them were caught, marked, and released back. Next time, d_2 of them were caught. Assuming independence, compute the probability that $m \leq d_2$ marked species were caught second time.*

Problem 503. *A gentleman has 10 dress shirts and 10 ties, one matching tie to every shirt. Preparing for a long meeting, he selects at random 2 shirts and 2 ties. What is the probability that he gets exactly one matching pair of a shirt and a tie? At least one matching pair? Two matching pairs? If he selects at random 5 shirts and 5 ties, what is the probability to have exactly 2 matching pairs?*

Problem 504. *The U.S. Senate consists of 100 senators, two from each state. If 10 senators are chosen at random, what is the probability that this cohort contains a senator from any particular state, say, from Alaska?*

Problem 505. *An urn contains 3 black, 3 white, and 3 yellow balls. n balls are taken at random without replacement. For each n, n = 1, 2, ..., 9, find the probability that among the n selected balls there are balls of all three colors?*

Problem 506. *An urn contains 8 balls with a letter of the word STALLION on every ball. If 4 balls are chosen at random without replacement, what is the probability that either the word TOLL or the word LION can be composed of these balls?*

Problem 507. *Two fair right indistinguishable tetrahedrons, with faces marked 1 through 4, are tossed.*

1) What is the probability that the sum of the numbers on the bottom faces is 7?

2) What is the probability that the sum of these numbers is greater than or equal to 7?

Problem 508. *Let m and n be natural numbers. Consider four points O(0, 0), A(m, 0), B(m, n), and C(0, n) in the coordinate plane, and choose a random rectangle R with sides parallel to the coordinate axes and with vertices at points with integer coordinates inside or at the boundary of the rectangle OABC. What is the probability that R is a square?*

Problem 509. *Let $S = \{1, 2, \ldots, n\}$, where n can be any natural number, and $f(s) = cs^2$, c is a constant. Find c so that f is a random variable on S.*

Problem 510. *Let $S = \{1, 2, \ldots, n\}$, where n can be any natural number, and $f(s) = c / s$, c is constant. Find c so that f is a random variable on S.*

Problem 511. *Let $S = \{1, 2, 3, \ldots\}$, that is, in this problem the sample space is infinite, and $f(s) = c / s^2$, c is a constant. Find c so that f is a random variable on S.*

Problem 512. *What is the probability to get at least one 3 in a roll of two dice if the sum is odd?*

Problem 513. *Two hunters simultaneously shoot a dragon. The probability to kill the dragon for each of them is 1/3. What is the probability for the dragon to survive?*

Problem 514. *Six cards are drawn at random from a regular deck of 52 cards. What is the probability that*

(1) The Queen of Spades was chosen among these six cards?
(2) All four suits will appear among these six cards?

Problem 515. *Several cards are drawn at random from a regular deck of 52 cards. We want to guarantee, with the probability more than 1/2, that at least 2 cards of the same kind appear among the cards chosen. What is the smallest number of cards that must be drawn for that?*

Problem 516. *If $p(A) = 0.55$, $p(B) = 0.75$, and the events A and B are independent, what are the conditional probabilities $p(A \mid B)$ and $p(B \mid A)$?*

Problem 517. *In a certain population, 30% of men and 35% of women have a college degree; it is also known that 52% of the population are women. If a person chosen at random in this population has a college degree, what is the probability that the person is a woman?*

Problem 518. *The Student Government at The Game College sold 250 lottery tickets worth $1 each. There are one $100 prize, one $50, and three $10 prizes. If a student bought 2 tickets, what is the expected value of her net gain?*

Problem 519. *Among 10 000 coins all but one are fair, and one has tails on both sides. A randomly chosen coin was thrown 12 times.*

(1) What is the probability that this coin was false, if a tail was observed in all 12 trials?

(2) What is the probability that this coin was fair, if a tail was observed in all 12 trials?

Problem 520. *Every juror makes a right decision with the probability p. In a jury of three people two jurors follow their instincts, but the third juror flips a fair coin, and then the verdict follows the majority of jurors. What is the probability that the jury makes the right decision? Does this probability change if the jury consists of four jurors, and only one among them flips a coin?*

Problem 521. *Assuming independence, what is the probability that at least two of the first 43 Presidents of the USA have the same birthday?*

Problem 522. *We roll a fair die 6 times. If a 1 occurs first, or if a 2 occurs second, or if a 3 occurs third, ..., or if a 6 occurs sixth, we get $1. What is the expected value of our gain?*

Remark 23. *This result illustrates an important theorem that the expected value of the finite sum of random variables is equal to the sum of their expected values.*

Problem 523. *The President of the Discreet Club introduced the following game. A participant pays $1 and selects at random an integer number between 1 and 1 000 000 inclusive. If the decimal representation of the number contains a 1, the participant gets $2, otherwise the participant looses the game. What is the expected value of the game?*

Problem 524. *State the Inclusion–Exclusion Theorem 16 in probabilistic terms of this section.*

Problem 525. *A student scored 76, 81, 89, and 92 at four class tests. At the fifth test, she can equally likely get scores from 75 through 95 inclusive. What is the probability that her average will be 85? At least 85? What is the probability that her average will be 85 if her fifth score is 85?*

Problem 526. *In a class of 20, every two students have a common grandfather. Prove that among them at least 14 students have the same grandfather.*

Problem 527. *The spring in an old-fashioned watch breaks at a random moment. What is the probability that the hour hand will show time after 2 a.m. but before 4 a.m.?*

Problem 528. *The Discreet Club has a round roulette table with a rotating pointer. The table is divided in three sectors, one half-circle marked by the digit 3, and two quarter-circles marked by 1 and by 2, respectively. You pay $1 to make a single spin and get back the reward in $ equal to the number in the sector the pointer stops. Using negative numbers for a loss, find a probability distribution for the net gain if the game was played once. What is the average (expected) gain for each play? What would be the fair price of the game? Solve this problem if you pay $2 for a spin; $1.50 for a spin. Explore the similar problem if the table is divided in 6 equal sectors marked by the digits 1, 2, 3, 4, 5, 6.*

Problem 529. *A multiple-choice exam consists of 20 questions, with 4 possible answers for every question. If a student randomly guesses the answer to each question, find the probabilities that she gets zero, one, two, three, four correct answers. What is the expected number of questions guessed correct? Solve the same problem if each question has 5 possible answers.*

Problem 530. *Suppose that a shop makes on average 10% defective items. The supervisor has chosen 3 items at random. What is the probability that none among them is defective? One among them is defective? Two are defective? All the three are defective?*

Problem 531. *Assuming that any day has equal chances to be the birthday of a person chosen at random from a very big population, find the probability that two randomly selected people both have birthdays on Sunday; find the probability that at least one of two randomly selected people will have a birthday on Sunday; find the probability that two randomly selected people have birthdays on the same day of a week.*

Problem 532. *Each member of the Discreet Club is required to take a course in Discrete Mathematics (DM) and in Probability Theory (PT). The Registrar Office reports that 80% pass course DM, 75% pass course PT, and 90% pass at least one of the two courses. Find the probability of passing both courses. What is the probability that a person who passes course DM will also pass course PT? Are the passing course DM and the passing course PT independent events? What about passing one course and failing another?*

Problem 533. *The probability that an event E occurs in one trial, is p, the probability of the complementary event is $q = 1 - p$. Find the value of p, such that the outcome has the biggest indeterminacy?*

Solution. By formula (21.1), we must maximize the sum

$$f(p) = -p\log_2 p - (1-p)\log(1-p)$$

for $0 < p \le 1$. The stationary point of the function f occurs at $p = 1/2$ when $f(1/2) = 1$.

22.5 EXERCISES

Exercise 22.1. *Define the sample space in the lottery experiment in Example 69, if*

(1) You have two tickets
(2) You have one ticket that costs $1 and are interested in the net income.

Exercise 22.2. *This weekend Kathy either goes to the movies, with the probability of this event 0.7, or goes to the restaurant with the probability 0.5.*

(1) Given this information, is it possible to conclude that these two events are mutually exclusive?
(2) What is the smallest and the largest possible probability that this weekend Kathy will have both these pleasures?
(3) How can we change the problem to be able to determine precisely the probability that this weekend Kathy gets at least one of these pleasures? Both these pleasures?

Exercise 22.3. *Two dice are rolled simultaneously. What is the probability to get at least one number greater than 4?*

Exercise 22.4. *Find the expected value of the net gain in Problem 486, if the jackpot was $100 000 000.*

Exercise 22.5. *Describe the sample space if we simultaneously flip a coin and roll a die.*

Exercise 22.6. *Describe the sample space if we flip a coin six times. What is the probability that at least one head and at least two tails will appear in the six tosses? What is the probability that a streak of at least four consecutive tails will appear in the six tosses?*

Exercise 22.7. *Simultaneously toss a coin and roll a die. What is the probability to get a head and a multiple of 3?*

Exercise 22.8. *Simultaneously roll a die and draw a card from a standard deck of 52 cards. What is the probability to get an even number and a red face card?*

Exercise 22.9. *A lottery ticket contains six boxes–two for letters followed by four for digits. If there is only one winning ticket, what is the probability to win the lottery?*

Exercise 22.10. *A 9-digit natural number is chosen at random. What is the probability that all its digits are different?*

Exercise 22.11. *Three hunters can shoot the dragon with the probabilities of 0.1, 0.4, 0.8, respectively. Each shoots once, and the dragon was killed by one bullet. What is the probability that the dragon was killed by the first, the second, the third hunter, respectively?*

Exercise 22.12. *There are n students, who should attend the class in Discrete Mathematics. Yesterday's lecture was missed by r students. To check the attendance, the lector calls r names at random. What is the probability that the lecturer calls all the absentees?*

Exercise 22.13. *A fair coin is tossed three times. The event E_1 occurs when the first coin comes up tails. The event E_2 occurs when exactly two heads in a row come up. Are the events E_1 and E_2 independent?*

Exercise 22.14. *What is the probability to get (at least) two consecutive tails if a fair coin is tossed 12 times?*

Exercise 22.15. *What is the probability that in a random permutation of numbers 1, 2, 3, ..., 1000 at least one number occupies its own place (for example, the 5 is the 5th number in the permutation)?*

Exercise 22.16. *Among the whole numbers, some can be written down without the digit 1, like 527, while the others contain a 1, like 21 345. If you* *randomly pick a whole number from 0 through 999,999 inclusive, what is more probable, to pick a number with or without a 1 in its decimal representation? Does the probability change if we consider numbers from 1 to 999, or from 1 through 999 999 999?*

Exercise 22.17. *Seven people get in an elevator on the first floor of an 11-story building. What is the probability that no two of them get out of the elevator at the same floor?*

Exercise 22.18. *A $5 \times 5 \times 10$ parallelepiped with red sides cut into 250 unit cubes. What is the probability that a randomly chosen unit cube has no red face? One red face? Two red faces? Three red faces? Has four or more red faces?*

Exercise 22.19. *A die has 4 blue and 2 red faces. What is the probability that the two red faces have a common edge? What is the probability that the two blue faces have a common edge?*

TURING MACHINES, P AND NP CLASSES, AND OTHER MODELS OF COMPUTATION

23.1 TURING MACHINES

In this chapter, we return to the major theme of the book, which is *Models of Computations* – what and how should we compute, what can and cannot be computed. People were thinking about these issues since ancient times, and the word "algorithm" is the distorted name of Arab scientist Al Khwarizmi from Khorezm (c. 780–c. 850). But only in 19th century, the very idea appeared that computations must be defined. In the first half of the 20th century, several definitions were given, such as Church's thesis, Markov normal algorithms, some results of Post and Gödel (1906–1978), and first of all, Turing machines. All these concepts are primary notions, which are, as we understand today, equivalent to each other in the sense that they describe the same family of computational processes. We consider in more detail only the Turing machines. The latter are *abstract, thoughtful* machines,[1] modeling mental computational processes of human beings.[2]

---

[1] Even though acting and touchable Turing machines can be seen on the Internet and in some museums.

[2] Notwithstanding and without taking sides, we believe that the crucial question of Alan Turing, "Can a machine think?" still remains a question, even though the world chess champion was defeated by the computer. Maybe, this is because we cannot satisfactory answer the question "What is thinking?"

In this model, we do not try to emulate a fantastic speed or huge memory capacities of modern computers. Quite the contrary, we want to model computational processes in the human brains at the most elementary level, as sequences of the simplest possible operations. Therefore, we go down to the elementary processes, and perform the arithmetic operations not even in the binary but in the unary system.

The binary system has two symbols for the two digits 0 and 1. The *unary system* represents any natural number by making use of only one symbol, the stroke |; hence, every natural number is written as a sequence of these strokes. For example, 1 is written as ||, 2 as |||, 3 as ||||, etc., and all our arithmetic is reduced to manipulating these finite sequences of the vertical bars |. We also need a carrier for these bars, and some device to perform arithmetic operations with them.

Problem 534. *Why is it more convenient to represent a natural number $n \geq 0$ in the unary number system as not n but $n + 1$ vertical strokes?*

Hence, we arrive at the following description of this model of computations, that is, the Turing machine. It is useful here to refresh the definition of the finite-state automaton, see Chapter 19, because the new model is more capable than the finite automata. The model, Turing machine, includes the three alphabets and two procedures (functions) for computing the next internal state and the next output of the machine.

The machine works in discrete time. At each time moment, the machine executes certain instructions (for this moment) and prepares itself for the next time moment. The carrier is a tape, consisting of the cells and infinite in both directions. Every cell can either carry a stroke |, or be empty. At the start and at the every next moment, all but finitely many cells are empty. The initial string of the symbols on the tape is called the *initial command*. The machine also has an arithmetic device, called a read-write head, which can move along the tape in both directions, can read the content of the cell observed, can change this content or leave it unchanged, and then either stay at the same cell, or move along the tape one step to the right or to the left and stop there, according to the controlling program. It is important to mention that at every given time moment, the tape is finite, but it can be extended as long to the left and to the right, as the computational process requires. This is the crucial distinction between the Turing machines and the finite automata.

Certain modifications are possible, for example, one can define a machine with several tapes, or a machine with a semi-infinite tape. However, all these variants are equivalent to one another and to the basic model in the sense that all of them represent the same class of algorithms, and we do not consider them here.

The algorithms/functions, which can be represented by the Turing machine, are called *computable functions*. They are convenient to write but cumbersome to deal with. Consider an example of a Turing machine.

Problem 535. *Design the Turing machine, that is, compose its program, that adds two natural numbers m + n.*

Solution. We represent this computable function, or the algorithm as a program, namely, a series of commands to be placed on the tape of a Turing machine. Let us say, $m = 2 = |||$, and $n = 1 = ||$. On tape, these numbers m and n must be initially separated with at least one empty cell; if they have no separator, then no Turing machine can distinguish them. If they are separated with more than one empty cell, this just slightly prolongs the program.

Thus, consider the initial configuration in Table 23.1, as the head observes the leftmost non-empty cell. In addition to the symbol |, this cell is marked also with the symbol of the head H. Then the read-write head deletes the symbol | in the cell it observes and gradually, step-by-step moves to the right, until it sees the empty cell, see Table 23.2.

TABLE 23.1 Turing Machine that Adds 2 + 1. Initial Configuration.

		H I	I	I		I	I		

TABLE 23.2 Turing Machine that Adds 2 + 1. The Head Deleted the Leftmost Symbol | and Moved to the Right to the Empty Separator between ||| and ||.

			I	I	H	I	I		

After that, the head writes into this cell the symbol |, which it remembers from the start.

TABLE 23.3 Turing Machine that Adds 2 + 1. The Machine Transferred the Unity Symbol from the Leftmost Cell into the Cell-Separator.

			I	I	H I	I			

Finally, the head returns to the leftmost occupied cell. The program has been executed, and the machine stops.

TABLE 23.4 Turing Machine that Adds 2 + 1. The Machine Transferred the Stroke from the Leftmost Cell into the Cell-Separator.

			H I	I	I	I			

Problem 536. *Design a program, that is, Turing machine, which adds two natural numbers, separated with three empty cells.*

Problem 537. *Design a Turing machine, which, given the natural numbers m and n ≥ m, computes, that is, prints out the natural number n − m if n − m ≥ 0, and prints 0 if n < m; thus, it computes the function $(n − m)^+$.*

Everyone who had to do computer programing, would probably saw computer programs, which seemingly worked *endlessly*, they computed something without interruption, and there was no indication, whether the program will ever stop. An important question in Computer Science is whether we can design an algorithm, a Turing machine that answers this question for any program. Of course, if we want to solve the question *mathematically*, we must, first of all, give relevant definitions and specify certain statements. We assume hereafter that there exists a generic alphabet, such that any program can be written in that alphabet. Moreover, there is a certain Turing machine, such that any program can be written on its tape.

The question now occurs, if we initialize this machine with that program on its tape, is there some algorithm, that is, another Turing machine, which can answer the question without running the initial machine, and whether this machine will *halt* after finitely many steps? This question is called the *solvability problem*, that is, can we design a Turing machine that *solves* this problem? We repeat again, that this is a *mass problem* – the question is not whether we can solve that specific problem, the question is, whether we can solve any problem from a precisely described class of problems?

The answer was found by Alan Turing almost a century ago: namely, the halting problem is insoluble (unsolvable), that is, there is no Turing machine that for any problem from a given class tells whether there is the Turing machine, which solves *every* problem from this class. We do not doubt that a specific problem can be solved; for instance, we believe that every particular Diophantine equation eventually can be solved, but the solutions for two different equations may be different. The claim states that no universal algorithm does exist; there is no algorithm solving every such an equation.

It is worth mentioning that a solution to the halting problem is similar to the Russell paradox, see Chapter 3.

To put some flesh onto these dry bones, we show a couple of programs for Turing machines. In the next problems, we assume that both external alphabets are binary {0, 1}, the initial state of the Turing machine is q_1 and the final state is q_0, and at the initial time moment, the machine head looks at the leftmost unity 1 on the type. The symbols L, R, S in the program indicate that after executing this command, the head moves one cell to the left, or one cell to the right, or does not move.

Problem 538. A *Turing machine works according to the program*

$$\begin{cases} q_1 0 q_2 1R \\ q_1 1 q_3 0R \\ q_2 0 q_3 1L. \\ q_2 1 q_2 1S \\ q_3 1 q_1 1R \end{cases}$$

Determine whether the machine can be applied to the words (1) $w = 1^4 01$; (2) $w = 1^3 01^2$ (3) $w = 1^6$. If the answer is in the affirmative, compute the result.

Problem 539. *The Turing machine has the initial configuration* $K = 1^2 q_1 1^3 01$, *the program below and the concluding state* q_0. *Find the final configuration.*

K	q_1	q_2
0	$q_0 1S$	$q_1 0R$
1	$q_2 0R$	$q_2 1L$

The word *algorithm* has been repeated in this book several times, in particular, recently in Chapter 19 and in Chapter 8, where we discussed the Euclidean algorithm, but it has never been defined. What is more, it has never been defined during almost a millennia since Al Khwarizmi was born. Algorithms penetrate both mathematics and computer science. It is not uncommon that there are several algorithms solving the same problem, and we want to compare their performance, speed, and other properties. These features constitute a separate part of Computer Science, called *Complexity of Algorithms*.

The complexity theory mostly studies two features of algorithms, the speed of algorithms, i.e., its *running time*, and the required computer memory. Any algorithm, any problem usually has certain singled out parameters, maybe just one, which are often run through the natural numbers. The set of natural numbers is *infinite*; therefore, it is "bigger" than any finite set. However, we also know from Analysis, that there are *different infinities* with different cardinalities; in particular, the set of real numbers has bigger cardinality than any countable set. We understand "smaller" or "bigger" here in the sense that if a set $\mathbb{A}$ has a smaller cardinality than a set $\mathbb{B}$, then there is an injective map $\mathbb{A} \to \mathbb{B}$. In particular, there exists an injective map $\mathbb{N} \to \mathbb{R}$, but not vice versa.

The sets which have the same cardinality as the set of natural numbers $\mathbb{N}$, are called *countable sets*. The sets of natural numbers, of even or of odd numbers, of the integers, of prime numbers are all countable. The set of real numbers is not countable. It is said that the set $\mathbb{R}$ and all the sets of the same cardinality, like the sets of complex numbers, the Euclidean spaces $\mathbb{R}^n$, the sets of the polynomials of any degree with real or complex coefficients, all have the cardinality of *continuum*.

The sets, we encounter in practical problems in computer science, are finite, and we do not discuss these "memory issues" here, but we look at the velocity of algorithms.

Let us begin with the following simple example.

Example 73. *Analyze the following algorithm, where "while loop" is similar to that loop in any programming language. This algorithm must print the factorials $n!$ in the range $1 \leq n \leq n_0$, where $n_0 > 0$ is a specified natural number. In the example, $n_0 = 3$:*

```
n₀ := 3
while n < n₀ do
n := n + 1
print n!
```

Solution. The initial value of the parameter is $n = 0$; thus the program (algorithm!) enters the loop and checks the condition $n < 3$, which is $0 < 3$. This is obviously true, and the algorithm executes the commands in the body of the algorithm, that is, assigns $n := 0 + 1 = 1$, prints $0! = 1$, and returns to the condition of the *while loop* $n < 3$. Now, however, $n = 1$, and the program verifies the inequality $1 < 3$, which is also true. Hence the program increases the counter and sets $n := 1 + 1 = 2$, prints $2! = 2$ and returns to the condition $n < 3$, i.e., $2 < 3$. Since this inequality is still true, the algorithm returns to the body of the program and executes the assignment $n := 2 + 1 = 3$, prints $3! = 6$, and returns into the cycle again. However, the condition $n < 3$, that is, $3 < 3$, now fails; thus, the algorithm skips the body of the loop and returns to the external program. So that, this program prints three numbers, 1, 2, 6, and continues execution of the external program.

Problem 540. *Let the body of the loop to be written in the changed order, as* print n! *followed by* n := n + 1 *with the same condition; execute this algorithm and compare with Example 73.*

We have mentioned a few crucial features of the algorithms. Now in Example 73, we definitely see that this is a *mass problem*, since we can ask the same question for any natural n_0. Moreover, for every $n_0 > 0$, it requires a *finitely many* precisely defined steps. Therefore, Example 73, indeed, describes an algorithm.

Tracing the execution of this algorithm, we see also that when $n_0 = 3$, the algorithm makes 4 comparisons, 3 additions, and prints 3 numbers. There are many similar algorithms, such that the number of steps is comparable with a certain critical parameter, like n_0 here. The next notations, called "*Big O*" and "*small o*," are convenient in many instances; we consider here sequences, that is, functions of the discrete argument $n = 0, 1, 2, ...$

Definition 137. *Given two sequences $a(n)$ and $b(n)$, it is said that $a(n) = O(b(n))$, as $n \to \infty$, iff the ratio $|a(n) / b(n)|$ is bounded as $n \to \infty$, that is, there exists a constant $C > 0$, which does not depend upon n_0, such that $|a(n) / b(n)| \le C$ for all $n \ge n_0$.*

It is said that $a(n) = o(b(n))$ as $n \to \infty$, iff the limit $\lim_{n \to \infty} a(n) / b(n) = 0$, that is, for every $\varepsilon > 0$ there exists an $n = n(\varepsilon)$ such that $\left| \dfrac{a(n)}{b(n)} \right| < \varepsilon$ *for all* $n > n(\varepsilon)$.

Problem 541. *Compare Definition 137 with a modified definition, which says that "for every $n_0 > 0$ there exists a $C > 0$, such that...." Are these two definitions equivalent?*

Problem 542. *Prove that $\log n = O(n^a)$ as $n \to \infty$ for any $a > 0$ but not for $a = 0$.*

Problem 543. *Prove that $n^b = O(n^a)$ as $n \to \infty$ for any $0 \le b \le a$.*

Problem 544. *State the Big O and small o notations for functions of the real argument $x \to \infty$.*

Power functions with a positive exponent make a very limited family of functions. There are also logarithmic functions $f(x) = (\log_b x)^p$, which grow slower than any power function.

Problem 545. *Prove that $|\log_b x|^p = O(x^a), x \to \infty$, where all the parameters a, b, p are positive and $b \ne 1$.*

On the other hand, any exponential function grows faster than any power function (all the parameters are to be positive).

Problem 546. *Prove that $x^a = O(e^{px^b})$, $x \to \infty$, where all the parameters a, b, p are positive.*

These facts are proved in mathematical analysis.

Problem 547. *Prove that Stirling formula in Chapter 1 implies that $n! = o(n^n)$ as $n \to \infty$.*

The crucial property of an algorithm is its performance. For instance, the question of whether a natural number n is prime, looks trivial – just check every natural number not exceeding $\sqrt{n}$, see Problem 250. However, if n is large, this answer can arrive many years *after* it was needed. That is why certain cryptography systems, like RSA, are considered to be reliable now – until people would discover a feasible quick algorithm for checking the primality of natural numbers.

In particular, this observation can explain, at least, partially why the problem "P versus NP" for classes of algorithms is important in Computer Science and Mathematics. We introduce these problems without being completely rigorous.

If we look back at Example 73, we observe that the number of operations, when one applies this algorithm, grows with n as a linear function – see the solution above. A linear function is a polynomial of degree 1. Current computers perform computations with polynomials of any degree without a noticeable delay, and P in "P versus NP" stands for problems, which can be solved in *polynomial* time.

Informally though, the class P consists of deterministic algorithms, containing a natural parameter n, whose solution time equals $O(p(n))$, where p is any polynomial. These problems are called *tractable* problems. For instance, the linear programming problem to maximize a linear function subject to linear inequality constraints, is in class P.

Then, what problems are currently *intractable*? Those are the problems, such that we either do not know now, that a tractable solution does exist, or we definitely know that any solution grows in some regards *faster* than any polynomial function. These problems do not have to be deterministic, though, and make the class of NP problems. Again informally, these problems have solutions, which can be *checked*, but not necessarily found, in polynomial time. For example, the Knapsack problem and Hamiltonian path problem, see Chapter 8, are known to be in NP class.

Finally, we want to mention in passing that an important distinction is whether we study the *worst case* scenario, or the *average case*. The former possibility means that the estimation is given over all the possible cases; the result could not be worse than that. The average caase means that we assume a certain probability distribution over all the possible outcomes, and we compute the expectation of the possible outcomes. In this case, we mostly neglect the outliers, and the result can be much better than in the worst case.

23.2 EXERCISE

Exercise 23.1. *Design a Turing machine that executes the algorithm of Example 73.*

CHAPTER 24

ANSWERS AND SOLUTIONS TO SELECTED EXERCISES

Chapter 1

Exercise 1.18. Give the recursive definitions of the following sequences.

(a) 1, 3, 5, 1, 3, 5, 1, ... (b) 1, 4, 9, 16, 25, ... (c) 1, 2, 3, 5, 8, 13, 21, ...

Solution. If we start the numbering with 0, then $a_{3k} = 1$ for $k = 0$, 1, 2, 3, ..., $a_{3k+1} = 3$, and $a_{3k+2} = 5$. If we start the numbering with 1, then there must be an obvious shift.

Exercise 1.19. Compute a_0, a_{13}, a_{k-1} and a_{k2}, if a_n are terms of the sequence with the general term $a_n = 4n - 3$.

Solution. $a_0 = -3$, $a_{13} = 49$, $a_{k-1} = 4k - 7$, $a_{k2} = 4k^2 - 3$.

Exercise 1.22. Find the inverses of the functions $f(x) = 1/x : (0, \infty) \to (0, \infty)$, $f(x) = 1/x$ and $g(x) = 2 - x$.

Solution. $f^{-1}(x) = 1/x$; $g^{-1}(x) = 2 - x$.

Chapter 2

Exercise 2.1, 2)

x	y	z	$x \to yz \equiv (x \vee y) \wedge (x \vee z)$
0	0	0	0
0	0	1	0
0	1	0	0
0	1	1	1
1	0	0	0

(Continued)

x	y	z	$x \to yz \equiv (x \vee y) \wedge (x \vee z)$
1	0	1	0
1	1	0	0
1	1	1	1

Exercise 2.3 Which of the following are propositions? Simple propositions? Compound propositions? Determine the truth values of the propositions.

1. $2 \times 2 = 4$

2. $2 \times 2 = 5$

3. Long Live Discrete Mathematics!

4. $x \times 3 = 7$

5. Discrete Mathematics is Indiscreet

6. $3 \leq 1 + 2$

7. Is it true that you are a lier?

8. I am a lier only when I say truth.

Solution. (1) and (6) are true propositions, (2) is a false proposition, the others are not propositions.

Exercise 2.5 What are the negations of the following propositions: (1) $2 + 2 = 4$; (2) $+ 2 \neq 4$; (3) *If a whole number is a multiple of 4, then it is a multiple of 2*; (4) *If a whole number is a multiple of 2, then it is a multiple of 4*; (5) *Our Earth is flat like a pancake, and Sun is cold.*

Solution. (1) $2 + 2 \neq 4$; (2) $2 + 2 = 4$.

Exercise 2.7 Determine whether the conditions (A) and (B) in each pair are necessary and sufficient, or only necessary, or only sufficient for one another:

1. A) A natural number m is a multiple of 6 ... (B) Each digit of m is 6

2. (A) $\frac{1}{x} > 1$ ... (B) $x > 1$

3. (A) $\sin x > 0$ ... (B) $0 < x < \pi$

Exercise 2.9 Is the formula $(((p \to q) \wedge b) \to a)$ a tautology, a contradiction, or neither?

Solution. Neither.

Exercise 2.11 Rewrite the next sentences in the standard implication form "If p, then q."

1. Eat those sweets if you care.

2. Eat those sweets and you will be sorry.

3. You leave or the robot comes to you.

4. I will do if you will do.

5. I will go unless you will go.

Solution. (1) If you care, eat those sweets. (4) If you will do, I will do.

Exercise 2.13. Prove that 2 and 3, and also 31 and 37 are mutually prime numbers. Are 14 and 49 mutually prime?

Solution. Check the prime numbers not exceeding 7 only.

Exercise 2.15. Prove that a natural number is divisible by 10 iff it is even and is a multiple by 5.

Solution. Verify that 2 and 5 are mutually prime.

Chapter 3
Exercise 3.1. Prove that 2 and 3, and also 31 and 37 are mutually prime numbers. Are 14 and 49 mutually prime?

Solution. Check the prime numbers not exceeding 7 only.

Exercise 3.3. Prove that a natural number is divisible by 10 iff it is even and is a multiple by 5.

Solution. Verify that 2 and 5 are mutually prime.

Exercise 3.5. Prove that any integer can be written as a linear combination $3k + 7l$ with some integer coefficients k and l.

Solution. Any integer can be written as $3k$, or $3k + 1$, or $3k + 2$. In the first case, $l = 0$. In the second case we write $3k + 1 = 3(k - 2) + 7$, that is, $l = 1$. In the third case, $3k + 2 = 3(k - 4) + 14$, i.e., $l = 2$.

Chapter 4
Exercise 4.1. Solve the DLP for the values $g = 12$ and $h = 2, 3, 4$.

Solution. For instance, $\log_{12} 2(\pmod 7)) = 2$, since $12^4 = 20736$ $\pmod 7) = 2 \pmod 7$.

Exercise 4.3. Prove that $(g^{-1})^{-1} = g$ and $(gh)^{-1} = h^{-1}g^{-1}$.

Exercise 4.5. Prove that the set $B(0, 1)$ with the binary addition is an additive group, where every element is its own inverse.

Solution. Indeed, in the group, $0 \oplus 0 \equiv 0$ and $1 \oplus 1 \equiv 0$.

Chapter 5
Exercise 5.3. Are the following formulas identically true with respect to the predicate variables P and Q, or identically false, or satisfiable?

Solution. (1) $\forall x(P(x) \vee Q(x)) \equiv \forall x P(x) \vee \forall x Q(x)$ – Satisfiable (2) $\exists x(P(x) \wedge Q(x)) \equiv \exists x P(x) \wedge \exists x Q(x)$ – Satisfiable (3) $\forall y \exists x P(x, y) \to \exists x \forall y P(x, y)$ – Identically False (4) $\forall x(Q(x) \to P(x)) \to (\forall x Q(x) \to \forall x P(x))$ – Satisfiable.

Chapter 6

Exercise 6.1. Compute the Cartesian product of the sets $A = \{a, b\}$ and $P = \{p, q, r\}$, and the product $P \times A$.

Solution. By the definition, the product $A \times P$ is

$$A \times P = \{(a, p), (a, q), (a, r), (b, p), (b, q), (b, r)\},$$
$$\text{while } P \times A = \{(p, a), (q, a), (r, a), (p, b), (q, b), (r, b)\}.$$

Exercise 6.6. Find the negation, conjunction, disjunction, and conditional of the strings $S_1 = (1, 1, 1, 1, 1)$ and $S_2 = (0, 0, 0, 0, 0)$.

Solution. $\neg S_1 = S_2$, $S_1 \vee S_2 = S_1$, $S_1 \wedge S_2 = S_2$.

Chapter 7

Exercise 7.1. Every student in a section studies at most one foreign language, 13 students take French, 12 students take German, 5 students take Chinese, and 3 students do not take any foreign language. How many students are there in the section?

Solution. To answer the question, we have to add $13 + 12 + 5 + 3 = 33$.

Exercise 7.14. In how many ways can 12 identical balls be placed into 7 distinguishable boxes?

Solution. Put the balls in one line, since they are identical, there is only one way to do that. We must split the balls into 7 parts, for that we need 6 partitions; hence, we arrive at the permutations with repetitions. The answer is $C(12 + 7 - 1, 7 - 1) = C(18, 6)$.

Exercise 7.20. How many paths are there in the Hasse diagram, Figure. 6.2, from vertex 1 to vertex 6?

Solution. There are 3 paths.

Exercise 7.21. Prove that the quantity $(p!)! \cdot (p!)^{-(p-1)!}$ is an integer for every natural p.

Solution. This is a multinomial coefficient.

Chapter 8

Exercise 8.1. Compute the $gcd(72, 17)$ and find the integers s, t, such that $gcd(72, 17) = 72s + 17n$.

Solution. $gcd(72, 17) = 1$; $1 = 72 \times (-4) + 17 \times (17)$.

Exercise 8.5. Are the integers 128 and 125 mutually prime?

Solution $128 = 2^7$, $125 = 5^3$, therefore, these integers have no common factor and are mutually prime.

Exercise 8.7. Compute 29×31 (mod 3).

Solution. $29 = 3 \cdot 9 + 2$, $31 = 3 \cdot 10 + 1$, thus, 29×31 (mod 3) $= 2 \cdot 1 = 2$.

Exercise 8.11. Check whether the integers 12, 49, 73, 11 are multiplicative inverses of 7 (mod 61) and of 11 (mod ( )61).

Solution. Compute 12×7 (mod 61) $\equiv 84$ (mod 61) $\equiv 23$ (mod 61), but not 1. Hence, 12 and 7 are not inverses in this problem. However, 11×11 (mod 60) $\equiv 1$ (mod 60). Thus, modulo 60, $11^2 \equiv 1$, or 11 is multiplicative inverse to itself, or $11 \equiv \sqrt{60}$ (mod 60).

Exercise 8.12. Compute the multiplicative inverses of the integers $n = 1, 2, 3, 4, 5, 6, 7$ in Z_8.

Solution. Now the even numbers do not have multiplicative inverses, while $1^{-1} = 1$; $3^{-1} = 3$, since $3 \cdot 3 = 1 \cdot 8 + 1$; $5^{-1} = 5$; $7^{-1} = 7$.

Exercise 8.17. Find the smallest positive multiple of 7, such that after dividing over 2 the remainder is 1, after dividing over 3 the remainder is 2, after dividing over 4 the remainder is 3, after dividing over 5 the remainder is 4, after dividing over 6 the remainder is 5, and it is a multiple of 7. Find the smallest positive integer, if after dividing over 2 the remainder is 1, after dividing over 3 the remainder is 2, after dividing over 4 the remainder is 3, after dividing over 5 the remainder is 4, and after dividing over 6 the remainder is 5.

Solution. Setting out the system of congruencies and excluding step-by-step natural parameters, we arrive at the answer $x = 119$.

Chapter 9

Exercise 9.1. Construct normal forms of the proposition (or the Boolean function) a) $P \equiv a \to (b \to c)$

Solution. $P \equiv \bar{a} \vee \bar{b} \vee c$, which is both a DNF (of one elementary disjunction) and a CNF (of three elementary conjunctions).

Exercise 9.2. Find the Boolean polynomial of the Boolean function $Q \equiv (a \to b)(b \leftrightarrow c)$.

Solution. $Q \equiv a \wedge b + a \wedge c + 1$.

Chapter 12

Exercise 12.3. An ice hockey team has 9 forwards. The team plays four games a week. Prove that the coach can set up the triples of field players so that no two forwards play twice in the same triple in the week.

Solution. Replace professors with forwards in Problem 320.

Chapter 13

Exercise 13.1. For any element x in a Boolean Algebra, prove that $x + x = x$. Hence, in any Boolean Algebra, all the coeffcients in equations are equal to 1.

Solution. Directly follows from the axioms.

Exercise 13.4. Prove that $A \subset B$ iff $A \cdot B = A$, and iff $A \cup B = B$.

Solution. For example, if $A \cup B = B$, then $A \subset B$. All the other implications are proved similarly.

Chapter 14

Exercise 14.1. Do arithmetic operations with binary numbers

Solution. $(1, 0, 1, 1, 0)_2 + (0, 0, 1, 1, 1)_2 = (1, 1, 1, 0, 1)_2$; $(1, 0, 1, 1, 0)_2 - (0, 0, 1, 1, 1)_2 = (0, 1, 1, 1, 1)_2$; $(1, 0, 1, 1, 0)_2 \times (0, 0, 1, 1, 1)_2 = (1,0,0,1,1,0,1,0)_2$.

Exercise 14.2. Find the binary representations of the (decimal) numbers, and verify your computations by returning back from the binary numbers to the decimal system.

Solution. $78_{10} = (1, 0, 0, 1, 1, 1, 0)_2$.

Exercise 14.3. Change $(6304)_7$ to the decimal system.

Solution. $(6304)_7 = 2209_{10}$.

Chapter 15

Exercise 15.2. Prove that $\{\neg, \vee\}$, $\{1, \oplus, \wedge\}$, $\{0, 1, \vee, x \oplus y \oplus z\}$ are bases in the set of all Boolean functions. Can we exclude any function from these systems?

Solution. For example, the negation preserves neither 0, nor 1, is not a monotone boolean function, but is linear; while the disjunction is not self-dual and is not linear, but is self-dual. Any constant can be excluded from the last system in this example.

Exercise 15.5. Give an example of a linear and non-self-dual Boolean function. What linear Boolean functions are self-dual?

Solution. For example, $f(x + y + 1)$. Compare the boolean functions with an even and with an odd numbers of arguments.

Chapter 16

Exercise 16.2. Does there exist a graph with three vertices of degree 5 each, two vertices of degree 4, one vertex of degree 1, and the total of 11 edges?

Solution. By Lemma 23, no.

Exercise 16.3. Prove that a complete graph K_n has $n(n-1)/2$ edges. What is the degree of any vertex of K_n?

Solution. The degree of any vertex of K_n is $n-1$. Every one of the n vertices has $n-1$ incident edges, but the product $n(n-1)$ counts every edge twice.

Exercise 16.6. A connected planar graph has 6 vertices, among them there are 5 vertices of degree 3 and a vertex of degree 1. In how many regions does the graph split the plane? Draw the diagram.

Solution. For this graph, $v = 6$ and $e = 8$, thus, by the Euler formula, $f = 4$.

Exercise 16.8. Prove that the complete bipartite graph $K_{m,n}$ has $m \cdot n$ edges.

Solution. Every vertex has either m or n incident edges.

Chapter 17

Exercise 17.1. Prove that any tree of order $p \geq 2$ is bipartite.

Solution. Fix any vertex of a tree and fold the tree "up and down."

Exercise 17.7. A forest has 67 vertices and 35 edges. How many trees, i.e., connected components does it have?

Solution. $67 - 35 = 32$ trees.

Exercise 17.10. What graphs coincide with their spanning trees?

Solution. Spanning trees.

Exercise 17.12. How many non-isomorphic spanning trees does the bipartite graph $K_{3,3}$ have?

Solution. A spanning tree of $K_{3,3}$ must have 5 edges. Fix any vertex, now we can choose any of its 3 incident edges. The second vertex of this edge has 2 incident edges, and after that the choice is unique. Thus, there are $3 \times 2 = 6$ spanning trees.

Exercise 17.13. n towns are connected by highways without intersections, such that a driver can reach every town from each other town, and there is the only route between any two towns. Prove that the number of highways is $n - 1$.

Solution. If there are two different ways between any two towns, they make a cycle. Now one can apply the major theorem on the trees.

Exercise 17.15. Calculate the maximal number of vertices at level three in a full binary tree.

Solution. $2^3 = 8$, if we start numbering at the zero level.

Exercise 17.16. Calculate the number of layers in a full ternary tree with 44 vertices.

Solution. The number of vertices is $1 + 3 + 3^2 + 3^3 + 4 = 44$, hence, the number of layers is 4, if we do not count the root as a layer, or $4 + 1 = 5$, if we do.

Chapter 19

Exercise 19.2.

Solution. See equation (19.3).

Chapter 21

Exercise 21.3. Determine, how many errors can detect the encoding function

$$c : \mathbb{B}^3 \to \mathbb{B}^4,$$

$c((0, 0, 0)) = (0, 0, 0, 1)$
$c((0, 0, 1)) = (0, 0, 1, 0)$
$c((0, 1, 0)) = (1, 0, 0.0)$
$c((0, 1, 1)) = (1, 1, 0, 0).$
$c((1, 0, 0)) = (0, 1, 0, 0)$
$c((1, 0, 1)) = (0, 1, 1, 1)$
$c((1, 1, 0)) = (1, 1, 1, 1)$
$c((1, 1, 1)) = (0, 1, 0, 1)$

Solution. The minimum distance in the example is $\rho((0, 0, 0, 1), (0, 1, 0, 1)) = 1 = \kappa + 1$, therefore, $\kappa = 0$ and this code in general, cannot detect any error with certainty.

Chapter 22

Exercise 22.3. Two dice are rolled simultaneously. What is the probability to get at least one number greater than 4?

Solution. $1 - (4/6)(4/6) = 5/9$.

Exercise 22.7. Simultaneously toss a coin and roll a die. What is the probability to get a head and a multiple of 3?

Solution. 1/6

Exercise 22.19. A $5 \times 5 \times 10$ parallelepiped with red sides cut into 250 unit cubes. What is the probability that a randomly chosen unit cube has no red face? One red face? Two red faces? Three red faces? Has four or more red faces?

Solution. No red: $3 \times 3 \div 250 = 72/250$. One red: $(2 \times 3 \times 4 + 4 \times 3 \times 8) \div 250$.

BIBLIOGRAPHY

[1] Aleksandrov A.D., Kolmogorov A.N., Lavrent'ev M.A. Ed-s, *Mathematics, Its Content, Methods and Meaning*, Moscow, 1977.

[2] Aumasson J.-P., *Serious Cryptography*. No starch press, San Francisco, 2018.

[3] Beale, M., Monaghan, M. F., *Encryption Using Random Boolean Functions*, in "Cryptography and Coding" (Cirencester, 1986), 219–230. Oxford Univ. Press, New York, 1989.

[4] Berlekamp E., *Algebraic Coding Theory*, Revised 2nd Ed. World Sci. 2015.

[5] Birkhoff G., Th. C. Bartee, *Modern Applied Algebra*, McGraw-Hill, New York, 1970.

[6] Bollobás B., *Modern Graph Theory*, Springer-Verlag, New York, 1998.

[7] Borovik A. *Implementation of the Kid Krypto Concept, MSOR Connections*. Vol. 2, No. 3, 23–25.

[8] Bourbaki N., *Eléments d'histoire des mathématiques*, Hermann, Paris, 1960.

[9] Brown, E., *Saints and Scoundrels and Two Theorems That Are Really The Same, Coll. Math. J.*, Vol. 46 (2015), No. 5, 326–334.

[10] Cameron Peter J., *Combinatorics*, Cambridge Univ. Press, New York, Melbourne, 1994.

[11] Carlet C., Joyner d., Stănică P., Tang D., Cryptographic Properties of Monotone Boolean Functions, *J. Math. Cryptol.*, Vol. 10(2016), No. 1, 1–14.

[12] Courant R., Robbins H., *What Is Mathematics*. Oxford Univ. Press, London, New York and Toronto, 1941.

[13] Cozzens M., Miller S., J., *The Mathematics of Encryption*. AMS, Providence, RI, 2013.

[14] Cusick T. W., Stănică P., *Cryptographic Boolean functions and Applications*, 2nd Ed., Acad. Press, 2017.

[15] Epp S.S., *Discrete Mathematics With Applications*, 2nd Ed., Brooks/Cole, Belmont, CA, 1995.

[16] Feferman S., *The Number Systems. Foundations of Algebra and Analysis*. Addison-Wesley, 1963.

[17] Feller W., *An Introduction to Probability Theory and Its Applications*, Vol. I (1957), Vol. II (1966), J. Wiley, New York.

[18] Fomin S. V., *Number Systems*, Univ. Chicago Press, Transl. from 2nd Russian Ed. Chicago, 1974.

[19] Gavrilov G. P., Sapozhenko A. A., Problems and Exercises in Discrete Mathematics, Dordrecht: Springer Netherlands, 2010.

[20] Gindikin S., Algebraic Logic, Springer.

[21] Glushkov V. M., *The Abstract Automata Theory*, Russian Mathematical Surveys, vol 16, No. 5 (1961) 3–62.

[22] Golovina L.I., Yaglom I.M., *Induction in Geometry*. GITTL, Moscow, 1956.

[23] Gottschalk L., *Understanding History. A Primer of Historical Method*, A. A. Knopp, New York, 1963.

[24] Halmos P., *Naïve Set Theory*, Van Nostrand, 1960.

[25] Halmos P., Applied Mathematics Is Bad Mathematics. In Steen L.A. (ed.) *Mathematics Tomorrow*, pp. 9–20. Springer, New York (1981) (Reprinted in [Halmos 1983]).

[26] Harary F., Palmer E. M., *Graphical Enumeration*, Academic Press, New York, 1973.

[27] Jacobson N., *Lecturers in Abstract Algebra*, Vol. 1, D. Van Nostrand Co., Princeton, etc., 1951.

[28] Kheyfits A., *A Primer in Combinatorics*, De Gruyter, Berlin/New York, 2010.

[29] Kleene S.C., *Representation of Events in Nerve Nets and Finite Automata*, C. Shannon and J. McCarthy, (eds.) Automata Studies, Princeton University Press, NJ, 1956, pp. 3–41.

[30] Kleene S. C., *Mathematical Logic*, John Wiley & Sons, Inc., New York, 1967.

[31] Knuth D.E., *The Art of Computer Programming*, Vol. 1, Addison-Wesley, MA 1973.

[32] Kolman B., Busby R.C., Ross S.C., *Discrete Mathematical Structure*, 4th Ed. Prentice-Hall, NJ, 2000.

[33] Kolmogorov A.N., Dragalin A.G., *Introduction to Mathematical Logic* (In Russian) Moscow, 1982.

[34] Lax P. D., *Linear Algebra and Its Applications*, 2nd Ed. Wiley-Interscience, 2007.

[35] Lovasz L., Pelikan J., Vesztergombi K., *Discrete Mathematics: Elementary and Beyond*. Cham: Springer International Publishing, 2003.

[36] Marion W., Discrete Mathematics A Mathematics Course or a Computer Science Course? *PRIMUS*, Vol. 1, No. 2, (1991) 314–324.

[37] Maurer S. B., Ralston A., *Discrete Algorithmic Mathematics*, A. K. Peters, Natick, MA, 1998.

[38] Von Neumann J., Morgenstern O., *Theory of Games and Economic Behavior*, Princeton Univ. Press, Princeton, 1953.

[39] O'Connor L., *The Inclusion-Exclusion Principle and its Applications to Cryptography, Cryptologia,* Vol. 17 (1993), No. 1, 63–79.

[40] Paar C., Pelzl J., *Understanding Cryptology*, Springer-Verlag, Berlin Heidelberg 2010.

[41] Rasmusen E., *Games and Information: An Introduction to Game Theory*, 3rd Ed. Blackwell, Oxford. 2001.

[42] Regan G., *Guide to Discrete Mathematics*, Springer, 2016.

[43] Rosen K.H., *Discrete Mathematics and Its Applications*, 8th Ed., New York, McGraw-Hill, 2019.

[44] Savage C., A Survey of Combinatorial Gray Codes, *SIAM Review*, Vol. 39(1997) No. 4, 605-629.

[45] Schroeder M., *Number Theory in Science and Communication*, 5th Ed., Springer, 2009.

[46] Shannon C. E., A Mathematical Theory of Communication, *Bell System Technical Journal*, Vol. 27 (1948), pp. 379–423, 623–656, July and October.

[47] Sipser M., *Introduction to the Theory of Computation*, 3rd Ed., Thomson, 2012.

[48] Turocy T.L., von Stengel B., Game Theory. In *Encyclopedia of Information Systems*, Vol. 2. San Diego. Elsevier Science, pp. 403–420, 2002.

[49] Washburn Sh., Marlowe T., Ryan, C.T. *Discrete Mathematics*, Addison-Wesley, Reading, 2000.

[50] Weil A., *Number Theory for Beginners*, Springer-Verlag, New York-Heidelberg-Berlin, 1979.

[51] Weil G., Mathematics and Logic, *Amer. Math. Mon.* Vol. 53 (1946), 2–13.

[52] Wiener N., *A Life in Cybernetics*, The MIT Press, Cambridge, MA, 2017.

[53] Wilson R.J., *Introduction to Graph Theory*, AP, New York, 1972.

INDEX

www.ingramcontent.com/pod-product-compliance
Lightning Source LLC
Chambersburg PA
CBHW061922190326
41458CB00009B/2621